1 8 0 7

WILEY

2 0 0 7

BICENTENNIAL · BICENTENNIAL · BICENTENNIAL · BICENTENNIAL

THE WILEY BICENTENNIAL—KNOWLEDGE FOR GENERATIONS

*E*ach generation has its unique needs and aspirations. When Charles Wiley first opened his small printing shop in lower Manhattan in 1807, it was a generation of boundless potential searching for an identity. And we were there, helping to define a new American literary tradition. Over half a century later, in the midst of the Second Industrial Revolution, it was a generation focused on building the future. Once again, we were there, supplying the critical scientific, technical, and engineering knowledge that helped frame the world. Throughout the 20th Century, and into the new millennium, nations began to reach out beyond their own borders and a new international community was born. Wiley was there, expanding its operations around the world to enable a global exchange of ideas, opinions, and know-how.

For 200 years, Wiley has been an integral part of each generation's journey, enabling the flow of information and understanding necessary to meet their needs and fulfill their aspirations. Today, bold new technologies are changing the way we live and learn. Wiley will be there, providing you the must-have knowledge you need to imagine new worlds, new possibilities, and new opportunities.

Generations come and go, but you can always count on Wiley to provide you the knowledge you need, when and where you need it!

WILLIAM J. PESCE
PRESIDENT AND CHIEF EXECUTIVE OFFICER

PETER BOOTH WILEY
CHAIRMAN OF THE BOARD

Visualizing
ENVIRONMENTAL
SCIENCE

Linda R. Berg

Mary Catherine Hager

BICENTENNIAL
1807
WILEY
2007
BICENTENNIAL

In collaboration with
THE NATIONAL GEOGRAPHIC SOCIETY

CREDITS

EXECUTIVE PUBLISHER Kaye Pace
MANAGING DIRECTOR Helen McInnis
DIRECTOR OF DEVELOPMENT Barbara Heaney
MARKETING MANAGER Jeffrey Rucker
PRODUCTION MANAGER Kelly Tavares
PRODUCTION ASSISTANT Courtney Leshko
ASSOCIATE EDITOR Merillat Staat
CREATIVE DIRECTOR Harry Nolan
COVER DESIGNER Harry Nolan
INTERIOR DESIGN Vertigo Design
PHOTO RESEARCHER Tara Sanford/Mary Ann Price/
Stacy Gold, National Geographic Society
ILLUSTRATION COORDINATOR Anna Melhorn
COVER CREDIT Bill Curtsinger/NG Image Collection

This book was set in Times New Roman by PrePress Company, Inc., printed and bound by Quebecor World. The cover was printed by Phoenix Color.

To order books or for customer service please, call 1-800-CALL WILEY (225-5945).

ISBN-13 978-0-471-69702-2
ISBN-10 0-471-69702-8

Printed in the United States of America

10 9 8 7 6 5 4 3

Visualizing Environmental Science is designed to help your students learn effectively. Created in collaboration with the National Geographic Society (NGS) and our Wiley Visualizing Consulting Editor, Professor Jan Plass of New York University, *Visualizing Environmental Science* integrates rich visuals and media with text to direct student's attention to important information. This approach represents complex processes, organizes related pieces of information, and integrates information into clear representations. Beautifully illustrated, *Visualizing Environmental Science* shows your students what the discipline is all about, its main concepts and applications, while also instilling an appreciation and excitement about the richness of the subject.

Visuals, as used throughout this text, are instructional components that display facts, concepts, processes, or principles. They create the foundation for the text and do more than simply support the written or spoken word. The visuals include diagrams, graphs, maps, photographs, illustrations, schematics, animations, and videos.

Why should a textbook based on visuals be effective? Research shows that we learn better from integrated text and visuals than from either medium separately. Beginners in a subject benefit most from reading about the topic, attending class, and studying well-designed and integrated visuals. A visual, with good accompanying discussion, really can be worth a thousand words!

Well-designed visuals can also improve the efficiency with which information is processed by a learner. The more effectively we process information, the more likely it is that we will learn. This processing of information takes place in our working memory. As we learn we integrate new information in our working memory with existing knowledge in our long-term memory.

Have you ever read a paragraph or a page in a book, stopped, and said to yourself: "I don't remember one thing I just read?" This may happen when your working memory has been overloaded, and the text you read was not successfully integrated into long-term memory. Visuals don't automatically solve the problem of overload, but well-designed visuals can reduce the number of elements that working memory must process, thus aiding learning.

You, as the instructor, facilitate your student's learning. Well-designed visuals, used in class, can help you in that effort. Here are six methods for using the visuals in *Visualizing Environmental Science* in classroom instruction.

1. **Assign students to study visuals in addition to reading the text.**

   Instead of assigning only one medium of presentation, it is important to make sure your students know that the visuals are just as essential as the written material.

2. **Use visuals during class discussions or presentations.**

   By pointing out important information as the students look at the visuals during class discussions, you can help focus students' attention on key elements of the visuals and help them begin to organize the information and develop an integrated model of understanding. The verbal explanation of important information combined with the visual representation is highly effective.

3. **Use visuals to review content knowledge.**

   Students can review key concepts, principles, processes, vocabulary, and relationships displayed visually. Better understanding results when new information in working memory is linked to prior knowledge.

4. **Use visuals for assignments or when assessing learning.**

   Visuals can be used for comprehension activities or assessments. For example, students could be asked to identify examples of concepts portrayed in visuals. Higher-level thinking activities that require critical thinking, deductive and inductive reasoning, and prediction can also be based on visuals. Visuals can be very useful for drawing inferences, for predicting, and for problem solving.

5. **Use visuals to situate learning in authentic contexts.**

   Learning is made more meaningful when a learner can apply facts, concepts, and principles to realistic situations or examples. Visuals can provide that realistic context.

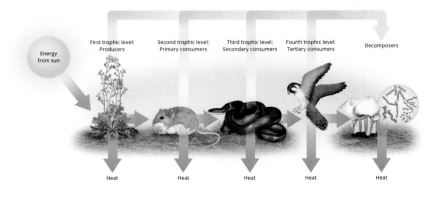

6. Use visuals to encourage collaboration.

Collaborative groups often are required to practice interactive processes such as giving explanations, asking questions, clarifying ideas, and arguing points of view. These interactive, face-to-face processes provide the information needed to build a verbal mental model. Learners also benefit from collaboration over visuals that require decision making or problem solving.

Visualizing Environmental Science not only aids student learning with extraordinary use of visuals, but it also offers an array of remarkable photos, media, and film from the National Geographic Society collections. Students using *Visualizing Environmental Science* also benefit from the long history and rich, fascinating resources of National Geographic.

National Geographic has also performed an invaluable service in fact-checking *Visualizing Environmental Science*: They have verified every fact in the book with two outside sources, ensuring the accuracy and currency of the text.

Given all of its strengths and resources, *Visualizing Environmental Science* will immerse your students in the discipline, its main concepts and applications, while also instilling an appreciation and excitement about the richness of the subject area.

Additional information on learning and instructional design is provided in a special guide to using this book, *Learning from Visuals: How and Why Visuals Can Help Students Learn*, prepared by Matthew Leavitt, of Arizona State University. This article is available at the Wiley web site: www.wiley.com/college/visualizing. The online *Instructor's Manual* also provides guidelines and suggestions on using the text and visuals most effectively.

Visualizing Environmental Science also offers a rich selection of visuals in the supplementary materials that accompany the book. To complete this robust package the following materials are available: Test Bank with visuals used in assessment, PowerPoints, Image Gallery to provide you with the same visuals used in the text, and web-based learning materials for homework and assessment, including images, video, and media resources from National Geographic.

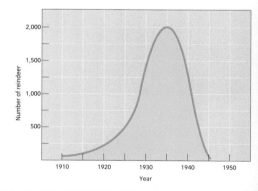

PREFACE

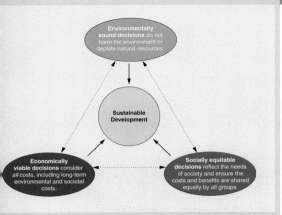

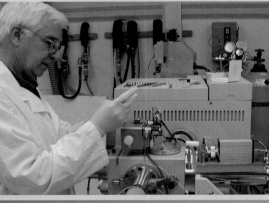

This debut edition of *Visualizing Environmental Science* offers students a valuable opportunity to identify and connect the central issues of environmental science through a visual approach. As students explore the critical topics of environmental science, their study of the role of humans on Earth must be interwoven with basic concepts of ecology, geography, chemistry, economics, ethics, policy, and many other disciplines. *Visualizing Environmental Science* reinforces these interacting components and, with its premier art program, vividly illustrates the overarching role that humans play in our planet's environmental problems and successes.

We begin *Visualizing Environmental Science* with an introduction of the environmental dilemmas we face in our world today, emphasizing particularly how unchecked population growth and economic inequity complicate our ability to solve these problems. We stress that solutions rest in creativity and diligence at all levels, from individual commitment to international cooperation. We revisit this theme throughout the text, offering concrete suggestions that students can adopt to make their own difference in solving environmental problems, and explaining the complications that arise when solutions are tackled on a local, regional, national, or global scale.

Yet *Visualizing Environmental Science* is not simply a check-list of "to do" items to save the planet. In the context of an engaging visual presentation, we offer solid discussions of such critical environmental concepts as sustainability, conservation and preservation, and risk analysis. We weave the threads of these concepts throughout our treatment of ecological principles and their application to various ecosystems, the impacts of human population change, and the problems associated with our use of the world's resources. We particularly instruct students in the importance of ecosystem services to a functioning world, and the threats that restrict our planet's ability to provide such services.

This book is intended to serve as an introductory text primarily for nonscience undergraduate students. The accessible format of *Visualizing Environmental Science*, coupled with our assumption that students have little prior knowledge of ecosystem ecology, allows students to easily make the transition from jumping-off points in the early chapters to the more complex concepts they encounter later. With its interdisciplinary presentation, which mirrors the nature of environmental science itself, this book is appropriate for use in one-semester environmental science courses offered by a variety of departments, including environmental studies, biology, agriculture, geography, and geology.

ORGANIZATION

Visualizing Environmental Science is organized around the premise that humans have caused many of the world's environmental dilemmas and now must address these issues as we use Earth's resources and seek to avoid the future disasters so often predicted in the media. The book's first four chapters lay the groundwork for creating an understanding of what environmental issues we face, how environmental sustainability and human values play a critical role in addressing these issues, how the environmental movement developed over time, how economics shapes environmental policy, and how environmental threats sometimes cause health hazards..

Chapters 5, 6, and 7 of *Visualizing Environmental Science* present the intricacies of ecological concepts in a human-dominated world, including energy flow and the cycling of matter through ecosystems, and the various ways that species interact and partition resources. Gaining familiarity with these concepts allows students to appreciate the variety of terrestrial and aquatic ecosystems that we introduce. Students also develop a richer understanding of the implications to the environment that rise from human population change.

The remaining 11 chapters of *Visualizing Environmental Science* all deal in some way with our world's resources, as we use them today and as we assess their availability and impacts for the future. These issues cover a broad spectrum that includes the sources and effects of air pollution, climate and global atmospheric change, freshwater resources and causes and effects of water pollution, the ocean and fisheries, mineral and soil resources, land resources, agriculture and food resources, biological resources, solid and hazardous waste, and nonrenewable and renewable energy resources. We are particularly pleased to dedicate an entire chapter to a discussion of ocean processes and resources, recognizing the importance of the global ocean to environmental issues.

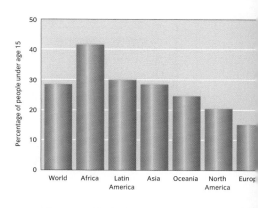

ILLUSTRATED BOOK TOUR

Many visual pedagogical features have been developed specifically for *Visualizing Environmental Science*. Presenting the highly varied and often technical concepts woven throughout environmental science raises challenges for reader and instructor alike. The Illustrated Book Tour on the following pages provides a guide to the diverse features contributing to *Visualizing Environmental Science*'s pedagogical plan.

CHAPTER INTRODUCTIONS provide concise stories about some of today's most pressing environmental issues. These narratives are featured alongside striking accompanying photographs. The chapter openers also include illustrated **CHAPTER OUTLINES** that use thumbnails of illustrations from the chapter to refer visually to the content.

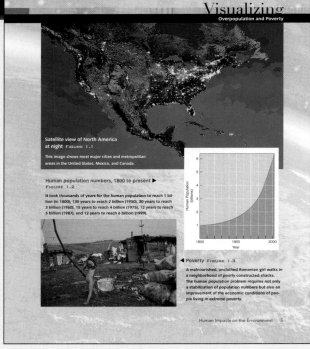

PROCESS DIAGRAMS present a series of figures or a combination of figures and photos that describe and depict a complex process, helping students to observe, follow, and understand the process.

VISUALIZING features are specially designed multi-part visual spreads that focus on a key concept or topic in the chapter, exploring it in detail or in broader context using a combination of photos and figures.

WHAT A SCIENTIST SEES features highlight a concept or phenomenon, using photos and figures that would stand out to a professional in the field, and helping students to develop observational skills.

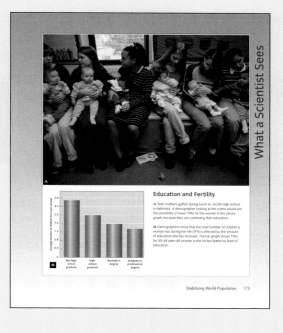

What a Scientist Sees

Education and Fertility

A Teen mothers gather during lunch at Lincoln High School in Nebraska. A demographer looking at this scene would see the possibility of lower TFRs for the women in this photograph, because they are continuing their education.

B Demographers know that the total number of children a woman has during her life (TFR) is affected by the amount of education she has received. The bar graph shows TFRs for 35–44 year-old women in the United States by level of education.

Stabilizing World Population 173

EnviroDiscovery   **Using Goats to Fight Fires**

California has about 6,000 wildfires each year, and they are becoming increasingly expensive and dangerous to manage because so many people are building homes and living in the fire-vulnerable chaparral. Yet the topography of chaparral is so steep that firefighters often cannot use mechanized equipment that must be transported to fires by helicopters. Afraid that prescribed burns will get out of control, local governments are increasingly trying an effective, low-tech method to reduce the fuel load. During the 6-month fire season, goats are clearing hills around Oakland, Berkeley, Monterey, and Malibu.

A herd of 350 goats can denude an entire acre of heavy brush in about a day, but their use entails a lot of advance organization and support. Before goats can clear hazardous dry fuels from surrounding hillsides, botanists must walk the terrain to put fences around any small trees or other plants that are rare or endangered; fencing keeps the goats from eating those plants. A portable home site for goatherds is then installed, as are electric fencing and water troughs for the goats. The goatherds typically use dogs to help herd the goats.

Goats are an excellent tool for fire management because they preferentially browse woody shrubs and thick undergrowth—the exact fuel that causes disastrous fires. Fires that have occurred in areas after goats have browsed there are much easier to contain.

Goats prefer woody and weedy species.

different parts of the world, even though the individual species differ by location. A dense thicket of evergreen shrubs—often short, drought-resistant pine or scrub oak trees that grow 1 to 3 m (3.3 to 9.8 ft) tall—usually dominates chaparral. During the rainy winter season, the environment may be lush and green, but during the hot, dry summer, the plants lie dormant. The hard, small, leathery leaves of trees and shrubs resist water loss. Many plants are also specifically fire-adapted and

grow best in the months following a fire. Such growth is possible because fire releases the nutrient minerals present in the aboveground parts of the plants that burned. These minerals are released into the soil. The seeds and underground parts of plants that survive fire make use of the newly available nutrient minerals and sprout vigorously during winter rains. Mule deer, wood rats, chipmunks, lizards, and many species of birds are common animals of the chaparral.

134 CHAPTER 6 Ecosystems and Evolution

ENVIRODISCOVERY features provide additional topical material about relevant environmental issues.

CASE STUDY

Environmental Problems in Central and Eastern Europe

The fall of the Soviet Union and communist governments in Central and Eastern Europe during the late 1980s revealed a grim legacy of environmental destruction. Chemicals leaking out of dump sites contaminated agricultural soil and water. Buildings and statues eroded, and entire forests died because of air pollution and acid rain. One of the most polluted areas in the world is the "Black Triangle," which consists of the bordering regions of eastern Germany, the northern Czech Republic, and southwest Poland.

Massive pollution affects human health in the region. Many people suffer from asthma, emphysema, chronic bronchitis, and other respiratory diseases. Levels of cancer, miscarriages, and birth defects are extremely high. Life expectancies are still lower there than in other industrialized nations. In 2005, the average Eastern European lived to age 69—10 years less than the average Western European.

The economic assumption behind communism was one of high production and economic self-sufficiency. Governments supported heavy industry—power plants,

chemicals, metallurgy, and large machinery—at the expense of more environmentally benign service industries. As a result, Central and Eastern Europe became overindustrialized, and most of the region's plants lacked the pollution abatement equipment now required in factories in most industrialized countries.

Experts predict that eliminating the pollution legacy of communism will take decades. How much will it cost? The numbers are staggering. Improving the environment in eastern Germany alone will cost up to an estimated $300 billion.

While switching from communism to democracy with a free-market economy, current Central and Eastern European governments also face the responsibility of improving the environment. Hungary, Poland, and the Czech Republic have been relatively successful in moving toward a market economy, generating enough money to invest in environmental cleanup. From

Birth defects in Moscow

Since 1973, at least 90 children with terminal-limb defects were born in two neighborhoods of Moscow. Medical researchers have not investigated limb defects thoroughly and do not know if their cause is environmental, genetic, or both.

1990 to 2000, sulfur dioxide emissions in the Czech Republic declined 86 percent, nitrogen oxide emissions from industries declined 69 percent, and particulate emissions declined 91 percent. Economic recovery is slow in other countries, such as Bulgaria, Romania, and Russia, and severe budgetary problems there have forced the environment to take a back seat to political and economic reform.

Pollution problems in former communist countries

A former uranium processing plant produced radioactive waste that damaged this land in the Estonian town of Sillamae. Note the exceedingly high reading on the Geiger counter. Thousands of polluted sites exist in former communist countries, the result of rapid expansion of industrialization without regard for the environment. It will take decades for the contamination from these sites to be cleaned up.

66 CHAPTER 3 Environmental History, Politics, and Economics

The illustrated **CASE STUDIES** that cap off the text sections of each chapter offer a wide variety of in-depth examinations that address important issues in the field of environmental science.

KEY TERMS
- poverty p. 4
- highly developed countries p. 6
- moderately developed countries p. 6
- less developed countries p. 7
- nonrenewable resources p. 7
- renewable resources p. 7
- people overpopulation p. 9
- consumption overpopulation p. 9
- environmental sustainability p. 12
- environmental science p. 14
- scientific method p. 17

CRITICAL AND CREATIVE THINKING QUESTIONS

1. Criticize the following statement: "Population growth in developing countries is of much more concern then is population growth in highly developed countries."

2. Use the IPAT equation to calculate the environmental impact in terms of CO_2 emissions per year at the beginning of the 21st century, when there were 6 billion people, an average of 0.1 motor vehicles per person, and 5.4 tons of CO_2 emitted by each car per year. Then make a similar calculation for the year 2050, based on these projections: a population of 10 billion people, 0.4 cars per person, and CO_2 emissions per vehicle similar to what we have today (that is, no technological improvements). How might we hold global CO_2 emissions from motor vehicles to 2000 levels in the year 2050?

3. Give at least two examples of things you can do as an individual to promote environmental sustainability.

4. People want scientists to give them precise, definitive answers to environmental problems. Explain why this is not possible.

5. Make a three-column list of the material items you own or would like to own. Column 1 consists of items you have that meet your basic needs; column 2 contains items you have that

you wanted but didn't need; and column 3 contains items you would like to possess in the near future. Which of the items in columns 2 and 3 would you be willing to give up to reduce your personal impact on the environment?

6. Write a brief paragraph describing two or three ways your personal quality of life could be improved. Do any of these involve material items? If so, explain how these items would improve your life.

7–9. Your throat feels scratchy and you think you're coming down with a cold. You take a couple of vitamin C pills and feel better. You conclude that vitamin C helps prevent colds.

7. Is your conclusion valid from a scientific standpoint? Why or why not?

8. Develop a testable hypothesis to answer the question, "Does taking vitamin C pills help prevent colds?"

9. Describe how you would conduct a controlled experiment to test your hypothesis.

What is happening in this picture?

What are these people protesting?

How is this opposition an example of NIMBY?

What kinds of incentives might make these people more willing to consider locating a nuclear waste site near them?

PEOPLE OF NEW HAMPSHIRE
AGAINST THE NUCLEAR DUMP

Critical and Creative Thinking Questions 23

WHAT IS HAPPENING IN THIS PICTURE? is an end-of-chapter feature that presents students with a photograph relevant to chapter topics but that illustrates a situation students are not likely to have encountered previously. The photograph is paired with questions designed to stimulate creative thinking.

OTHER PEDAGOGICAL FEATURES

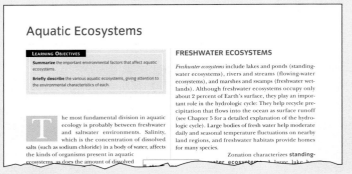

Aquatic Ecosystems

LEARNING OBJECTIVES

Summarize the important environmental factors that affect aquatic ecosystems.

Briefly describe the various aquatic ecosystems, giving attention to the environmental characteristics of each.

The most fundamental division in aquatic ecology is probably between freshwater and saltwater environments. Salinity, which is the concentration of dissolved salts (such as sodium chloride) in a body of water, affects the kinds of organisms present in aquatic ecosystems, as does the amount of dissolved

FRESHWATER ECOSYSTEMS

Freshwater ecosystems include lakes and ponds (standing-water ecosystems), rivers and streams (flowing-water ecosystems), and marshes and swamps (freshwater wetlands). Although freshwater ecosystems occupy only about 2 percent of Earth's surface, they play an important role in the hydrologic cycle: They help recycle precipitation that flows into the ocean as surface runoff (see Chapter 5 for a detailed explanation of the hydrologic cycle). Large bodies of fresh water help moderate daily and seasonal temperature fluctuations on nearby land regions, and freshwater habitats provide homes for many species.

Zonation characterizes **standing-water ecosystems**. A large lake

LEARNING OBJECTIVES at the beginning of each section head indicate in behavioral terms what the student must be able to do to demonstrate mastery of the material in the chapter.

---

in physical factors like temperature, salinity, and depth of light penetration, estuarine organisms must have a high tolerance for changing conditions.

Temperate estuaries usually feature **salt marshes**, shallow wetlands in which salt-tolerant grasses grow (**FIGURE 6.14A**). Salt marshes perform many ecosystem services, including providing biological habitats, trapping sediment and pollution, supplying groundwater, and buffering storms by absorbing their energy, which prevents flood damage elsewhere.

Mangrove forests, the tropical equivalent of salt marshes, cover perhaps 70 percent of tropical coastlines (**FIGURE 6.14B**). Like salt marshes, mangrove forests provide valuable ecosystem services. Their interlacing roots are breeding grounds and nurseries for several commercially important fishes and shellfish, such as crabs, shrimp, mullet, and spotted sea trout. Mangrove branches are nesting sites for many species of birds, such

as pelicans, herons, egrets, and roseate spoonbills. Mangrove roots stabilize the submerged soil, thereby preventing coastal erosion and providing a barrier against the ocean during storms.

CONCEPT CHECK STOP

How are flowing-water ecosystems and standing-water ecosystems alike? How are they different?

What are some ecosystem services of salt marshes and mangrove swamps?

What is a freshwater wetland? How does a freshwater wetland differ from an estuary?

CONCEPT CHECK questions at the end of each section give students the opportunity to test their comprehension of the learning objectives.

GLOBAL LOCATOR MAPS accompany figures addressing issues encountered in a particular geographic region. These hemispheric locator maps help students visualize where the area discussed is situated on a continent.

---

risk The probability of harm (such as injury, disease, death, or environmental damage) occurring under certain circumstances.

MARGINAL GLOSSARY TERMS (in green boldface) introduce each chapter's most important terms. Other important terms appear in black boldface and are defined in the text.

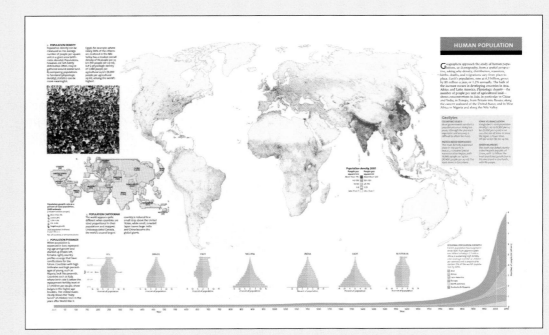

NATIONAL GEOGRAPHIC SOCIETY MAPS are featured in the special section following Chapter 1.

Probability of risk of dying by selected causes, 1998 TABLE 4.1		
Cause of death	*U.S. deaths in 1998*	*Probability of risk*
Cardiovascular disease	940,600	3.5 of every 1,000 people
Cancer (all types)	541,500	2.0 of every 1,000 people
Accidents (including motor vehicle)	97,800	3.6 of every 10,000 people
Suicide	30,600	1.1 of every 10,000 people
Homicide	18,300	0.7 of every 10,000 people
Accidental falls	16,274	0.6 of every 10,000 people
Accidental poisonings by drugs	9,838	3.6 of every 100,000 people
Accidental drownings	3,964	1.5 of every 100,000 people
Fire	3,255	1.2 of every 100,000 people
Accidents by firearms	726	2.6 of every 1,000,000 people
Accidents (airplane)	692	2.5 of every 1,000,000 people

TABLES AND GRAPHS, with data sources cited at the end of the text, summarize and organize important information.

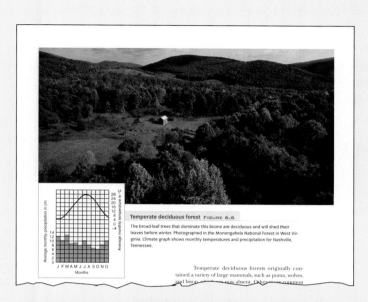

Temperate deciduous forest FIGURE 6.6
The broad-leaf trees that dominate this biome are deciduous and will shed their leaves before winter. Photographed in the Monongahela National Forest in West Virginia. Climate graph shows monthly temperatures and precipitation for Nashville, Tennessee.

Temperate deciduous forests originally contained a variety of large mammals, such as puma, wolves, and bison, which are now absent. Other more common

ILLUSTRATIONS AND PHOTOS support concepts covered in the text, elaborate on relevant issues, and add visual detail. Many of the photos originate from National Geographic's archives.

The **CHAPTER SUMMARY** revisits each learning objective and redefines each marginal glossary term, featured in boldface here, and included in a list of Key Terms. Students are thus able to study vocabulary words in the context of related concepts.

CRITICAL AND CREATIVE THINKING QUESTIONS, some of which include visuals, highlight each chapter's important concepts and applications.

MEDIA AND SUPPLEMENTS

Visualizing Environmental Science is accompanied by a rich array of media and supplements that incorporate the visuals from the textbook extensively to form a pedagogically cohesive package. For example, a Process Diagram from the book appears in the Instructor's Manual with suggestions for using it as a PowerPoint in the classroom; it may be the subject of a short video or an online animation; and it may appear with questions in the Test Bank, as part of the chapter review, homework assignment, assessment questions, and other online features.

INSTRUCTOR SUPPLEMENTS

VIDEOS

A collection of videos from the award-winning National Geographic Film Collection have been selected to accompany the text. Each chapter includes two to three video clips, available online as digitized streaming video, that illustrate and expand on a concept or topic.

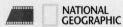 This logo in the text indicates that video from the archives of the National Geographic Society is available.

Contextualized commentary and questions to further develop student understanding accompany each of the videos.

The videos are available on the main website: (www.wiley.com/college/berg).

POWERPOINT PRESENTATIONS AND IMAGE GALLERY

A complete set of visual PowerPoint presentations by Dawn Keller of Hawkeye Community College is available online to enhance classroom presentations. Tailored to the text's topical coverage and learning objectives, these presentations are designed to convey key text concepts, illustrated by embedded text art.

All photographs, figures, maps, and other visuals from the text are online and can be used as you wish in the classroom. These online electronic files allow you to easily incorporate them into your PowerPoint presentations as you choose, or to create your own overhead transparencies and handouts.

As noted previously, by using visuals in class discussions, you can help focus students' attention on key elements of the visuals and help them begin to organize the information and develop greater understanding. The classroom explanation of important information combined with the visual representation is highly effective.

TEST BANK (AVAILABLE IN WILEYPLUS AND ELECTRONIC FORMAT)

The visuals from the textbook are also included in the Test Bank by Madelyn Logan of Northshore Community College and Sandy Buczynski of San Diego University (assisted by biologist Jessica O'Leary). The Test Bank has approximately 1200 test items, with at least 25 percent incorporating visuals from the book. The test items include multiple choice and essay questions testing a variety of comprehension levels. The test bank is available online in two formats: MS Word files and a Computerized Test Bank. The easy-to-use test-generation program fully supports graphics, print tests, student answer sheets, and answer keys. The software's advanced features allow you to create an exam to meet your exact specifications.

INSTRUCTOR'S MANUAL (AVAILABLE IN WILEYPLUS AND ELECTRONIC FORMAT)

The Manual begins with a special introduction on Using Visuals in the Classroom, prepared by Matthew Leavitt of the Arizona State University, in which he provides guidelines and suggestions on how to use the visuals in teaching the course. For each chapter, materials by Barry Vroginday of Devry University include suggestions and directions for using web-based learning modules in the classroom and for homework assignments, as well as over 50 creative ideas for in-class activities by David Hassenzahl of the University of Nevada-Las Vegas (assisted by biologist Joy Sales Colquitt).

WEB-BASED LEARNING MODULES

A robust suite of multi-media learning resources have been designed for *Visualizing Environmental Science*, again focusing on and using the visuals from the book. Delivered via the web, the content is organized into tutorial animations.

www.wiley.com/college/berg

Tutorial Animations: Animations visually support the learning of a difficult concept, process, or theory, many of them built around a specific feature such as a Process Diagram, Visualizing piece, or key visual in the chapter. The animations go beyond the content and visuals presented in the book, providing additional visual examples and descriptive narration.

BOOK COMPANION SITE (WWW.WILEY.COM/COLLEGE/BERG)

Instructor Resources on the book companion site include the Test Bank, Instructor's Manual, all illustrations and photos in the textbook in jpeg format, as well as select flash animations for use in classroom presentations.

WILEYPLUS

Visualizing Environmental Science is available with WileyPlus, a powerful online tool that provides instructors and students with an integrated suite of teaching and learning resources in one easy-to-use Web site.

Wiley PLUS Helping Teachers Teach and Students Learn

www.wiley.com/college/berg

This title is available with *Wiley PLUS*, a powerful online tool that provides instructors and students with an integrated suite of teaching and learning resources in one easy to use Web site. *Wiley PLUS* is organized around the activities you and your students perform in class.

For Instructors

Prepare & Present: Create class presentations using a wealth of Wiley-provided resources—such as an online version of the textbook, PowerPoint slides, animations, overviews, and visuals from the Wiley Image Gallery—making your preparation time more efficient. You may easily adapt, customize, and add to this content to meet the needs of your course.

Create Assignments: Automate the assigning and grading of homework or quizzes by using Wiley-provided question banks, or by writing your own. Student results will be automatically graded and recorded in your gradebook. *Wiley PLUS* can link the pre-lecture quizzes and test bank questions to the relevant section of the online text.

Track Student Progress: Keep track of your students' progress via and instructor's gradebook, which allows you to analyze individual and overall class results to determine students' progress and level of understanding.

Administer Your Course: *Wiley PLUS* can easily be integrated with another course management system, gradebook, or other resources you are using in your class, providing you with the flexibility to build your course, your way.

For Students

Wiley PLUS provides immediate feedback on student assignments and a wealth of support materials. This powerful study tool will help your students develop their conceptual understanding of the class material and increase their ability to answer questions.

A "**Study and Practice**" area links directly to text content, allowing students to review the text while they study and answer. Resources include National Geographic videos, concept animations and tutorials, visual learning interactive exercises, and links to websites that offer opportunities for further exploration.

An "**Assignment**" area keeps all the work you want your students to complete in one location, making it easy for them to stay "on task." Students will have access to a variety of interactive self- assessment tools, as well as other resources for building their confidence and understanding. In addition, all of the pre-lecture quizzes contain a link to the relevant section of the multimedia book, providing students with context-sensitive help that allows them to conquer problem-solving obstacles as they arise.

A **Personal Gradebook** for each student will allow students to view their results from past assignments at any time.

Please view our online demo at **www.wiley.com/college/wileyplus.** Here you will find additional information about the features and benefits of *Wiley PLUS*, how to request a "test drive" of *Wiley PLUS* for this title, and how to adopt it for class use.

ACKNOWLEDGMENTS

CLASS TESTING AND STUDENT FEEDBACK

To make certain that *Visualizing Environmental Science* meets the needs of today's students, we asked several instructors to class test chapters. The feedback that we received from students and instructors confirmed our belief that the visualizing approach taken in this book is highly effective in helping students to learn. We wish to thank the following instructors and their students who provided us with helpful feedback and suggestions:

Hernan Aubert
Pima Community College

Keith Hench
Kirkwood Community College

Dawn Keller
Hawkeye Community College

Dale Lambert
Tarrant Community College

Janice Padula
Clinton College

Ashok Malik
Evergreen Valley College

PROFESSIONAL FEEDBACK

Throughout the process of writing and developing this text and the visual pedagogy, we benefited from the comments and constructive criticism provided by the instructors listed below. We offer our sincere appreciation to these individuals for their helpful reviews:

Mark Anderson
The University of Maine

Raymond Beiersdorfer
Youngstown State University

Richard Bowden
Allegheny College

Scott Brame
Clemson University

James A. Brenneman
University of Evansville

Huntting W. Brown
Wright State University

Michael S. Dann
Penn State University

R. Laurence Davis
Northeastern Cave Conservancy, Inc.

Brad C. Fiero
Pima Community College

Michael Freake
Lee University

Jennifer Frick-Ruppert
Brevard College

Todd G. Fritch
Northeastern University

Marcia L. Gillette
Indiana University, Kokomo

Arthur Goldsmith
Hallandale High

Cliff Gottlieb
Shasta College

Syed E. Hasan
University of Missouri-Kansas City

Dawn Keller
Hawkeye Community College

Martin G. Kelly
Buffalo State College

David Kitchen
University of Richmond

Paul Kramer
SUNY Farmingdale

Dale Lambert
Tarrant County College

Meredith Gooding Lassiter
Wiona State University

Ernesto Lasso de la Vega
Edison College

Madelyn E. Logan
North Shore Community College

Linda Lyon
Frostburg State University

Timothy F. Lyon
Ball State University

Matthew H. McConeghy
Johnson & Whales University

Rick McDaniel
Henderson State University

Leslie Nesbitt
Niagara University

Ken Nolte
Shasta College

Natalie Osterhoudt
Broward Community College

Barry Perlmutter
Community College of Southern Nevada

Thomas E. Pliske
Florida International University

Katherine Prater
Texas Wesleyan University

Sabine Rech
San Jose State University

Howie Scher
University of Rochester

Nan Schmidt
Pima Community College

Richard B. Schultz
Elmhurst College

Bo Sosnicki
Florida Community College at Jacksonville

Jerry Skinner
Keystone College

Ravi Srinivas
University of St. Thomas

David Steffy
Jacksonville State University

Andrew Suarez
University of Illinois

Charles Venuto
Brevard Community College

Margaret E. Vorndam
Colorado State University Pueblo

Maud M. Walsh
Louisiana State University

Arlene Westhoven
Ferris State University

Susan M. Whitehead
Becker College

John Wielichowski
Milwaukee Area Technical College

FOCUS GROUPS AND TELESESSION PARTICIPANTS

A number of instructors and students participated in focus groups and telesessions, providing feedback on the text, visuals, and pedagogy. Our thanks to the following instructors for their helpful comments and suggestions:

Sylvester Allred
Northern Arizona University

David Bastedo
San Bernardino Valley College

Ann Brandt-Williams
Glendale Community College

Natalie Bursztyn
Bakersfield College

Stan Celestian
Glendale Community College

O. Pauline Chow
Harrisburg Area Community College

Diana Clemens-Knott
California State University, Fullerton

Mitchell Colgan
College of Charleston

Linda Crow
Montgomery College

Smruti Desai
Cy-Fair College

Charles Dick
Pasco-Hernando Community College

Donald Glassman
Des Moines Area Community College

Mark Grobner
California State University, Stanislaus

Michael Hackett
Westchester Community College

Gale Haigh
McNeese State University

Roger Hangarter
Indiana University

Michael Harman
North Harris College

Terry Harrison
Arapahoe Community College

Javier Hasbun
University of West Georgia

Hasiotis, Stephen
University of Kansas

Adam Hayashi
Central Florida Community College

Laura Hubbard
University of California, Berkeley

James Hutcheon
Georgia Southern University

Scott Jeffrey
Community College of Baltimore County, Catonsville Campus

Matthew Kapell
Wayne State University

Arnold Karpoff
University of Louisville

Dale Lambert
Tarrant County College NE

Arthur Lee
Roane State Community College

Harvey Liftin
Broward Community College

Walter Little
University at Albany, SUNY

Mary Meiners
San Diego Miramar College

Scott Miller
Penn State University

Jane Murphy
Virginia College Online

Bethany Myers
Wichita State University

Terri Oltman
Westwood College

Keith Prufer
Wichita State University

Ann Somers
University of North Carolina, Greensboro

Donald Thieme
Georgia Perimeter College

Kip Thompson
*Ozarks Technical Community
College*

Judy Voelker
Northern Kentucky University

Arthur Washington
Florida A&M University

Stephen Williams
Glendale Community College

Feranda Williamson
Capella University

SPECIAL THANKS

We are extremely grateful to the many members of the editorial and production staff at John Wiley and Sons who guided us through the challenging steps of developing this book. Their tireless enthusiasm, professional assistance, and endless patience smoothed the path as we found our way. We thank in particular: Executive Editor Rebecca Hope, who expertly launched and directed our process; Barbara Heaney, Director of Product and Market Development, who prepared book maps for most of the chapters and provided valuable suggestions for page layout; Helen McInnis, Managing Director, Wiley Visualizing, who oversaw the concept of the book; Kelly Tavares, Production Manager, who stepped in whenever we needed expert advice; Kaye Pace, Vice President and Publisher, who oversaw the entire project; and Jeffrey Rucker, Executive Marketing Manager for Wiley Visualizing, who adeptly represents the Visualizing imprint.

We wish also to acknowledge the contributions of Vertigo Design for the interior design concept and Harry Nolan, Wiley's Creative Director who gave art direction, refined the design and other elements and the cover. We appreciate the efforts of Hilary Newman and Tara Sanford in obtaining some of our text photos, and Sandra Rigby, Senior Illustration Editor, for her expertise in managing the illustration program. We also wish to thank those who worked on the media and ancillary materials: Tom Kulesa, Senior Media Editor, for his expert work in developing the video and electronic components; and Merillat Staat, Associate Editor, for carefully guiding the completion of the ancillary materials.

We thank Mary Ann Price for her unflagging, always swift work in researching and obtaining many of our text images; Stacy Gold, Research Editor and Account Executive at the National Geographic Image Collection, for her valuable expertise in selecting National Geographic photos; and Charity Robey, who offered early input on the content and design of the text.

Many other individuals at National Geographic offered their expertise and assistance in developing this book: Francis Downey, Vice President and Publisher, and Richard Easby, Supervising Editor, National Geographic School Division; Mimi Dornack, Sales Manager, and Lori Franklin, Assistant Account Executive, National Geographic Image Collection; Dierdre Bevington-Attardi, Project Manager, and Kevin Allen, Director of Map Services, National Geographic Maps; and Devika Levy, Jim Burch, and Michael Garrity of the National Geographic Film Library. We appreciate their contributions and support.

A number of individuals at Pre-Press Company, Inc.—particularly Beth Kluckhohn, Assistant Project Manager; Abigail Greshik, Project Manager; and Gordon Laws, Development Manager—provided unfailing dedication to shaping all drafts of each chapter of our book.

CONTENTS *in Brief*

Foreword   v
Preface   viii

1   The Environmental Dilemmas We Face   2

2   Environmental Sustainability and Human Values   24

3   Environmental History, Politics, and Economics   46

4   Risk Analysis and Environmental Hazards   70

5   How Ecosystems Work   94

6   Ecosystems and Evolution   124

7   Human Population Change and the Environment   154

8   Air and Air Pollution   184

9   Global Atmospheric Changes   210

10 Freshwater Resources and Water Pollution   234

11 The Ocean and Fisheries   262

12 Environmental Geology: Mineral and Soil Resources   286

13 Environmental Geography: Land Resources   310

14 Agriculture and Food Resources   336

15 Biological Resources   360

16 Solid and Hazardous Waste: An Unrecognized Resource   384

17 Nonrenewable Energy Resources   406

18 Renewable Energy Resources   428

Glossary   453
Credits   459
Index   464

CONTENTS

1 The Environmental Dilemmas We Face 2

Introduction: A World in Crisis 2

Human Impacts on the Environment 4
  The Gap Between Rich and Poor Countries 6
  Population, Resources, and the Environment 6
    ■ ENVIRODISCOVERY: THE FLOOD OF REFUGEES 8

Sustainability and Earth's Capacity to Support Humans 12

Environmental Science 14
  The Goals of Environmental Science 14
  Science as a Process 15
    ■ WHAT A SCIENTIST SEES: THE SCIENTIFIC METHOD 16

How We Handle Environmental Problems 18
    ■ ENVIRODISCOVERY: NIMBY AND NIMTOO 20

Case Study: Tropical Rainforests Versus Agricultural Land 21

2 Environmental Sustainability and Human Values 24

Introduction: The Global Commons 24

Human Use of the Earth 26
  Sustainable Consumption 27

Human Values and Environmental Problems 29
  Worldviews 30
  Voluntary Simplicity 33

Environmental Justice 34
  Environmental Justice and Ethical Issues 35

An Overall Plan for Sustainable Living 35 
  Control Human Population Growth and Improve the Quality of Human Life 36

  Respect and Care for Earth's Biological Diversity 37
  Work on Development and Conservation at Local, National, and Global Levels 39
    ■ ENVIRODISCOVERY: RELIGION AND THE ENVIRONMENT 40
  What Kind of World Do You Want? 40

Case Study: The 2002 World Summit 42

NATIONAL GEOGRAPHIC MAPS:

Plate 1. The Political World
Plate 2. Human Population
Plate 3. Water
Plate 4. Energy
Plate 5. Environment

3 Environmental History, Politics, and Economics   46

Introduction: Old-Growth Forests of the Pacific Northwest 46

Conservation and Preservation of Resources 48

Environmental History 49
  Protecting Forests 49
  Establishing National Parks and Monuments 50
  Conservation in the Mid-20th Century 52
  The Environmental Movement 53
  ■ ENVIRODISCOVERY: ENVIRONMENTAL LITERACY 56

Environmental Legislation 57
  Environmental Regulations 58
  Accomplishments of Environmental Legislation 58

Environmental Economics 60
  National Income Accounts and the Environment 60
  An Economist's View of Pollution 63
  Economic Strategies for Pollution Control 65

Case Study: Environmental Problems in Central and Eastern Europe 66

4 Risk Analysis and Environmental Hazards   70

Introduction: Pesticides and Children 70

A Perspective on Risks 72

Environmental Health Hazards 75
  Disease-Causing Agents in the Environment 75
  Environmental Changes and Emerging Diseases 76

Movement and Fate of Toxicants 79
  Mobility in the Environment 80
  The Global Ban of Persistent Organic Pollutants 82

How We Determine the Health Effects of Pollutants 83
  Cancer-Causing Substances 84
  Risk Assessment of Chemical Mixtures 86
  Children and Chemical Exposure 86
  ■ ENVIRODISCOVERY: SMOKING: A SIGNIFICANT RISK 87

The Precautionary Principle 88

Case Study: Endocrine Disrupters 90

5 How Ecosystems Work 94

Introduction: Lake Victoria's Ecological Imbalance 94

What is Ecology? 96

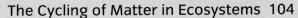

The Flow of Energy Through Ecosystems 99
- The First and Second Laws of Thermodynamics 99
- Producers, Consumers, and Decomposers 100
- The Path of Energy Flow in Ecosystems 102

The Cycling of Matter in Ecosystems 104
- The Carbon Cycle 104
- The Hydrologic Cycle 106
- The Nitrogen Cycle 107
- The Phosphorus Cycle 108

Ecological Niches 110
- ▦ WHAT A SCIENTIST SEES: RESOURCE PARTITIONING 112

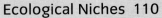

Interactions Among Organisms 113
- Symbiosis 113
- Predation 115
- Competition 118
- Keystone Species 119

Case Study: Global Warming: Is There an Imbalance in the Carbon Cycle? 120

6 Ecosystems and Evolution 124

Introduction: The Florida Everglades 124

Earth's Major Biomes 126
- Tundra 128
- Boreal Forest 129
- Temperate Rain Forest 130
- Temperate Deciduous Forest 131
- Tropical Rain Forest 132
- Chaparral 133
- ■ ENVIRODISCOVERY: USING GOATS TO FIGHT FIRES 134
- Temperate Grassland 135
- Savanna 136
- Desert 137

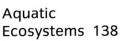

Aquatic Ecosystems 138
- Freshwater Ecosystems 138
- ▦ WHAT A SCIENTIST SEES: ZONATION IN A LARGE LAKE 139
- Freshwater Wetlands 141
- Brackish Ecosystems: Estuaries 142

Population Responses to Changing Conditions over Time: Evolution 143
- Natural Selection 144
- The Domains and Kingdoms of Life 146

Community Responses to Changing Conditions Over Time: Succession 147

Case Study: Wildfires 150

7 Human Population Change and the Environment 154

Introduction: Slowing Population Growth in China 154

Population Ecology 156
How Do Populations Change in Size? 157
Maximum Population Growth 157
Environmental Resistance and Carrying Capacity 158

Human Population Patterns 161
Projecting Future Population Numbers 162

Demographics of Countries 164
The Demographic Transition 164
Age Structure of Countries 166

Stabilizing World Population 169
Culture and Fertility 169

The Social and Economic Status of Women 171
Family Planning Services 172
■ WHAT A SCIENTIST SEES: EDUCATION AND FERTILITY 173
■ ENVIRODISCOVERY: REPRODUCTIVE RIGHTS IN DIFFERENT COUNTRIES 174
Government Policies and Fertility 174

Population and Urbanization 175
Environmental Problems of Urban Areas 176
Environmental Benefits of Urbanization 176
Urbanization Trends 177

Case Study: Urban Planning in Curitiba, Brazil 180

8 Air and Air Pollution 184

Introduction: Long-Distance Transport of Air Pollution 184

The Atmosphere 186
Atmospheric Circulation 188

Types and Sources of Air Pollution 190
Major Classes of Air Pollutants 190
■ WHAT A SCIENTIST SEES: AIR POLLUTION FROM NATURAL SOURCES 193
Sources of Outdoor Air Pollution 193

Effects of Air Pollution 194
Air Pollution and Human Health 195
Urban Air Pollution 195
How Weather And Topography Affect Air Pollution 197
Urban Heat Islands and Dust Domes 198

Controlling Air Pollutants 200
The Clean Air Act 201
Air Pollution in Developing Countries 202

Indoor Air Pollution 203
Radon 204

Case Study: Curbing Air Pollution in Chattanooga 206

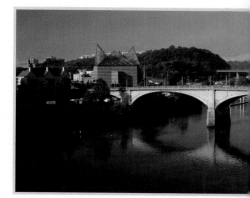

9 Global Atmospheric Changes 210

Introduction: Melting Ice and Rising Sea Levels 210

The Atmosphere and Climate 212
Solar Radiation and Climate 213
Precipitation 214
- WHAT A SCIENTIST SEES: RAIN SHADOW 215

Global Warming 216
Causes of Global Warming 217
Effects of Global Warming 219
Specific Ways to Deal with Global Warming 222

Ozone Depletion in the Stratosphere 224
Causes of Ozone Depletion 224
Effects of Ozone Depletion 224
Helping the Ozone Layer to Recover 226
- ENVIRODISCOVERY: LINKS BETWEEN CLIMATE AND ATMOSPHERIC CHANGE 227

Acid Deposition 227
How Acid Deposition Develops 228
Effects of Acid Deposition 228
The Politics of Acid Deposition 228
Facilitating Recovery from Acid Deposition 228

Case Study: International Implications of Global Warming 230

10 Freshwater Resources and Water Pollution 234

Introduction: The Missouri River: A Battle Over Water Rights 234

The Importance of Water 236
The Hydrologic Cycle and Our Supply of Fresh Water 236
Properties of Water 238

Water Resource Problems 238
Aquifer Depletion 240
Overdrawing Surface Waters 241
Salinization of Irrigated Soil 242
Global Water Issues 243

Water Management 244
Dams and Reservoirs: Managing the Columbia River 245
Water Conservation 246

Water Pollution 248
Types of Water Pollution 248
- WHAT A SCIENTIST SEES: OLIGOTROPHIC AND EUTROPHIC LAKES 250
Sources of Water Pollution 251
Groundwater Pollution 253

Improving Water Quality 254
Purification of Drinking Water 254
Municipal Sewage Treatment 255
Controlling Water Pollution 256

Case Study: Water Pollution in the Great Lakes 258

11 The Ocean and Fisheries 262

Introduction: Closing the Georges Bank Fishery 262

The Global Ocean 264
Patterns of Circulation in the Ocean 264
Ocean—Atmosphere Interaction 266

Major Ocean Life Zones 268
The Intertidal Zone: Transition between Land and Ocean 268
The Benthic Environment 270

■ ENVIRODISCOVERY: OTTERS IN TROUBLE 272
The Neritic Province: From the Shore to 200 Meters 273
The Oceanic Province: Most of the Ocean 273

Human Impacts on the Ocean 274
Marine Pollution and Deteriorating Habitat 274
World Fisheries 274
■ WHAT A SCIENTIST SEES: MODERN COMMERCIAL FISHING METHODS 276
Shipping, Ocean Dumping, and Plastic Debris 278
Coastal Development 278
Human Impacts on Coral Reefs 279
■ WHAT A SCIENTIST SEES: OCEAN WARMING AND CORAL BLEACHING 279
Offshore Extraction of Mineral and Energy Resources 280
Climate Change, Sea-Level Rise, and Warmer Temperatures 280

Addressing Ocean Problems 281
Future Actions 281

Case Study: Humans and the Antarctic Food Web 283

12 Environmental Geology: Mineral and Soil Resources 286

Introduction: Copper Basin, Tennessee 286

Plate Tectonics and Shifting Continents 288

Volcanoes 289
Earthquakes 290

Economic Geology: Useful Minerals 292
How Minerals Are Extracted and Processed 294

Environmental Implications of Mineral Use 296
Mining and the Environment 296
Environmental Impacts of Refining Minerals 296
■ ENVIRODISCOVERY: NOT-SO-PRECIOUS GOLD 297
Restoration of Mining Lands 298

Soil Properties and Processes 299
Soil Formation and Composition 299
■ WHAT A SCIENTIST SEES: SOIL PROFILE 300
Soil Organisms 301

Soil Problems and Conservation 302
Soil Erosion 303
Soil Pollution 303
Soil Conservation and Regeneration 304

Case Study: Industrial Ecosystems 307

13 Environmental Geography: Land Resources 310

Introduction: Korup National Park 310

Land Use in the United States 312

Forests 314
  Forest Management 315
  ▫ WHAT A SCIENTIST SEES: HARVESTING TREES 317
  ■ ENVIRODISCOVERY: ECOLOGICALLY CERTIFIED WOOD 318
  Deforestation 318
  Forests in the United States 322

Rangelands 323
  Rangeland Degradation and Desertification 323
  Rangeland Trends in the United States 325

National Parks and Wilderness Areas 326
  National Parks 326
  Wilderness Areas 328
  Management of Federal Lands 330

Conservation of Land Resources 331

Case Study: The Tongass Debate Over Clear-Cutting 332

14 Agriculture and Food Resources 336

Introduction: Maintaining Grain Stockpiles for Food Security 336

World Food Problems 338
  Population and World Hunger 338
  Poverty and Food 340

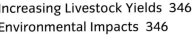

The Principle Types of Agriculture 341

Challenges of Agriculture 343
  Loss of Agricultural Land 343
  Global Decline in Domesticated Plant and Animal Varieties 344
  Increasing Crop Yields 344
  Increasing Livestock Yields 346
  Environmental Impacts 346

Solutions to Agricultural Problems 348
  Moving to Sustainable Agriculture 348
  ■ ENVIRODISCOVERY: A NEW WEAPON FOR LOCUST SWARMS 349
  Genetic Engineering: A Solution or a Problem? 350

Controlling Agricultural Pests 352
  Benefits of Pesticides 352
  Problems with Pesticides 353
  ▫ WHAT A SCIENTIST SEES: PESTICIDE USE AND NEW PEST SPECIES 354
  Alternatives to Pesticides 355
  Integrated Pest Management 356

Case Study: DDT and the American Bald Eagle 357

15 Biological Resources 360

Introduction: Disappearing Frogs 360

Species Richness and
Biological Diversity 362

How Many Species Are
There? 362

Why We Need
Biodiversity 363

Importance of Genetic
Diversity 365

Endangered and Extinct
Species 366

Endangered and
Threatened
Species 366

Areas of Declining Biological Diversity 368

■ ENVIRODISCOVERY: IS YOUR COFFEE BIRD
FRIENDLY®? 368

Earth's Biodiversity Hotspots 369

■ WHAT A SCIENTIST SEES: WHERE IS DECLINING
BIOLOGICAL DIVERSITY THE GREATEST
PROBLEM? 369

Human Causes of Species Endangerment 370

Conservation Biology 374

Protecting Habitats 375

Restoring Damaged or Destroyed Habitats 376

Conserving Species 376

Conservation Policies and Laws 378

The Endangered Species Act 378

International Conservation Policies and Laws 380

Case Study:
Reintroducing
Wolves to
Yellowstone 381

16 Solid and Hazardous Waste:
An Unrecognized Resource 384

*Introduction: Reusing and Recycling Old
Automobiles 384*

Solid Waste 386

Types of Solid Waste 386

Disposal of Solid Waste 387

□ WHAT A SCIENTIST SEES:
SANITARY LANDFILL 389

■ ENVIRODISCOVERY:
COMPOSTING FOOD
WASTE AT RIKER'S
ISLAND 392

Reducing Solid
Waste 392

Source Reduction 393

Reusing Products 393

Recycling Materials 393

Integrated Waste
Management 396

Hazardous Waste 397

Types of Hazardous
Waste 398

Managing Hazardous
Waste 400

Chemical Accidents 400

Public Policy and Toxic Waste Cleanup 400

Managing Toxic Waste Production 401

Case Study: Hanford Nuclear Reservation 403

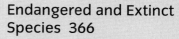

17 Nonrenewable Energy Resources   406

Introduction: Oil Prices and the World Oil Market 406

Energy Consumption 408

Coal 409
- Coal Mining 409
- Environmental Impacts of Coal 410
- Making Coal Cleaner 411

Oil and Natural Gas 412
- Reserves of Oil and Natural Gas 413
- Environmental Impacts of Oil and Natural Gas 414

Nuclear Energy 416
- Conventional Nuclear Fission 417
- Is Nuclear Energy Cleaner Than Coal? 418
- Can Nuclear Energy Decrease Our Reliance on Foreign Oil? 420

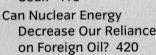
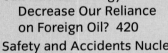

- Safety and Accidents Nuclear Power Plants 420
- The Link Between Nuclear Energy and Nuclear Weapons 421
- Radioactive Wastes 422
  - WHAT A SCIENTIST SEES: YUCCA MOUNTAIN 423
  - ENVIRODISCOVERY: A NUCLEAR WASTE NIGHTMARE 424

Decommissioning Nuclear Power Plants 424

Case Study: The Arctic National Wildlife Refuge 425

18 Renewable Energy Resources   428

Introduction: Cleaner Cars, Cleaner Fuels 428

Direct Solar Energy 430
- Heating Buildings and Water 430
- Photovoltaic Solar Cells 432
  - WHAT A SCIENTIST SEES: PHOTOVOLTAIC CELLS 432
- Solar Thermal Electric Generation 434
- Solar-Generated Hydrogen 434

Indirect Solar Energy 436
- Biomass Energy 436
- Wind Energy 438
- Hydropower 440

Other Renewable Energy Sources 442
- Geothermal Energy 442
- Tidal Energy 443

Energy Solutions: Conservation and Efficiency 444
- Energy Consumption Trends and Economics 444
- Energy-Efficient Technology 445
- Electric Power Companies and Energy Efficiency 447
- Energy Conservation at Home 448

Case Study: Meeting The Challenge: Green Architecture 449

Glossary 453

Credits 459

Index 467

VISUALIZING FEATURES

PROCESS DIAGRAMS

Multi-part visual presentations that focus on a key concept or topic in the chapter:

A series or combination of figures and photos that describe and depict a complex process:

Chapter 1
Overpopulation and Poverty
Environment Exploitation

Chapter 3
Economics and the Environment

Chapter 4
Bioaccumulation and Biomagnification

Chapter 7
Demographics of
    Countries

Chapter 8
The Atmosphere

Chapter 9
The Effects of Global
    Warming
The Ozone Layer
The Effects of Acid
    Depletion

Chapter 10
Water Conservation

Chapter 11
Ocean Currents

Chapter 12
Soil Conservation

Chapter 13
Tropical Deforestation
National Parks

Chapter 14
World Hunger
Environmental Impacts
    of Agriculture

Chapter 15
Habitat Fragmentation
Efforts to Conserve Species

Chapter 16
Recycling

Chapter 17
The *Exxon Valdez* Oil Spill

Chapter 18
Wind Energy

Chapter 1
Addressing Environmental Problems

Chapter 4
The Four Steps of Risk Assessment for Adverse
    Health Effects

Chapter 5
The Carbon Cycle
The Hydrologic Cycle
The Nitrogen Cycle
The Phosphorous Cycle

Chapter 6
Primary Succession on
Glacial Moraine
Secondary Succession on an
    Abandoned Field in
    North Carolina

Chapter 8
Coriolis Effect
Temperature, Topography, and Air Pollution

Chapter 10
Water Treatment
    for Municipal
    Use
Primary and
    Secondary
    Sewage
    Treatment

Chapter 11
Upwelling

Chapter 12
Nutrient Cycling

Chapter 13
Role of Forests in the Hydrologic Cycle

Chapter 14
Energy Inputs In Industrialized Agriculture

Chapter 16
Mass Burn, Waste-to-Energy Incinerator
Integrated Waste Management

Chapter 17
Fluidized-bed Combustion of Coal
Petroleum Refining

Chapter 18
Active Solar Water Heating

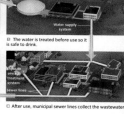

Contents   xxxi

FOR INSTRUCTORS

WileyPLUS is built around the activities you perform in your class each day. With WileyPLUS you can:

Prepare & Present

Create outstanding class presentations using a wealth of resources such as PowerPoint™ slides, image galleries, interactive simulations videos, and more. You can even add materials you have created yourself.

Create Assignments

Automate the assigning and grading of homework or quizzes by using the provided question banks, or by writing your own.

Track Student Progress

Keep track of your students' progress and analyze individual and overall class results.

Now Available with WebCT and Blackboard!

"It has been a great help, and I believe it has helped me to achieve a better grade."

Michael Morris,
Columbia Basin College

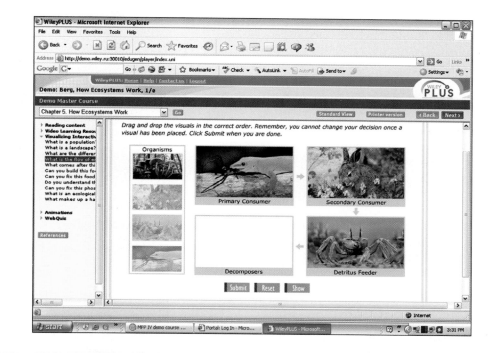

FOR STUDENTS

You have the potential to make a difference!

WileyPLUS is a powerful online system packed with features to help you make the most of your potential and get the best grade you can!

With WileyPLUS you get:

- A complete online version of your text and other study resources.

- Problem-solving help, instant grading, and feedback on your homework and quizzes.

- The ability to track your progress and grades throughout the term.

For more information on what *WileyPLUS* can do to help you and your students reach their potential, please visit www.wiley.com/college/*wileyplus*.

76% of students surveyed said it made them better prepared for tests. *

*Based on a survey of 972 student users of *WileyPLUS*

The Environmental Dilemmas We Face

A WORLD IN CRISIS

Of the millions of species inhabiting Earth, only one—*Homo sapiens*—has a truly superior intellectual capacity. This special ability has made it possible for humans to venture into space, allowing us a view of the uniqueness of our planet in the solar system.

Earth's abundant natural resources have provided the backdrop for a parade of living things to evolve. Although early Earth was inhospitable by modern standards, it provided the raw materials and energy needed for early life forms to arise and adapt. Today, millions of species inhabit the planet.

About 100,000 years ago—a mere blip in Earth's 4.5-billion-year history—an evolutionary milestone began with the appearance of modern humans in Africa. Over time, our population grew, we expanded our range throughout the planet, and we increasingly impacted the environment with our presence and our technologies. In many ways these technologies have made life better, at least for those of us who live in the United States and other highly developed nations.

At the same time, much in our world indicates that we are headed for environmental changes that could harm human well-being. Today the human species is the most significant agent of environmental change on our planet. We are overpowering the planet with our burgeoning population; transforming forests, prairies, and deserts to meet our needs and desires; and consuming ever-increasing amounts of Earth's abundant but finite resources—rich topsoil, clean water, and breathable air. We are eradicating thousands upon thousands of unique species as we destroy or alter their habitats. Evidence continues to accumulate that human-induced climate change is putting the natural environment at risk.

This book introduces the major environmental problems that humans have created and considers ways to address these issues. Most importantly, it explains why we must minimize human impact on our planet. We can't afford to ignore the environment because our lives, as well as those of future generations, depend on it. As the wise proverb says, "We have not inherited the world from our ancestors; we have only borrowed it from our children."

CHAPTER OUT

■ **Human Impacts on the Environment** p. 4

■ **Sustainability and Earth's Capaci to Support Humans** p. 12

■ **Environmental Science** p. 14

■ **How We Handle Environmental Problems** p. 18

■ **Case Study: Tropical Rain Forest Versus Agricultural Land** p. 21

Human Impacts on the Environment

www.wiley.com/
college/berg

LEARNING OBJECTIVES

Distinguish among highly developed countries, moderately developed countries, and less developed countries.

Relate human population size to natural resources and resource consumption.

Distinguish between people overpopulation and consumption overpopulation.

Describe the three factors that are most important in determining human impact on the environment and solve a problem using the IPAT equation.

The satellite photograph in **FIGURE 1.1** is a portrait of about 436 million people. The tiny specks of light represent cities, whereas the great metropolitan areas, such as New York City along the northeastern seacoast, are ablaze with light.

Earth's central environmental problem, which links all others together, is that there are many people, and the number, both in North America and worldwide, continues to grow. In 1999 the human population as a whole passed a significant milestone: 6 billion individuals. Not only is this number incomprehensibly large, but our population has grown this large in a very brief span of time. In 1960 the human population was only 3 billion (**FIGURE 1.2**). By 1975 there were 4 billion people, and by 1987 there were 5 billion. The 6.5 billion people who currently inhabit our planet consume vast quantities of food and water, use a great deal of energy and raw materials, and produce much waste.

Despite most countries' involvement with family planning, population growth rates won't change overnight. Several billion people will be added to the world in the 21st century, so even if we remain concerned about overpopulation and even if our solutions are very effective, the coming decades may very well present us with many problems.

The World Bank, which makes loans to developing countries for projects it thinks will encourage development, estimates that 2.8 billion people—about 43 percent of the world's population—are living in extreme **poverty** (**FIGURE 1.3**). By one measure, living in poverty is defined as having a per capita income of less than $2 per day, expressed in U.S. dollars ad-justed for purchasing power. Using this measure, more than half of the world's population currently lives at this level of poverty. Poverty is associated with a low life expectancy, illiteracy, and inadequate access to health services, safe water, and balanced nutrition.

The world population may stabilize toward the end of the 21st century, given the family planning efforts that are currently under way. U.N. population experts have noticed a decrease in the fertility rate worldwide to a current average of about three children per family, and the fertility rate is projected to continue to decline in coming decades. Population experts have made various projections for the world population at the end of the 21st century, from about 7.9 billion (their low projection) to 10.9 billion (their high projection), based on how fast the fertility rate decreases.

No one knows whether Earth can support so many people indefinitely. Finding ways for it to do so represents one of the greatest challenges of our times. Among the tasks to be accomplished is feeding a world population considerably larger than the present one without destroying the biological communities that support us. The quality of life available to our children and grandchildren will depend to a large extent on our ability to achieve this goal.

A factor as important as population size is a population's level of **consumption**, which is the human use of material and energy. Consumption is intimately connected to a country's **economic growth**, the expansion in output of a nation's goods and services. The world's economy is growing at an enormous rate, yet this growth is unevenly distributed across the nations of the world.

> **poverty** A condition in which people are unable to meet their basic needs for adequate food, clothing, shelter, education, or health.

Satellite view of North America at night FIGURE 1.1

This image shows most major cities and metropolitan areas in the United States, Mexico, and Canada.

Human population numbers, 1800 to present ▶
FIGURE 1.2

It took thousands of years for the human population to reach 1 billion (in 1800), 130 years to reach 2 billion (1930), 30 years to reach 3 billion (1960), 15 years to reach 4 billion (1975), 12 years to reach 5 billion (1987), and 12 years to reach 6 billion (1999).

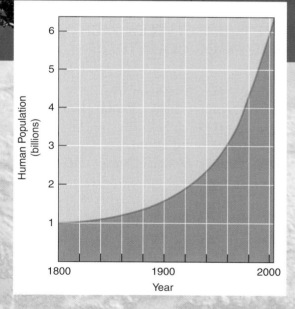

◀ Poverty FIGURE 1.3

A malnourished, unclothed Romanian girl walks in a neighborhood of poorly constructed shacks. The human population problem requires not only a stabilization of population numbers but also an improvement of the economic conditions of people living in extreme poverty.

Human Impacts on the Environment   5

Consumption of natural resources FIGURE 1.4

A A typical Japanese family, from Tokyo, Japan, with their possessions. People in highly developed countries consume a disproportionate share of natural resources.

B A typical Mexican family, from Guadalajara, with their possessions. Economic development in this moderately developed country has allowed many people to enjoy a middle-class lifestyle. Other Mexicans live in poverty, however.

THE GAP BETWEEN RICH AND POOR COUNTRIES

Generally speaking, countries are divided into rich (the "haves") and poor (the "have-nots"). Rich countries are known as **highly developed countries**. The United States, Canada, Japan, and most of Europe, which represent about 20 percent of the world's population, are highly developed countries (FIGURE 1.4A).

Poor countries, in which about 80 percent of the world's population live, fall into two subcategories: moderately developed and less developed. Mexico, Turkey, South Africa, and Thailand are examples of **moderately developed countries** (FIGURE 1.4B). **Less developed countries** (LDCs) include Bangladesh, Mali, Ethiopia, and Laos (FIGURE 1.4C). Cheap, unskilled labor is abundant in LDCs but capital for investment is scarce. Most economies of LDCs are agriculturally based, often on only one or a few crops. As a result, crop failure or a lower world market value for that crop is catastrophic to the economy. Hunger, disease, and illiteracy are common in LDCs.

highly developed countries Countries with complex industrialized bases, low rates of population growth, and high per capita incomes.

moderately developed countries Developing countries with a medium level of industrialization and per capita incomes lower than those of highly developed countries.

POPULATION, RESOURCES, AND THE ENVIRONMENT

Inhabitants of the United States and other highly developed countries consume many more resources per person than do citizens of developing countries. This high rate of resource consumption affects the environment at least as much as the explosion in population that is occurring in other parts of the world.

We can make two useful generalizations about the relationships among population growth, consumption of natural resources, and environmental degradation. One, the amount of resources essential to an individual's survival is small, but rapid population growth (often found in developing countries) tends to overwhelm and deplete a country's soils, forests, and other

NATIONAL GEOGRAPHIC

C A typical family from Kouakourou, Mali, with all their possessions. The rapidly increasing number of people in less developed countries overwhelms their natural resources, even though individual resource requirements may be low.

less developed countries Developing countries with a low level of industrialization, a very high fertility rate, a very high infant mortality rate, and a very low per capita income relative to highly developed countries.

natural resources. Two, in highly developed nations, individual demands on natural resources are far greater than the requirements for mere survival. To satisfy their desires as well as their basic needs, many people in more affluent nations deplete resources and degrade the global environment through increased consumption of nonessential items like televisions, jet skis, and cellular phones.

Types of resources

When examining the effects of population on the environment, it is important to distinguish between nonrenewable and renewable natural resources. **Nonrenewable resources** include minerals (such as aluminum, tin, and copper) and fossil fuels (coal, oil, and natural gas). Natural processes do not replenish nonrenewable resources within a reasonable duration on the human timescale. Fossil fuels, for example, take millions of years to form.

In addition to a nation's population and its level of resource use, several other factors affect how nonrenewable resources are used—including how efficiently the resource is extracted and processed and how much of it is required or consumed. Nonetheless, the inescapable fact is that Earth has a finite supply of nonrenewable resources that sooner or later will be exhausted. In time, technological advances may help find or develop substitutes for nonrenewable resources. Slowing the rate of population growth and resource consumption will help us buy time to develop such alternatives.

Some examples of **renewable resources** are trees, fishes, fertile agricultural soil, and fresh water. Nature replaces these resources fairly rapidly, on a scale of days to decades. Forests, fisheries, and agricultural land are particularly important renewable resources in developing countries because they provide food. Indeed, many people in developing countries are subsistence farmers who harvest just enough food for their families to survive.

Rapid population growth can cause renewable resources to be overexploited. For example, large numbers of poor people must grow crops on land—such as mountain slopes or tropical rain forests—that is inappropriate for farming. Although this practice may provide a short-term solution to the need for food, it does not work in the long run, because when these lands are cleared for farming, their agricultural productivity declines rapidly and severe environmental deterioration occurs. Renewable resources, then, are *potentially* renewable. They must be used in a manner that gives them time to replace or replenish themselves.

The effects of population growth on natural resources are particularly critical in developing countries. The economic growth of developing countries is

nonrenewable resources Natural resources that are present in limited supplies and are depleted as they are used.

renewable resources Resources that are replaced by natural processes and that therefore can be used forever, provided they are not overexploited in the short term.

Sometimes economic, political, or environmental problems force people to leave their homelands *en masse*. According to the United Nations, in 2003 there were 9.7 million international refugees—people who moved from one country to another. This number does not include internally displaced persons—in 2003, about 24 million people who were forced to move but did not cross a border into another country. For example, 5 million Sudanese were internally displaced in 2003.

Environmental degradation is recognized as a major cause of refugee flight. For the past several decades, deterioration of agricultural lands, coupled with extended droughts and other natural disasters, has caused the displacement of perhaps 30 million people worldwide. The large number of refugees strains the economies and natural resources of the regions into which the refugees migrate.

Many people attempting to flee their homelands are not welcomed with open arms. The industrial world's reluctance to embrace refugees stems from several sources:

- Economic problems at home.
- The high costs of processing asylum applications.
- Skepticism over whether refugees are actually fleeing political persecution and are not simply "economic migrants."

Failure to address the high demand for asylum leads to a lower quality of life for prospective refugees:

- Refugee camps are growing more violent, with sexual and gender-based violence against women refugees particularly pronounced.
- Human trafficking (charging asylum-seekers money to illegally transport them across borders) is on the rise, with its accompanying financial exploitation, human rights abuses, and violence.
- Less developed nations take their cue from highly developed ones, refusing to accept refugees when more prosperous nations will not.

Refugees

This photograph, taken in 2000, shows the Moqol Qeshlaq Refugee Camp in Afghanistan.

frequently tied to the exploitation of their natural resources, often for export to highly developed countries. Developing countries are faced with the difficult choice of exploiting natural resources to provide for their expanding populations in the short term (that is, to pay for food or to cover debts), or conserving those resources for future generations. It is instructive to note that the economic growth and development of the United States and of other highly developed nations came about through the exploitation—and in some cases the destruction—of their resources. Continued growth and development in highly developed countries now relies significantly on the importation of these resources from less developed countries.

Poverty is tied to the effects of population pressures on natural resources and the environment. Poor people in developing countries find themselves trapped in a vicious cycle of poverty. They use environmental resources unwisely for short-term gain (that is, to survive), but this exploitation degrades the resources and diminishes long-term prospects of economic development.

Population size and resource consumption

Resource issues are clearly related to population size—more people use more resources. An equally important factor is a population's resource *consumption*. Consumption is both an economic and a social act. Consumption provides the consumer with a sense of identity as well as status among peers. The media, including the advertising industry, promote consumption as a way to achieve happiness. We are encouraged to spend, to consume.

People in highly developed countries are extravagant and wasteful consumers; their use of resources is greatly out of proportion to their numbers. A single child born in a highly developed country such as the United States causes a greater impact on the environment and on resource depletion than perhaps 20 children born in a developing country. Many natural resources are needed to provide the automobiles, air conditioners, disposable diapers, cell phones, DVD players, computers, clothes, newspapers, athletic shoes, furniture, boats, and other "comforts" of life in highly developed nations. Thus, the disproportionately large consumption of re-

sources by the United States and other highly developed countries affects natural resources and the environment as much as or more than the population explosion in the developing world.

People overpopulation and consumption overpopulation
A country is overpopulated if the level of demand on its resource base results in damage to the environment. In comparing human impact on the environment in developing and highly developed countries, we see that a country can be overpopulated in two ways. **People overpopulation** occurs when the environment is worsening because there are too many people, even if those people consume few resources per person. People overpopulation is the current problem in many developing nations. In contrast, **consumption overpopulation** results from the consumption-oriented lifestyles popular in highly developed countries. The effects of consumption overpopulation on the environment are the same as those of people overpopulation—pollution, resource depletion, and degradation of the environment.

Many affluent, highly developed nations, including the United States, suffer from consumption overpopulation: Highly developed nations represent only 20 percent of the world's population, yet they consume significantly more than half of its resources. According to the Worldwatch Institute, highly developed nations account for the lion's share of total resources consumed:

- 86 percent of aluminum used
- 76 percent of timber harvested
- 68 percent of energy produced
- 61 percent of meat eaten
- 42 percent of the fresh water consumed

> ■ **people overpopulation** A situation in which there are too many people in a given geographic area.
>
> ■ **consumption overpopulation** A situation that occurs when each individual in a population consumes too large a share of resources.

These nations also generate 75 percent of the world's pollution and waste (FIGURE 1.5).

Mathis Wackernagel and colleagues at the Universidad Anáhuac de Xalapa in Mexico compared the ecological impact of individuals living in 52 large nations. In this study, each person has an **ecological footprint**, which is the average amount of productive land and ocean needed to supply that person with food, energy, water, housing, transportation, and waste disposal. In developing nations, such as India and Nigeria, the ecological footprint is about 1 hectare (2.5 acres). In the United States, the ecological footprint is about 9.6 hectares (23.7 acres). If all 6.5 billion people in the world had the same lifestyle and level of consumption as an average American, and assuming no changes in technology, we would need four additional planets the size of Earth (FIGURE 1.6).

New Consumers in Developing Countries   As developing nations increase their economic growth and improve their standard of living, more and more people in those countries purchase consumer goods. By the early 2000s, more new cars were sold annually in Asia than in North America and Western Europe combined. These new consumers may not consume at the high level of the average consumer in a highly developed nation, but their consumption has increasingly adverse effects on the environment. For example, air pollution from traffic in urban centers in developing countries is bad and getting worse every year. Millions

of dollars are lost to health problems caused by air pollution in these cities.

Population, Consumption, and Environmental Impact   When you turn on the tap to brush your teeth in the morning, you probably do not think about where the water comes from or about the environmental consequences of removing it from a river or the ground. Likewise, most North Americans do not think about where the energy comes from when they flip on a light switch or start their car, van, or truck. Many of us do not realize that all of the materials that make up the products we use every day come from Earth, nor do we grasp that these materials eventually are returned to Earth, much of it in sanitary landfills.

Such human impacts on the environment are difficult to assess. One way to estimate them is to use the three factors most important in determining environmental impact (I):

- The number of people (P).

- The affluence per person, which is a measure of the consumption or amount of resources used per person (A).

- The environmental effects (resources needed and wastes produced) of the technologies used to obtain and consume the resources (T).

This method of assessment is usually referred to as an equation: $I = P \times A \times T$.

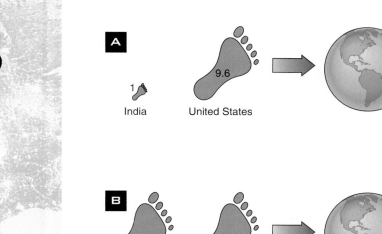

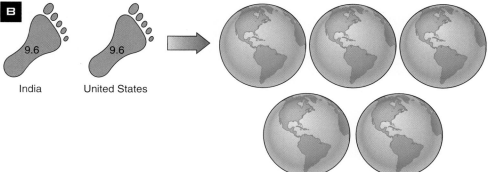

Ecological footprints FIGURE 1.6

A In LDCs such as India, about 1 hectare meets the resource requirements of an average person; the ecological footprint of each individual in a highly developed country such as the United States is almost 10 hectares.

B If everyone in the world had the same level of consumption as an average American, it would take the resources and land area of five Earths.

Biologist Paul R. Ehrlich and physicist John P. Holdren first proposed the IPAT model in the 1970s. It shows the mathematical relationship between environmental impacts and the forces that drive them. To determine the environmental impact of carbon dioxide (CO_2) emissions from motor vehicles, for example, multiply the population times the number of cars per person (affluence/consumption per person) times the average annual CO_2 emissions per year (technological impact). This model demonstrates that although improving motor vehicle efficiency and developing cleaner technologies will reduce pollution and environmental degradation, a larger reduction will result if population and per capita consumption are also controlled.

Although useful, the IPAT equation must be interpreted with care, in part because we often do not understand all of the environmental impacts of a particular technology. Motor vehicles, for example, are linked not only to global warming from CO_2 emissions but also to air pollution (tailpipe exhaust), water pollution (improperly disposed motor oil and antifreeze), stratospheric ozone depletion (from leakage of air conditioner coolants), and solid waste (disposal of automobiles in sanitary landfills). There are currently about 550 million cars on the planet, and the number is rising.

The three factors in the IPAT equation are always changing in relation to each other. For example, consumption of a particular resource may increase, but technological advances may decrease the environmental impact of the increased consumption. Consumer trends and choices may affect environmental impact. The average fuel economy of new cars and light trucks (sport utility vehicles, vans, and pickup trucks) in the United States has declined since 1988, in part because of the popularity of sport utility vehicles (SUVs). However, the projected increase in the sale of fuel-efficient hybrid cars may offset this decline in the next decade or so. In addition to being about 29 percent less fuel efficient than cars, SUVs release between 30 percent and 75 percent more emissions. Because of such trends and uncertainties, the IPAT equation is of limited usefulness when making long-term predictions.

The equation is valuable in that it helps to identify what we don't know or understand about consumption and its environmental impact. For example, which kinds of consumption have the greatest destructive impact on the environment? Which groups in society are responsible for the greatest environmental disruption? How can we alter the activities of these environmentally disruptive groups? It will take years to address such questions, but the answers should help decision-makers in business and government formulate policies that will alter consumption patterns in an environmentally responsible way. The ultimate goal should be to make consumption sustainable so that humanity's current practices do not compromise the ability of future generations to use and enjoy the riches of our planet.

To summarize, as human numbers and consumption increase worldwide, so does humanity's impact on Earth, posing new challenges to us all. Success in achieving sustainability in population size and consumption will require the cooperation of all the world's peoples.

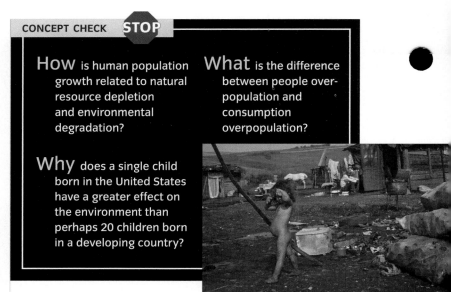

CONCEPT CHECK STOP

How is human population growth related to natural resource depletion and environmental degradation?

What is the difference between people overpopulation and consumption overpopulation?

Why does a single child born in the United States have a greater effect on the environment than perhaps 20 children born in a developing country?

Sustainability and Earth's Capacity to Support Humans

LEARNING OBJECTIVES

Define environmental sustainability.

Identify human behaviors that threaten environmental sustainability.

One of the most important concepts in this text is **environmental sustainability**. *Sustainability* implies that the environment will function indefinitely without going into a decline from the stresses that human society imposes on natural systems (such as fertile soil, water, and air) that maintain life. Environmental sustainability applies to many levels, including the individual, communal, regional, national, and global.

Environmental sustainability is based in part on the following ideas:

- We must consider the effects of our actions on the health and well-being of the natural environment, including all living things.

- Earth's resources are not present in infinite supply. We must live within limits that let renewable resources such as fresh water regenerate for future needs.

- We must understand all the costs to the environment and to society of products we consume.

- We must each share in the responsibility for environmental sustainability.

environmental sustainability
The ability to meet humanity's current needs without compromising the ability of future generations to meet their needs.

Many environmental experts think that human society is not operating sustainably because of the following human behaviors:

A **Bluefin tuna are overexploited in certain parts of the ocean.**

B **A clear-cut forest in the Gifford Pinchot National Forest in Washington State. Forest harvest is unsustainable in many parts of the world.**

C **An oil refinery produces noxious emissions. Air pollution control devices reduce these toxic emissions but often are not used because they are expensive. Photographed in Baton Rouge, Louisiana.**

- We are using nonrenewable resources such as fossil fuels as if they were present in unlimited supplies.

- We are using renewable resources such as fresh water and forests faster than they are replenished naturally (FIGURE 1.7A, B).

- We are polluting the environment—the land, rivers, ocean, and atmosphere—with toxins as if the capacity of the environment to absorb them were limitless (FIGURE 1.7C).

- Our numbers continue to grow despite Earth's finite ability to feed and sustain us and to absorb our wastes.

If left unchecked, these activities may threaten the life-support systems of Earth to the extent that recovery is impossible. Our first goal should be to critically evaluate which changes our society is willing to make.

At first glance, the issues may seem simple. Why don't we just stop the overconsumption, population growth, and pollution? The solutions are more complex than they may initially seem, in part because of various interacting ecological, societal, and economic factors.

Our inadequate scientific understanding of how the environment works and how different human choices affect the environment is a major reason that problems of environmental sustainability are difficult to resolve. Even for established environmental problems, political and social controversy often prevents widespread acceptance of the threats' reality. Because

the effects of many interactions between the environment and humans are unknown or difficult to predict, we generally don't know if corrective actions should be taken before our scientific understanding is more complete.

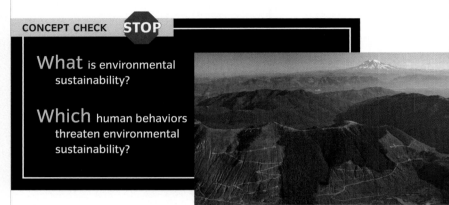

CONCEPT CHECK **STOP**

What is environmental sustainability?

Which human behaviors threaten environmental sustainability?

Environmental Science

Environmental science is an interdisciplinary field that combines information from many disciplines, such as biology, geography, chemistry, geology, physics, economics, sociology (particularly *demography*, the study of populations), cultural anthropology, natural resource management, agriculture, engineering, law, politics, and ethics.

Ecology, the discipline of biology that studies the interrelationships between organisms and their environment, is a basic tool of environmental science. Environmental scientists examine the complex relationships among human populations, their use of natural resources, and their production of pollution.

THE GOALS OF ENVIRONMENTAL SCIENCE

Environmental scientists try to establish general principles about how the natural world functions. They use these principles to develop viable solutions to environmental problems—solutions that are based as much as possible on scientific knowledge.

Environmental problems are generally complex, however, and scientific understanding of them is often less complete than we would like. Environmental scientists are often called on to reach a scientific consensus before the data are complete. As a result, they often make recommendations based on probabilities rather than precise answers.

environmental science The interdisciplinary study of humanity's relationship with other organisms and the non-living physical environment.

Many of the environmental problems considered in this book are serious ones that we must address. Yet environmental science is not simply a "doom and gloom" listing of problems, coupled with predictions of a bleak future. To the contrary, the focus of environmental science, and our focus as individuals and as world citizens, is on identifying, understanding, and solving problems that we as a society have generated. It is encouraging that individuals, businesses, and governments are already doing a great deal, although more must be done to address the problems of today's world.

SCIENCE AS A PROCESS

The key to successfully solving any environmental problem is rigorous scientific evaluation. It is important to understand clearly just what science is, as well as what it is not. Most people think of **science** as a body of knowledge—a collection of facts about the natural world. However, science is also a dynamic *process,* a systematic way to investigate the natural world. Science seeks to reduce the apparent complexity of our world to general principles, which are then used to make predictions, solve problems, or provide new insights.

Scientists collect objective **data** (singular, *datum*), the information with which science works (FIGURE 1.8). Data are collected by observation and experimentation and then analyzed or interpreted. Scientific conclusions are inferred from the available data and are not based on faith, emotion, or intuition. Scientists publish their findings in scientific journals, and other scientists examine and critique their work. A requirement of science is repeatability—that is, observations and experiments must produce consistent data when they are repeated. The scrutiny by other scientists reveals any inconsistencies in results or interpretation. These errors are discussed openly, and ways to eliminate them are developed.

There is no absolute certainty or universal agreement about anything in science. Science is an ongoing enterprise, and scientific concepts must be reevaluated in light of newly discovered data. Thus, scientists never claim to know the final answer about anything, because scientific understanding changes.

Data collection FIGURE 1.8

A biologist collects data from an experiment on the uptake of carbon dioxide (CO_2) by plants during photosynthesis. This research will contribute to our expanding knowledge of CO_2, a gas that is increasing in the atmosphere, primarily from the combustion of coal, oil, and natural gas. The buildup of CO_2 in the atmosphere is linked to climate warming. Photographed in the Center of Alpine Ecology in Trento, Italy.

Several areas of human endeavor are not scientific. Ethical principles often have a religious foundation, and political principles reflect social systems. Some general principles, however, derive not from religion or politics but from the physical world around us. If you drop an apple, it will fall, whether or not you wish it to, and despite any laws you may pass forbidding it to do so. Science aims to discover and better understand the general principles that govern the operation of the natural world.

The Scientific Method

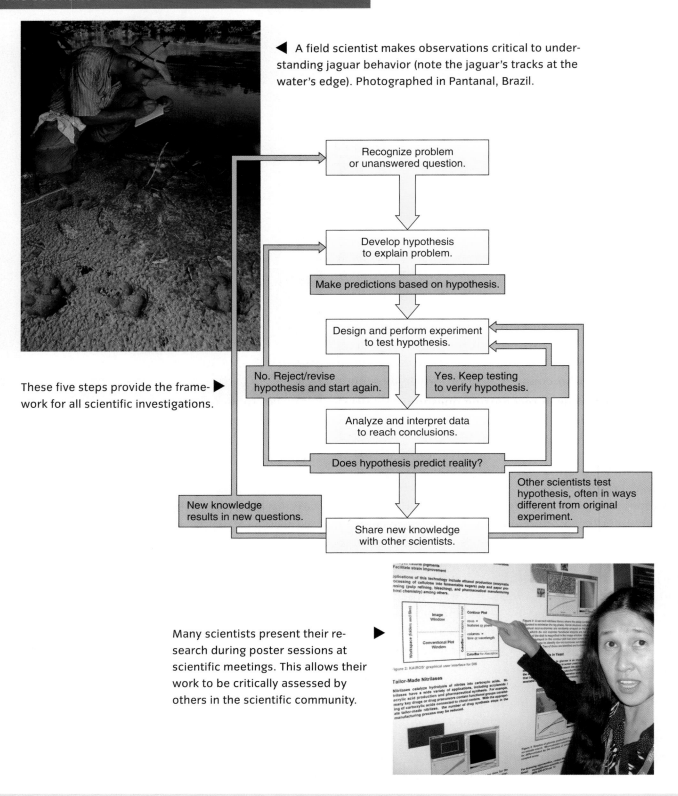

◀ A field scientist makes observations critical to understanding jaguar behavior (note the jaguar's tracks at the water's edge). Photographed in Pantanal, Brazil.

Recognize problem or unanswered question.

Develop hypothesis to explain problem.

Make predictions based on hypothesis.

Design and perform experiment to test hypothesis.

No. Reject/revise hypothesis and start again.

Yes. Keep testing to verify hypothesis.

These five steps provide the framework for all scientific investigations. ▶

Analyze and interpret data to reach conclusions.

Does hypothesis predict reality?

New knowledge results in new questions.

Other scientists test hypothesis, often in ways different from original experiment.

Share new knowledge with other scientists.

Many scientists present their research during poster sessions at scientific meetings. This allows their work to be critically assessed by others in the scientific community. ▶

The scientific method The established processes that scientists use to answer questions or solve problems are collectively called the **scientific method**. Although there are many variations of the scientific method, it basically involves five steps:

1. Recognize a question or unexplained occurrence in the natural world.

2. Develop a *hypothesis,* or educated guess, to explain the problem.

3. Design and perform an experiment to test the hypothesis.

4. Analyze and interpret the data to reach a conclusion.

5. Share new knowledge with the scientific community.

Although the scientific method is often portrayed as a linear sequence of events, science is rarely as straightforward or tidy as the scientific method implies (See "What a Scientist Sees"). Good science involves creativity, not only in recognizing questions and developing hypotheses but also in designing experiments. Because scientists try to expand our current knowledge, their work is in the realm of the unknown. Many creative ideas end up as dead ends, and there are often temporary setbacks or reversals of direction as scientific knowledge progresses. Scientific knowledge often expands haphazardly, with the "big picture" emerging slowly from confusing and sometimes contradictory details.

Scientific discoveries are often incorrectly portrayed in the media as "new facts" that have just come to light. At a later time, additional "new facts" that question the validity of the original study are reported. If you were to read the scientific papers on which such media reports are based, however, you would find that all the scientists involved probably made very tentative conclusions based on their data. Science progresses from uncertainty to less uncertainty, not from certainty to greater certainty. Thus, science is self-correcting over time, despite the fact that it never "proves" anything.

The importance of prediction A scientific **hypothesis** needs to be useful—it needs to tell you something you want to know. A hypothesis is most useful when it makes predictions, because the predictions provide a way to test the validity of the hypothesis. If your experiment refutes your prediction, then you must carefully recheck the entire experiment. If the prediction is still refuted, then you must reject the hypothesis. The more verifiable predictions a hypothesis makes, the more valid that hypothesis is.

Each of the many factors that influence a process is called a **variable**. To evaluate alternative hypotheses about a specific variable, it is necessary to hold all other variables constant so that they do not mislead or confuse us. To test a hypothesis about a variable, we carry out two forms of the experiment in parallel. In the **experimental group** the chosen variable is altered in a known way. In the **control group** that variable isn't altered. In all other respects the experimental group and the control group are the same. We then ask, "What is the difference, if any, between the outcomes for the two groups?" Any difference must be due to the influence of the variable we changed, because all other variables remained the same. Much of the challenge of science lies in designing control groups and in successfully isolating a single variable from all other variables.

Theories A **theory** is an integrated explanation of numerous hypotheses, each of which a large body of observations and experiments supports. A theory condenses and simplifies many data that previously appeared unrelated. Because a theory demonstrates the relationships among different data, it simplifies and clarifies our understanding of the natural world. A good theory grows as additional information becomes known. It predicts new data and suggests new relationships among a range of natural phenomena.

Theories are the solid ground of science, the explanations of which we are most sure. This definition contrasts sharply with the general public's use of the

> **scientific method** The way a scientist approaches a problem, by formulating a hypothesis and then testing it by means of an experiment.

word *theory,* implying lack of knowledge, or a guess. In this book, the word *theory* is always used in its scientific sense, to refer to a broadly conceived, logically coherent, and well-supported explanation.

Despite the fact that theories are generally accepted, there is no absolute truth in science, only varying degrees of uncertainty. Science is continually evolving as new evidence comes to light, and therefore its conclusions are always provisional or uncertain. It is always possible that the results of a future experiment will contradict a prevailing theory and show at least one aspect of it to be false.

Uncertainty, however, does not mean that scientific conclusions are invalid. For example, overwhelming evidence links cigarette smoking and incidence of lung cancer. We can't state with absolute certainty that every smoker will be diagnosed with lung cancer, but this uncertainty does not mean that there is no correlation between smoking and lung cancer. On the basis of the available evidence, we say that people who smoke have an increased risk of developing lung cancer.

In conclusion, the aim of science is to increase human comprehension by explaining the processes and events of nature. Scientists work under the assumption that all phenomena in the natural world have natural causes, and they formulate theories to explain these phenomena. The process of science as a human endeavor has shaped the world we live in and transformed our views of the universe and how it works.

CONCEPT CHECK STOP

What is environmental science? What are some of the disciplines involved in environmental science?

What are the five steps of the scientific method? Why is each important?

How We Handle Environmental Problems

Before examining the environmental problems discussed in the remaining chapters of this text, let's consider the elements that contribute to solving those problems. How, for example, can we handle water pollution in a river (**FIGURE 1.9**)? At what point are conclusions regarded as certain enough to warrant action? Who makes the decisions, and what are the trade-offs?

Monitoring water pollution FIGURE 1.9

This pollution control officer is measuring the oxygen level in the Severn River near Shrewsbury, England. When dissolved oxygen levels are high, pollution levels (of sewage, fertilizer, and such) are low.

Addressing environmental problems FIGURE 1.10

These five steps provide a framework for addressing environmental problems.

❶ Scientific Assessment: The problem is defined, hypotheses are tested, and models are constructed to show how the present situation developed and to predict the future course of events.

Scientists find higher-than-normal levels of bacteria are threatening a lake's native fish and determine the cause is human-produced pollution.

❷ Risk Analysis: The potential effects of various interventions—including doing nothing—are analyzed to determine the risks associated with each particular course of action.

If no action is taken, fishing resources—a major source of income in the region—will be harmed. If the pollution is reduced appreciably, the fishery will recover.

❺ Long-Term Evaluation: The results of any action taken should be carefully monitored to see if the environmental problem is being addressed.

Water quality in the lake is tested frequently, and the fish populations are monitored to ensure they do not decline.

❸ Public Education and Involvement: Changing public attitudes involves explaining the problem, presenting available alternatives for action, and revealing the probable risks, results, and the cost of each choice.

The public is informed of the ramifications—in this case, loss of income—if this problem is not addressed.

❹ Political Action: Elected officials, often at the urging of their constituencies, implement a course of action based on scientific evidence as well as economic and social considerations.

Elected officials, supported by the public, pass legislation to protect the lake and develop a lake cleanup plan.

Viewed simply, there are five stages in addressing an environmental problem (**FIGURE 1.10**). They are:

1. Scientific assessment.
2. Risk analysis.
3. Public education and involvement.
4. Political action.
5. Evaluation.

These five stages represent an ideal approach to systematically addressing environmental problems. In real life, seeking solutions to environmental problems is rarely so neat and tidy, particularly when the problem is exceedingly complex, of regional or global scale, or has high costs and unclear benefits for the money invested. Quite often, the public becomes aware of a problem and triggers discussion of remediation before the problem is clearly identified and scientifically assessed.

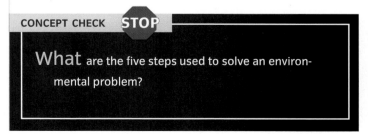

CONCEPT CHECK STOP

What are the five steps used to solve an environmental problem?

The NIMBY response is an unavoidable complication in addressing many environmental issues. NIMBY stands for "not in my backyard." As soon as people hear that a power plant, incinerator, or hazardous waste disposal site may be situated nearby, the NIMBY response rears its head. Part of the reason NIMBYism is so prevalent is that, despite the assurances experts give that a site will be safe, no one can guarantee complete safety, with no possibility of an accident.

A sister response to NIMBY is NIMTOO, which stands for "not in my term of office." Politicians who wish to get reelected are sensitive to their constituents' concerns and are not likely to support the construction of power plants or waste disposal sites in their districts.

Given human nature, the NIMBY and NIMTOO responses are not surprising. But emotional reactions, however reasonable they may be, do little to constructively solve complex environmental problems. Consider the disposal of radioactive waste from nuclear power plants. There is universal agreement that we need to safely isolate radioactive waste until it decays enough to cause little danger. But NIMBY and NIMTOO, with their associated demonstrations, lawsuits, and administrative hearings, prevent us from effectively dealing with radioactive waste disposal. Every potential disposal site is near someone's home, in some politician's state. Most people agree that our generation has the responsibility to dispose of hazardous waste. Only we want to put it in some other state, in someone else's backyard. Arguing against any disposal scheme that is proposed will simply result in letting the waste remain where it is now. Although this may be the only politically acceptable solution, it is unacceptable from a safety viewpoint.

Steam rises from the cooling towers of a nuclear power plant. Many nuclear power plants store highly radioactive spent fuel on site, because there is currently no place to safely dispose of it.

Tropical Rain Forests Versus Agricultural Land

This fire in a tropical rain forest in the Amazon Basin of Brazil was set deliberately to clear the forest and produce grazing land. Many developing countries clear forests to produce pasture or agricultural land. In addition, forests are cleared in many parts of the world to meet the demand for timber or to provide firewood. (The poor in many developing countries have no alternative to wood for fuel.)

What are some of the environmental consequences of this particular forest being destroyed? Most obvious is the loss of hundreds of species of plants, animals, and microorganisms that lived in the forest. The worldwide decrease in biological diversity is of great concern to scientists today.

Slash-and-burn agriculture, in which the forest is cut down, allowed to dry, and burned for cropland, also has adverse effects outside of the forest. The living trees no longer hold the forest soil in place, and so the soil is washed into surrounding streams, filling them with sediment.

The burning of the rain forest may affect the quality and flow of drinking water for people living far from the forest. **Watersheds** are areas that river systems drain. In a forest-covered watershed, the soil absorbs more water, which effectively filters out many contaminants. The forest regulates the flow of this water as it drains into rivers and their tributaries. With the clearing of the forest, most rainfall quickly washes into rivers without the benefits of filtration and flow regulation.

Rainforest clearing may contribute to global climate warming. Trees and other forest plants absorb CO_2 from the atmosphere and incorporate it into the carbon compounds found in wood and other plant tissues. The combustion of the wood, however, releases this stored CO_2 into the atmosphere.

Clearing of the forest reduces the long-term potential for people living in the area to earn livelihoods from the forest. If harvested using sustainable methods, tropical forests provide much more than timber, including latex for rubber, Brazil nuts, chicle for chewing gum, turpentine, and spices. The forest could be developed as a site for *ecotourism,* in which tourists pay to observe wildlife in natural settings. Sustainable timber practices and the promotion of nontimber products, including ecotourism, may provide people with more income than they would earn if the forest were cleared to grow crops or raise cattle.

Fire in a tropical rain forest.

Ranchers set this fire to clear land for cattle pasture. Photographed in Rondonia State, Brazil.

CHAPTER SUMMARY

1 Human Impacts on the Environment

1. **Highly developed countries** are countries that have complex industrialized bases, low rates of population growth, and high per capita incomes. **Moderately developed countries** are developing countries that have a medium level of industrialization and average per capita incomes lower than those of highly developed countries. **Less developed countries (LDCs)** are developing countries with a low level of industrialization, a very high fertility rate, a very high infant mortality rate, and a very low per capita income (relative to highly developed countries). **Poverty**, which is common in LDCs, is a condition in which people are unable to meet their basic needs for adequate food, clothing, shelter, education, or health.

2. The increasing global population is placing stresses on the environment, as humans consume ever-increasing quantities of food and water, use more and more energy and raw materials, and produce enormous amounts of waste and pollution. **Nonrenewable resources** are natural resources that are present in limited supplies and are depleted as they are used. **Renewable resources** are resources that natural processes replace and that therefore can be used forever, provided they are not exploited in the short term.

3. **People overpopulation** is a situation in which too many people live in a given geographic area. Developing countries have people overpopulation. **Consumption overpopulation** is a situation that occurs when each individual in a population consumes too large a share of resources. Highly developed countries have consumption overpopulation.

4. Environmental impact and the forces that drive it can be modeled by the mathematical equation, $I = P \times A \times T$. Environmental impact (I) has three factors: the number of people (P); the affluence per person (A), which is a measure of the consumption or amount of resources used per person; and the environmental effect of the technologies used to obtain and consume those resources (T).

2 Sustainability and Earth's Capacity to Support Humans

1. **Environmental sustainability** is the ability to meet humanity's current needs without compromising the ability of future generations to meet their needs. Sustainability implies that the environment can function indefinitely without going into a decline from the stresses that human society imposes on natural systems.

2. Human behaviors that threaten environmental sustainability include overuse of renewable and nonrenewable resources, pollution, and overpopulation.

3 Environmental Science

1. **Environmental science** is the interdisciplinary study of humanity's relationship with other organisms and the nonliving physical environment. Environmental science encompasses many problems involving human numbers, Earth's natural resources, and environmental pollution.

2. The **scientific method** is the way a scientist approaches a problem, by formulating a **hypothesis** and then testing it by means of an experiment. (1) A scientist recognizes and states the problem or unanswered question. (2) The scientist develops a hypothesis, or educated guess, to explain the problem. (3) An experiment is designed and performed to test the hypothesis. (4) **Data**, the results obtained from the experiment, are analyzed and interpreted to reach a conclusion. (5) The conclusion is shared with the scientific community.

4 How We Handle Environmental Problems

1. Addressing environmental problems ideally requires five stages. (1) Scientific assessment involves identifying a potential environmental problem and collecting data to construct a model. (2) Risk analysis evaluates the potential effects of intervention. (3) Public education and involvement occur when the results of scientific assessment and risk analysis are placed in the public arena. (4) In political action, elected or appointed officials implement a particular risk-management strategy. (5) Evaluation monitors the effects of the action that was taken.

- **poverty** p. 4
- **highly developed countries** p. 6
- **moderately developed countries** p. 6
- **less developed countries** p. 7

- **nonrenewable resources** p. 7
- **renewable resources** p. 7
- **people overpopulation** p. 9
- **consumption overpopulation** p. 9

- **environmental sustainability** p. 12
- **environmental science** p. 14
- **scientific method** p. 17

CRITICAL AND CREATIVE THINKING QUESTIONS

1. Criticize the following statement: "Population growth in developing countries is of much more concern than is population growth in highly developed countries."

2. Use the IPAT equation to calculate the environmental impact in terms of CO_2 emissions per year at the beginning of the 21st century, when there were 6 billion people, an average of 0.1 motor vehicles per person, and 5.4 tons of CO_2 emitted by each car per year. Then make a similar calculation for the year 2050, based on these projections: a population of 10 billion people, 0.4 cars per person, and CO_2 emissions per vehicle similar to what we have today (that is, no technological improvements). How might we hold global CO_2 emissions from motor vehicles to 2000 levels in the year 2050?

3. Give at least two examples of things that you can do as an individual to promote environmental sustainability.

4. People want scientists to give them precise, definitive answers to environmental problems. Explain why this is not possible.

5. Make a three-column list of the material items you own or would like to own. Column 1 consists of items you have that meet your basic needs; column 2 contains items you have that you wanted but didn't need; and column 3 contains items you would like to possess in the near future. Which of the items in columns 2 and 3 would you be willing to give up to reduce your personal impact on the environment?

6. Write a brief paragraph describing two or three ways your personal quality of life could be improved. Do any of these involve material items? If so, explain how these items would improve your life.

7–9. Your throat feels scratchy and you think you're coming down with a cold. You take a couple of vitamin C pills and feel better. You conclude that vitamin C helps prevent colds.

7. Is your conclusion valid from a scientific standpoint? Why or why not?

8. Develop a testable hypothesis to answer the question, "Does taking vitamin C pills help prevent colds?"

9. Describe how you would conduct a controlled experiment to test your hypothesis.

What is happening in this picture ?

- What are these people protesting?

- How is this opposition an example of NIMBY?

- What kinds of incentives might make these people more willing to consider locating a nuclear waste site near them?

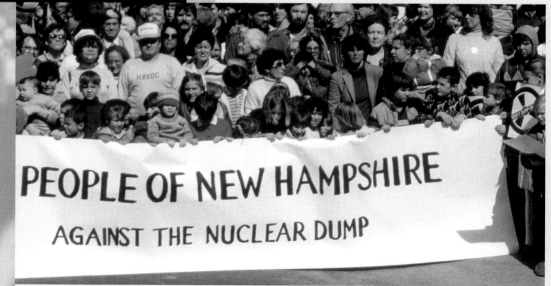

PEOPLE OF NEW HAMPSHIRE
AGAINST THE NUCLEAR DUMP

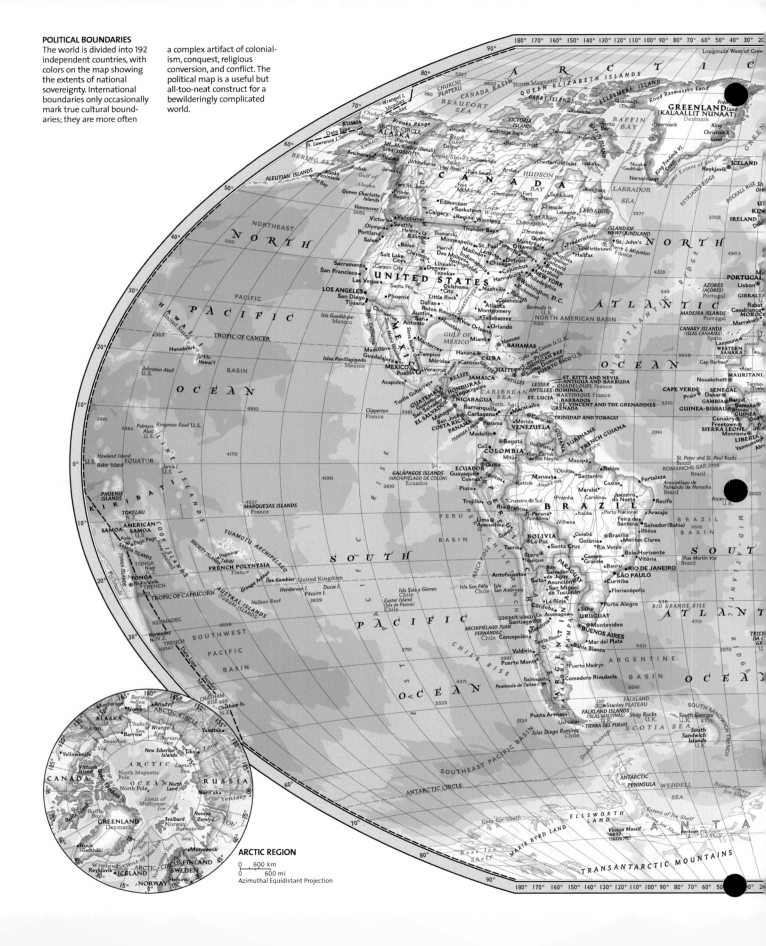

POLITICAL BOUNDARIES

The world is divided into 192 independent countries, with colors on the map showing the extents of national sovereignty. International boundaries only occasionally mark true cultural boundaries; they are more often a complex artifact of colonialism, conquest, religious conversion, and conflict. The political map is a useful but all-too-neat construct for a bewilderingly complicated world.

ARCTIC REGION

0   600 km
0   600 mi
Azimuthal Equidistant Projection

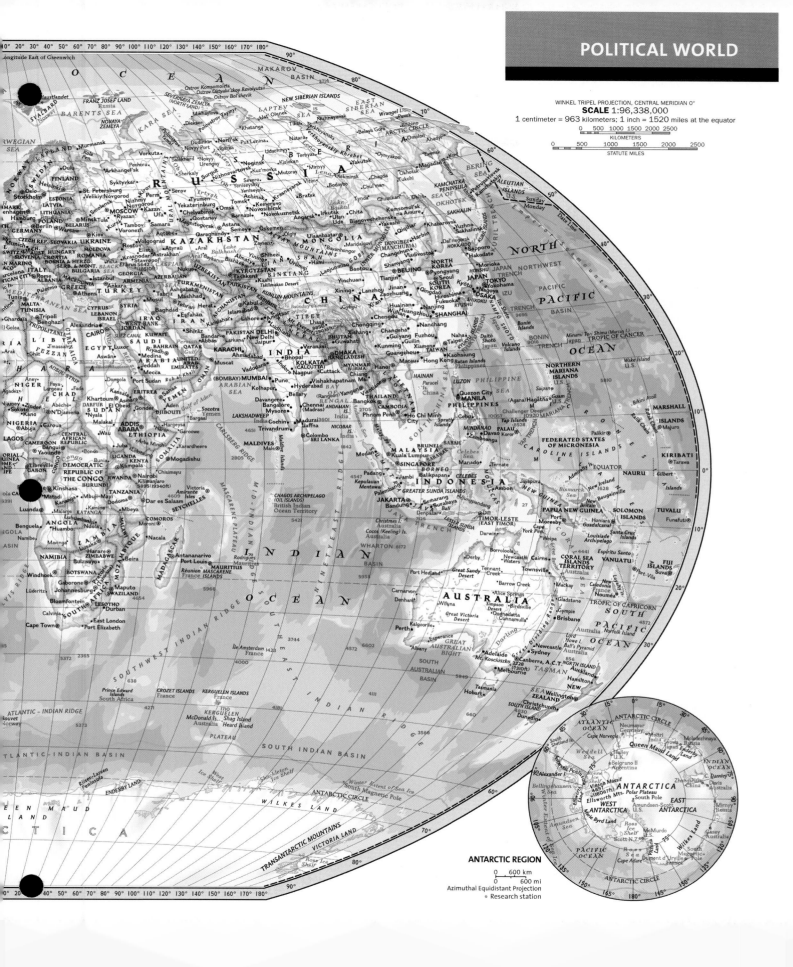

POLITICAL WORLD

WINKEL TRIPEL PROJECTION, CENTRAL MERIDIAN 0°
SCALE 1:96,338,000
1 centimeter = 963 kilometers; 1 inch = 1520 miles at the equator

0 500 1000 1500 2000 2500
KILOMETERS

0 500 1000 1500 2000 2500
STATUTE MILES

ANTARCTIC REGION

0 600 km
0 600 mi
Azimuthal Equidistant Projection
• Research station

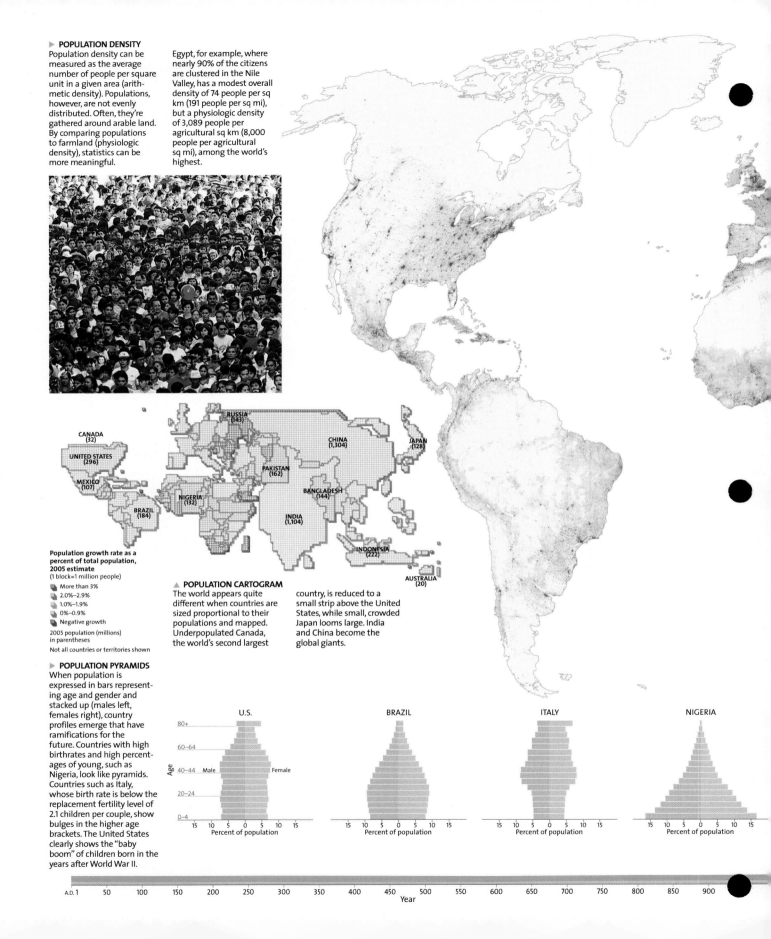

▶ POPULATION DENSITY

Population density can be measured as the average number of people per square unit in a given area (arithmetic density). Populations, however, are not evenly distributed. Often, they're gathered around arable land. By comparing populations to farmland (physiologic density), statistics can be more meaningful.

Egypt, for example, where nearly 90% of the citizens are clustered in the Nile Valley, has a modest overall density of 74 people per sq km (191 people per sq mi), but a physiologic density of 3,089 people per agricultural sq km (8,000 people per agricultural sq mi), among the world's highest.

Population growth rate as a percent of total population, 2005 estimate
(1 block=1 million people)

- More than 3%
- 2.0%–2.9%
- 1.0%–1.9%
- 0%–0.9%
- Negative growth

2005 population (millions) in parentheses

Not all countries or territories shown

▲ POPULATION CARTOGRAM

The world appears quite different when countries are sized proportional to their populations and mapped. Underpopulated Canada, the world's second largest country, is reduced to a small strip above the United States, while small, crowded Japan looms large. India and China become the global giants.

Cartogram labels:
CANADA (32), UNITED STATES (296), MEXICO (107), BRAZIL (184), RUSSIA (143), NIGERIA (132), PAKISTAN (162), BANGLADESH (144), INDIA (1,104), CHINA (1,304), JAPAN (128), INDONESIA (222), AUSTRALIA (20)

▶ POPULATION PYRAMIDS

When population is expressed in bars representing age and gender and stacked up (males left, females right), country profiles emerge that have ramifications for the future. Countries with high birthrates and high percentages of young, such as Nigeria, look like pyramids. Countries such as Italy, whose birth rate is below the replacement fertility level of 2.1 children per couple, show bulges in the higher age brackets. The United States clearly shows the "baby boom" of children born in the years after World War II.

Population pyramids: U.S., BRAZIL, ITALY, NIGERIA
(Age axis: 0–4, 20–24, 40–44, 60–64, 80+; Male left, Female right; x-axis: Percent of population, 15 10 5 0 5 10 15)

Timeline axis: A.D. 1, 50, 100, 150, 200, 250, 300, 350, 400, 450, 500, 550, 600, 650, 700, 750, 800, 850, 900 — Year

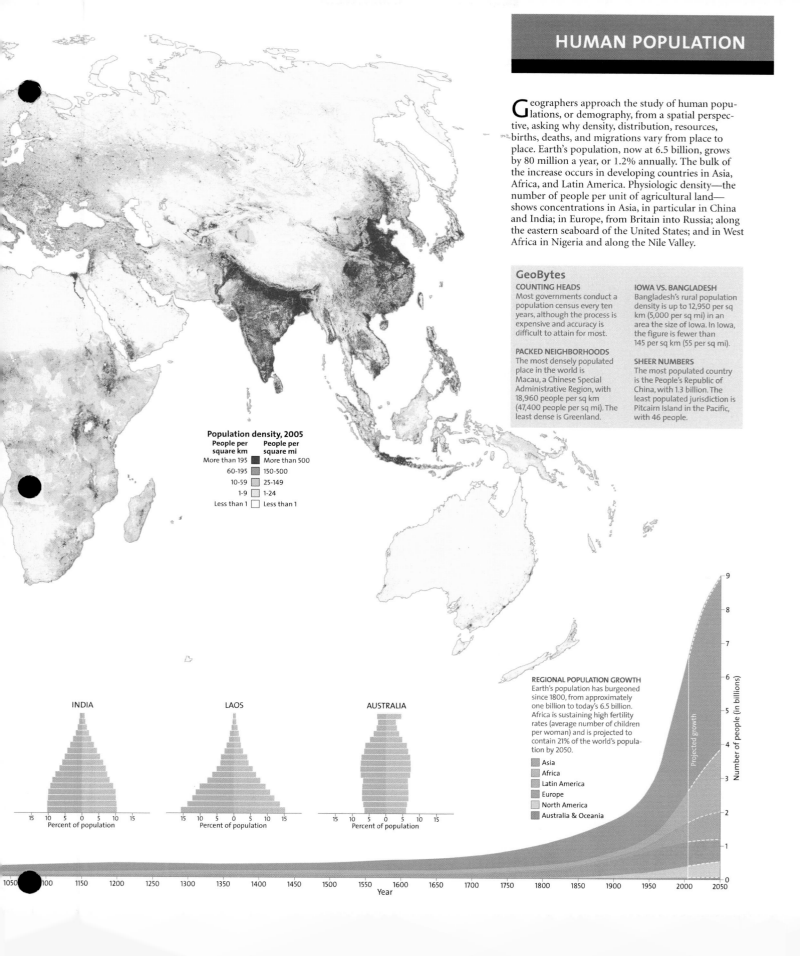

Geographers approach the study of human populations, or demography, from a spatial perspective, asking why density, distribution, resources, births, deaths, and migrations vary from place to place. Earth's population, now at 6.5 billion, grows by 80 million a year, or 1.2% annually. The bulk of the increase occurs in developing countries in Asia, Africa, and Latin America. Physiologic density—the number of people per unit of agricultural land—shows concentrations in Asia, in particular in China and India; in Europe, from Britain into Russia; along the eastern seaboard of the United States; and in West Africa in Nigeria and along the Nile Valley.

GeoBytes

COUNTING HEADS
Most governments conduct a population census every ten years, although the process is expensive and accuracy is difficult to attain for most.

PACKED NEIGHBORHOODS
The most densely populated place in the world is Macau, a Chinese Special Administrative Region, with 18,960 people per sq km (47,400 people per sq mi). The least dense is Greenland.

IOWA VS. BANGLADESH
Bangladesh's rural population density is up to 12,950 per sq km (5,000 per sq mi) in an area the size of Iowa. In Iowa, the figure is fewer than 145 per sq km (55 per sq mi).

SHEER NUMBERS
The most populated country is the People's Republic of China, with 1.3 billion. The least populated jurisdiction is Pitcairn Island in the Pacific, with 46 people.

Population density, 2005

People per square km	People per square mi
More than 195	More than 500
60-195	150-500
10-59	25-149
1-9	1-24
Less than 1	Less than 1

INDIA
Percent of population

LAOS
Percent of population

AUSTRALIA
Percent of population

REGIONAL POPULATION GROWTH
Earth's population has burgeoned since 1800, from approximately one billion to today's 6.5 billion. Africa is sustaining high fertility rates (average number of children per woman) and is projected to contain 21% of the world's population by 2050.

- Asia
- Africa
- Latin America
- Europe
- North America
- Australia & Oceania

Number of people (in billions)

Projected growth

Year

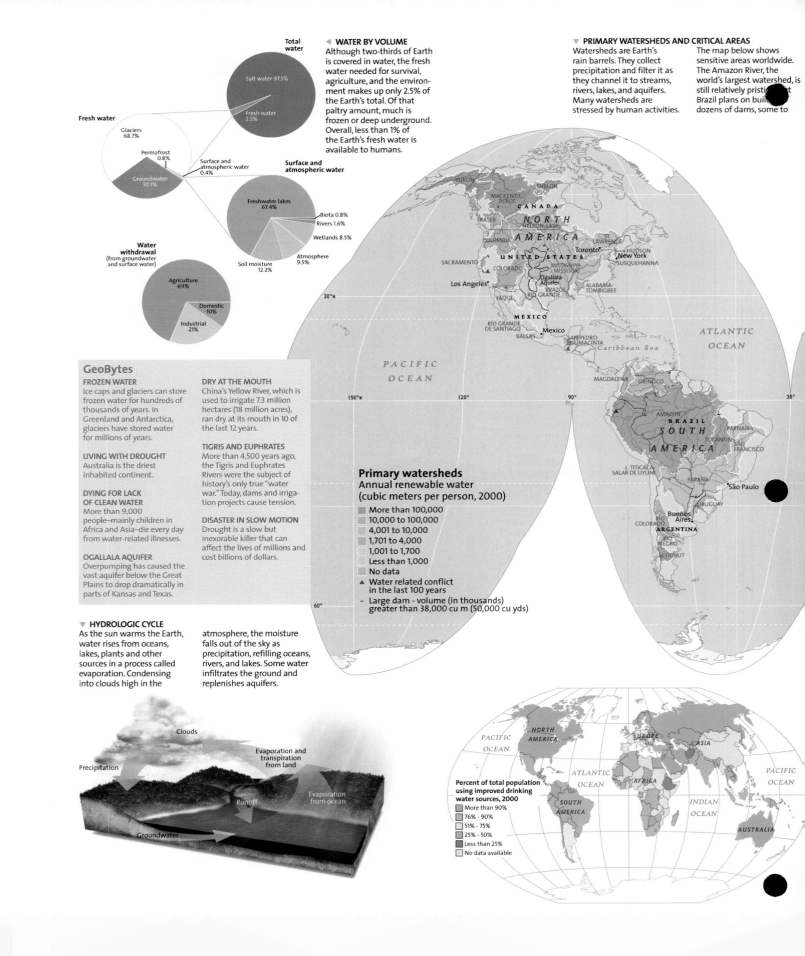

Total water

Salt water 97.5%

Fresh water 2.5%

◀ **WATER BY VOLUME**
Although two-thirds of Earth is covered in water, the fresh water needed for survival, agriculture, and the environment makes up only 2.5% of the Earth's total. Of that paltry amount, much is frozen or deep underground. Overall, less than 1% of the Earth's fresh water is available to humans.

▼ **PRIMARY WATERSHEDS AND CRITICAL AREAS**
Watersheds are Earth's rain barrels. They collect precipitation and filter it as they channel it to streams, rivers, lakes, and aquifers. Many watersheds are stressed by human activities.

The map below shows sensitive areas worldwide. The Amazon River, the world's largest watershed, is still relatively pristine, but Brazil plans on building dozens of dams, some to

Fresh water

Glaciers 68.7%

Permafrost 0.8%

Surface and atmospheric water 0.4%

Groundwater 30.1%

Surface and atmospheric water

Freshwater lakes 67.4%

Biota 0.8%

Rivers 1.6%

Wetlands 8.5%

Atmosphere 9.5%

Soil moisture 12.2%

Water withdrawal
(from groundwater and surface water)

Agriculture 69%

Domestic 10%

Industrial 21%

GeoBytes

FROZEN WATER
Ice caps and glaciers can store frozen water for hundreds of thousands of years. In Greenland and Antarctica, glaciers have stored water for millions of years.

LIVING WITH DROUGHT
Australia is the driest inhabited continent.

DYING FOR LACK OF CLEAN WATER
More than 9,000 people–mainly children in Africa and Asia–die every day from water-related illnesses.

OGALLALA AQUIFER
Overpumping has caused the vast aquifer below the Great Plains to drop dramatically in parts of Kansas and Texas.

DRY AT THE MOUTH
China's Yellow River, which is used to irrigate 7.3 million hectares (18 million acres), ran dry at its mouth in 10 of the last 12 years.

TIGRIS AND EUPHRATES
More than 4,500 years ago, the Tigris and Euphrates Rivers were the subject of history's only true "water war." Today, dams and irrigation projects cause tension.

DISASTER IN SLOW MOTION
Drought is a slow but inexorable killer that can affect the lives of millions and cost billions of dollars.

Primary watersheds
Annual renewable water
(cubic meters per person, 2000)

- More than 100,000
- 10,000 to 100,000
- 4,001 to 10,000
- 1,701 to 4,000
- 1,001 to 1,700
- Less than 1,000
- No data
- ▲ Water related conflict in the last 100 years
- – Large dam - volume (in thousands) greater than 38,000 cu m (50,000 cu yds)

▼ **HYDROLOGIC CYCLE**
As the sun warms the Earth, water rises from oceans, lakes, plants and other sources in a process called evaporation. Condensing into clouds high in the atmosphere, the moisture falls out of the sky as precipitation, refilling oceans, rivers, and lakes. Some water infiltrates the ground and replenishes aquifers.

Clouds

Evaporation and transpiration from land

Precipitation

Evaporation from ocean

Runoff

Groundwater

Percent of total population using improved drinking water sources, 2000
- More than 90%
- 76% - 90%
- 51% - 75%
- 25% - 50%
- Less than 25%
- No data available

power aluminum smelters. In Africa and Asia, lack of access to water and water-related diseases are the main problems. In Europe and the Middle East, overuse, pollution, and disagreement over diverting water are the major challenges. Hope rests in better planning and community-scale projects.

It's as vital to life as air. Yet fresh water is one of the rarest resources on Earth. Only 2.5% of Earth's water is fresh, and of that the usable portion for humans is less than 1% of all fresh water, or 0.01% of all water on Earth. Water is constantly recycling through Earth's hydrologic cycle. But population growth and pollution are combining to make less and less available per person per year, while global climate change adds new uncertainty.

Efficiency, conservation, and technology can help ensure that the water you absorb today will still be usable and clean hundreds of years from now.

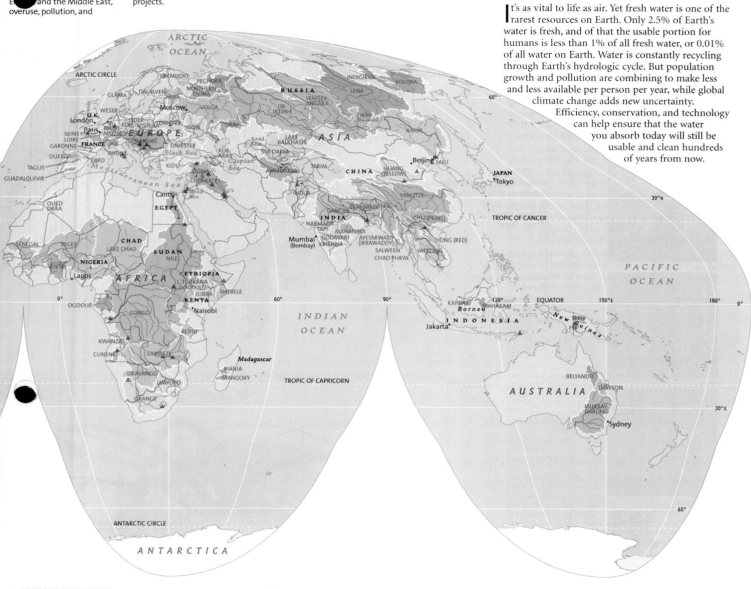

◀ **ACCESS TO FRESH WATER**
Access to clean fresh water is critical for human health. Yet, in many regions, potable water is becoming scarce because of heavy demands and pollution. Especially worrisome is the poisoning of aquifers—a primary source of water for nearly a third of the world—by sewage, pesticides, and heavy metals.

▼ **GLOBAL IRRIGATED AREAS AND WATER WITHDRAWALS**
Since 1970, global water withdrawals have correlated with the rise in irrigated area. Some 70% of withdrawals are for agriculture, mostly for irrigation that helps produce 40% of the world's food.

Freshwater withdrawal as a percentage of total water utilization, 2000

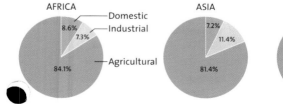

AFRICA
- Domestic 8.6%
- Industrial 7.3%
- Agricultural 84.1%

ASIA
- 7.2%
- 11.4%
- 81.4%

OCEANIA
- 17.6%
- 10%
- 72.4%

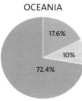

SOUTH AMERICA
- 19.3%
- 12.5%
- 68.2%

NORTH AMERICA
- 14.1%
- 44.5%
- 41.4%

EUROPE
- 15.2%
- 32.4%
- 52.4%

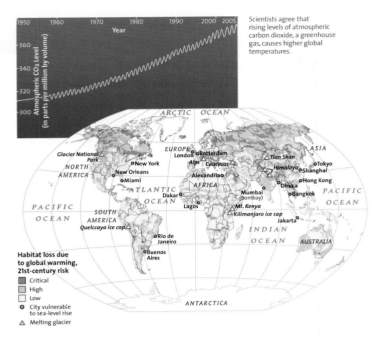

Scientists agree that rising levels of atmospheric carbon dioxide, a greenhouse gas, causes higher global temperatures.

Habitat loss due to global warming, 21st-century risk
- ■ Critical
- ■ High
- □ Low
- ◎ City vulnerable to sea-level rise
- △ Melting glacier

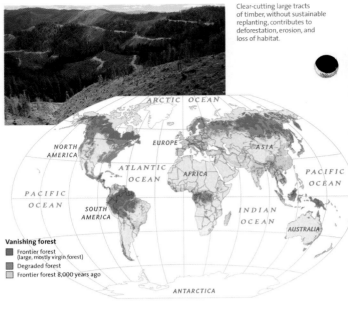

Clear-cutting large tracts of timber, without sustainable replanting, contributes to deforestation, erosion, and loss of habitat.

Vanishing forest
- ■ Frontier forest (large, mostly virgin forest)
- ▨ Degraded forest
- □ Frontier forest 8,000 years ago

▲ GLOBAL WARMING

Temperatures across the world are increasing at a rate not seen at any other time in the last 10,000 years. Although climate variation is a natural phenomenon, human activities that release carbon dioxide and other greenhouse gases into the atmosphere—industrial processes, fossil fuel consumption, deforestation, and land use change—are contributing to this warming trend. Scientists predict that if this trend continues, one-third of plant and animal habitats will be dramatically altered and more than one million species will be threatened with extinction in the next 50 years. And even small increases in global temperatures can melt glaciers and polar ice sheets, raising sea levels and flooding coastal cities and towns.

▲ DEFORESTATION

Of the 13 million hectares (32 million acres) of forest lost each year, mostly to make room for agriculture, more than half are in South America and Africa, where many of the world's tropical rain forests and terrestrial plant and animal species can be found. Loss of habitat in such species-rich areas takes a toll on the world's biodiversity. Deforested areas also release, instead of absorb, carbon dioxide into the atmosphere, contributing to global climate change. Deforestation can also affect local climates by reducing evaporative cooling, leading to decreased rainfall and higher temperatures.

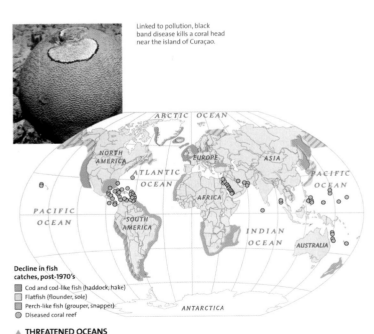

Linked to pollution, black band disease kills a coral head near the island of Curaçao.

Decline in fish catches, post-1970's
- ■ Cod and cod-like fish (haddock, hake)
- ■ Flatfish (flounder, sole)
- ■ Perch-like fish (grouper, snapper)
- ◎ Diseased coral reef

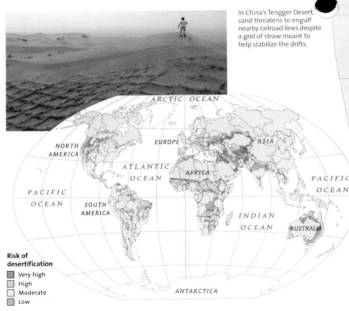

In China's Tengger Desert, sand threatens to engulf nearby railroad lines despite a grid of straw meant to help stabilize the drifts.

Risk of desertification
- ■ Very high
- ■ High
- □ Moderate
- □ Low

▲ THREATENED OCEANS

Oceans cover more than two-thirds of the Earth's surface and are home to at least half of the world's biodiversity, yet they are the least understood ecosystems. The combined stresses of overfishing, pollution, increased carbon dioxide emissions, global climate change, and coastal development are having a serious impact on the health of oceans and ocean species. Over 70% of the world's fish species are depleted or nearing depletion, and 50% of coral reefs worldwide are threatened by human activities.

▲ DESERTIFICATION

Climate variability and human activities, such as grazing and conversion of natural areas to agricultural use, are leading causes of desertification, the degradation of land in arid, semiarid, and dry subhumid areas. The environmental consequences of desertification are great—loss of topsoil, increased soil salinity, damaged vegetation, regional climate change, and a decline in biodiversity. Equally critical are the social consequences—more than 2 billion people live in and make a living off these dryland areas, covering about 41% of Earth's surfac

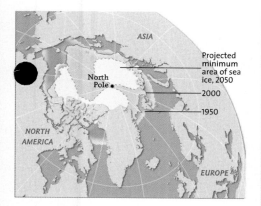

▲ POLAR ICE CAP
Over the last 50 years, the extent of polar sea ice has noticeably decreased. Since 1970 alone, an area larger than Norway, Sweden, and Denmark combined has melted. This trend is predicted to accelerate as temperatures rise in the Arctic and across the globe.

GeoBytes

ACIDIFYING OCEANS
Oceans are absorbing an unprecedented 20 to 25 million tonnes (22 to 28 millions tons) of carbon dioxide each day, increasing the water's acidity.

ENDANGERED REEFS
Some 95% of coral reefs in Southeast Asia have been destroyed or are threatened.

RECORD TEMPERATURES
The 1990s were the warmest decade on record in the last century.

WARMING ARCTIC
While the world as a whole has warmed nearly 0.6°C (1°F) over the last hundred years, parts of the Arctic have warmed 4 to 5 times as much in only the last 50 years.

DISAPPEARING RAIN FORESTS
Scientists predict that the world's rain forests will disappear within the next one hundred years if the current rate of deforestation continues.

OIL POLLUTION
Nearly 1.3 million tonnes (1.4 million tons) of oil seep into the world's oceans each year from the combined sources of natural seepage, extraction, transportation, and consumption.

ACCIDENTAL DROWNINGS
Entanglement in fishing gear is one of the greatest threats to marine mammals.

A FAREWELL TO FROGS?
Worldwide, almost half of the 5,700 named amphibian species are in decline.

With the growth of scientific record keeping, observation, modeling, and analysis, our understanding of Earth's environment is improving. Yet even as we deepen our insight into environmental processes, we are changing what we are studying. At no other time in history have humans altered their environment with such speed and force. Nothing occurs in isolation, and stress in one area has impacts elsewhere. Our agricultural and fishing practices, industrial processes, extraction of resources, and transportation methods are leading to extinctions, destroying habitats, devastating fish stocks, disturbing the soil, and polluting the oceans and the air. As a result, biodiversity is declining, global temperatures are rising, polar ice is shrinking, and the ozone layer continues to thin.

▼ POLLUTION
No corner of the earth is immune to pollution, be it in the air, soil, or water. Concentrations of pollution can be found in the industrial centers of North America, Europe, and, increasingly, Asia—and areas downwind or downstream from them. Shipping routes are sources of pollution, from oil spills to garbage dumpings.

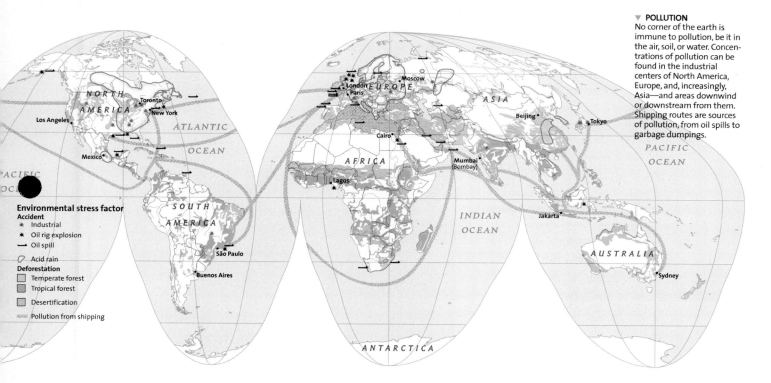

Environmental stress factor
Accident
* Industrial
* Oil rig explosion
— Oil spill
⟳ Acid rain
Deforestation
☐ Temperate forest
☐ Tropical forest
☐ Desertification
〜 Pollution from shipping

▶ OZONE DEPLETION
First noted in the mid-1980s, the springtime "ozone hole" over the Antarctic continues to grow. With sustained efforts to restrict chlorofluorocarbons (CFCs) and other ozone-depleting chemicals, scientists have begun to see what they hope is a leveling off in the rate of depletion. Stratospheric ozone shields the Earth from the sun's ultraviolet radiation. Thinning of this protective layer puts people at risk for skin cancer and cataracts. It can also have devastating effects on the Earth's biological functions.

1980

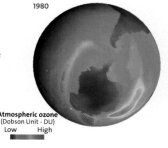

2000

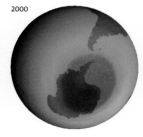

2004

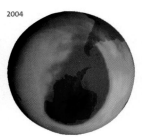

Atmospheric ozone
(Dobson Unit - DU)
Low   High

Environmental Sustainability and Human Values

2

THE GLOBAL COMMONS

Ecologist Garrett Hardin (1915–2003) is best known for his 1968 essay "The Tragedy of the Commons." In it he contended that our inability to solve complex environmental problems is the result of a struggle between short-term individual welfare and long-term *environmental sustainability* (the ability to meet humanity's current needs without compromising the ability of future generations to meet their needs).

Hardin used the commons to illustrate this struggle. In medieval Europe, the inhabitants of a village shared pastureland, called the commons, and each herder could bring animals onto the commons to graze. The more animals a herder brought onto the commons, the greater the advantage to that individual. When every herder in the village brought as many animals onto the commons as possible, however, the plants were killed from overgrazing.

In today's world, Hardin's parable has particular relevance at the global level. The commons are those parts of our environment that are available to everyone but for which no single individual has responsibility: the atmosphere, water, forests, wildlife (such as the green sea turtle), and fisheries. These modern-day commons, sometimes collectively called the *global commons,* are experiencing increasing environmental stress. The world needs effective legal and economic policies to prevent the degradation of our global commons. Clearly, all people, businesses, and governments must foster a strong sense of *stewardship*, or shared responsibility, for the sustainable care of our planet.

This chapter examines the role of ethics and values in environmental issues. As you read, keep in mind these words from the *Earth Charter*, formulated in 1992 by representatives from 178 countries. "Let ours be a time remembered for the awakening of a new reverence for life, the firm resolve to achieve sustainability, the quickening of the struggle for justice and peace, and the joyful celebration of life."

Human Use of the Earth p. 26

Human Values and Environmenta
Problems p. 29

Environmental Justice p. 34

**An Overall Plan for Sustainable
Living p. 35**

WORLD SUMMIT ON SUSTAINAB

**Case Study: The 2002 World
Summit p. 42**

Human Use of the Earth

LEARNING OBJECTIVES

Define sustainable development.

Outline some of the complexities associated with the concept of sustainable consumption.

Environmental sustainability is a concept that people have discussed for many years. *Our Common Future,* the 1987 report of the U.N. World Commission on Environment and Development, presented the closely related concept of **sustainable development** (FIGURE 2.1). The authors of *Our Common Future* pointed out that sustainable development includes meeting the needs of the world's poor. The report also linked the environment's ability to meet present and future needs to the state of technology and social organization existing at a given time and in a given place. The num-

> **sustainable development**
> Economic growth that meets the needs of the present without compromising the ability of future generations to meet their own economic needs.

ber of people, their degree of affluence (that is, their level of consumption), and their choices of technology all interact to produce the total effect of a given society, or of society at large, on the sustainability of the environment.

Even using the best technologies imaginable, Earth's productivity still has its limits, and our use of it can't be expanded indefinitely. *Sustainable development can occur only within the limits of the environment.*

To live within these limits, population growth must be held at a level that we can sustain, and the wealthy must first stabilize their use of natural resources and then reduce this use to a level that can be maintained. The world does not contain nearly enough resources to sustain everyone at the level of consumption that is enjoyed in the United States, Europe, and Japan. Suitable strategies, however, do exist to reduce these levels of consumption without concurrently reducing the real quality of life.

Sustainable development

FIGURE 2.1

Three factors—environmentally sound decisions, economically viable decisions, and socially equitable decisions—interact to promote sustainable development.

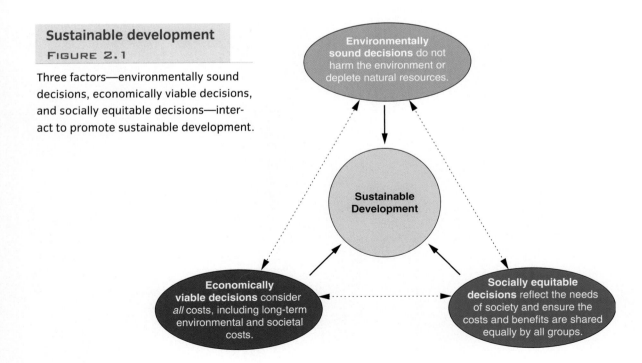

The challenge of eradicating poverty FIGURE 2.2

Slum in Tupac Amaru, Peru. Many of the world's people live in extreme poverty. *(Inset)* Desperately poor "pavement people" eat, sleep, and raise their families on a street in Calcutta, India. Homelessness is a serious problem in cities of highly developed nations and an even greater problem in cities of developing nations.

SUSTAINABLE CONSUMPTION

Consumption overpopulation is pollution and degradation of the environment that occurs when each individual in a population consumes too large a share of resources. Consumption overpopulation stems from the lifestyles of people living in highly developed nations. *Lifestyle* is interpreted broadly to include goods and services bought for food, clothing, housing, travel, recreation, and entertainment. In evaluating consumption overpopulation, all aspects of the production, use, and disposal of these goods and services are taken into account, including environmental costs. Such an analysis provides a sense of what it means to consume sustainably versus unsustainably.

■ **sustainable consumption**
The use of goods and services that satisfy basic human needs and improve the quality of life but that also minimize the use of resources.

Sustainable consumption, like sustainable development, forces us to address whether our present actions undermine the long-term ability of the environment to meet the needs of future generations. Factors that affect sustainable consumption include population, economic activities, technology choices, social values, and government policies.

At the global level, sustainable consumption requires the eradication of poverty, which in turn requires that poor people in developing countries *increase* their consumption of certain essential resources (**FIGURE 2.2**). For their increased consumption to be sustainable, however, the consumption patterns of people in highly developed countries must change.

Consumption in the United States FIGURE 2.3

This "typical" family of four lives in Pearland, Texas. All their material posses-sions were removed from the house and garage to emphasize the large amount of natural resources that people living in highly developed countries consume.

Widespread adoption of sustainable consumption will not be easy. It will require major changes in the consumption patterns and lifestyles of most people in highly developed countries (FIGURE 2.3). Some examples of promoting sustainable consumption include switching from motor vehicles to public transport and bicycles, and developing durable, repairable, recyclable products.

Sustainable consumption is not a popular idea with policymakers, politicians, and economists, or even with consumers such as you and me. It contains an inherent threat to "business as usual," but it also offers new and exciting opportunities. Many scientists and population experts increasingly advocate adoption of sustainable consumption before it is forced on us by an environmentally degraded, resource-depleted world.

As people adopt new lifestyles, they must be educated so that they understand the reasons for changing practices that may be highly ingrained or tradi-tional. Formal education and informal education are both important in bringing about change and in contributing to sustainable consumption. If people understand the way the natural world functions, they can appreciate their own place in it and value sustainable actions.

Any long-term involvement in the condition of the world must start with individuals—our values, attitudes, and practices. Each of us makes a difference, and it is ultimately our collective activities that make the world what it is.

CONCEPT CHECK **STOP**

What is sustainable development?

What is sustainable consumption? How is it linked to a reduction in world poverty?

Human Values and Environmental Problems

LEARNING OBJECTIVES

Define environmental ethics.

Discuss distinguishing features of the Western and deep ecology worldviews.

Define voluntary simplicity.

We now shift our attention to the views of different individuals and societies and how those views affect our ability to understand and solve sustainability problems. **Ethics** is the branch of philosophy that is derived through the logical application of human **values**. These values are the principles that an individual or society considers important or worthwhile. Values are not static entities but change as societal, cultural, political, and economic priorities change. Ethics helps us determine which forms of conduct are morally acceptable or unacceptable, right or wrong. Ethics plays a role in whatever types of human activities involve intelligent judgment and voluntary action. Whenever alternative, conflicting values occur, ethics helps us choose which value is better, or worthier, than other values.

Environmental ethics examines moral values to determine how humans should relate to the natural environment. Environmental ethicists consider such questions as what role we should play in determining the fate of Earth's resources, including other species, or how might we develop an environmental ethic that is acceptable in the short term for us as individuals but also in the long term for our species and the planet. These questions and others like them are difficult intellectual issues that involve political, economic, societal, and individual trade-offs.

Environmental ethics considers not only the rights of people living today, both individually and collectively, but also the rights of future generations (**FIGURE 2.4**). This aspect of environmental ethics is critical because the impacts of today's activities and technologies are changing the environment. In some cases these impacts may be felt for hundreds or even thousands of years. Addressing issues of environmental ethics puts us in a better position to use science and technology for long-term environmental sustainability.

environmental ethics A field of applied ethics that considers the moral basis of environmental responsibility and how far this responsibility extends.

Tomorrow's generation FIGURE 2.4

The choices made today will determine if future generations, such as these students from Bailey Elementary School in Falls Church, Virginia, will inherit a sustainable world.

WORLDVIEWS

Each of us has a particular **worldview**—that is, a commonly shared perspective based on a collection of our basic values that helps us make sense of the world, understand our place and purpose in it, and determine right and wrong behaviors. These worldviews lead to behaviors and lifestyles that may or may not be compatible with environmental sustainability.

Two extreme, competing **environmental worldviews** are the Western worldview and the deep ecology worldview. These two worldviews, admittedly broad generalizations, are at nearly opposite ends of a spectrum of worldviews relevant to global sustainability problems, and each approaches environmental responsibility in a radically different way.

The traditional **Western worldview**, also known as the *expansionist worldview*, is human-centered and utilitarian. It mirrors the beliefs of the 19th-century **frontier attitude**, a desire to conquer and exploit nature as quickly as possible (**FIGURE 2.5**). The Western worldview also advocates the inherent rights of individuals, accumulation of wealth, and unlimited consumption of goods and services to provide material comforts. According to the Western worldview, humans have a primary obligation to humans and are therefore responsible for managing natural resources to benefit human society. Thus, any concerns about the environment are derived from human interests.

environmental worldview A worldview that helps us make sense of how the environment works, our place in the environment, and right and wrong environmental behaviors.

Western worldview An understanding of our place in the world based on human superiority and dominance over nature, the unrestricted use of natural resources, and increased economic growth to manage an expanding industrial base.

A

Western worldview FIGURE 2.5

A Logging operations in 1884. This huge logjam occurred on the St. Croix River near Taylors Falls, Minnesota. B The Western worldview in action today. A logging company road cuts through an Indonesian forest, making the region's hardwoods available for logging.

Global Locator

B

The **deep ecology worldview** is a diverse set of viewpoints that dates from the 1970s and is based on the work of Arne Naess, a Norwegian philosopher, and others, including ecologist Bill Devall and philosopher George Sessions. The principles of deep ecology, as expressed by Naess in *Ecology, Community and Lifestyle* (1989), include:

1. Both human and nonhuman life has intrinsic value. The value of nonhuman life-forms is independent of the usefulness they may have for narrow human purposes.

2. Richness and diversity of life-forms contribute to the flourishing of human and nonhuman life on Earth (**FIGURE 2.6**).

3. Humans have no right to reduce this richness and diversity except to satisfy vital needs.

4. Present human interference with the nonhuman world is excessive, and the situation is rapidly worsening.

5. The flourishing of human life and cultures is compatible with a substantial decrease in the human population. The flourishing of nonhuman life requires such a decrease.

Preservation of biological diversity is an important part of the deep ecology worldview FIGURE 2.6

A California meadow. *(Inset)* A rain forest grasshopper in Indonesia. According to the deep ecology worldview, these plant and animal species have as much right to a place in the environment as humans do.

6. Significant change of life conditions for the better requires changes in economic, technological, and ideological structures.

7. The ideological change is mainly that of appreciating life quality rather than adhering to a high standard of living.

8. Those who subscribe to the foregoing points have an obligation to participate in the attempt to implement the necessary changes.

Compared to the Western worldview, the deep ecology worldview represents a radical shift in how humans relate themselves to the environment. The deep ecology worldview stresses that all forms of life have the right to exist, and that humans are not different or separate from other organisms. Humans have an obligation to themselves and to the environment. The deep ecology worldview advocates sharply curbing human population growth. It does not advocate returning to a society free of today's technological advances but instead proposes a significant rethinking of our use of current technologies and alternatives. It asks individuals and societies to share an inner spirituality connected to the natural world.

Most people today do not fully embrace either the Western worldview or the deep ecology worldview. The Western worldview is **anthropocentric** and emphasizes the importance of humans as the overriding concern in the grand scheme of things. In contrast, the deep ecology worldview is **biocentric** and views humans as one species among others. The planet's natural resources could not support its more than 6.5 billion humans if each consumed the high level of goods and services sanctioned by the Western worldview. On the other hand, the world as envisioned by the deep ecology worldview could support only a fraction of the existing human population (FIGURE 2.7).

These worldviews, while not practical for widespread adoption, are useful to keep in mind as you examine various environmental issues in later chapters. In the meantime, you should think about your own worldview and discuss it with others—whose worldviews will probably be different from your own. Thinking leads to actions, and actions lead to consequences. What are the short-term and long-term consequences of your particular worldview? We must develop and incorporate into our culture a long-lasting, environmentally sensitive worldview if the environment is to be sustainable for us, for other living organisms, and for future generations.

Embracing deep ecology FIGURE 2.7

At one time or another, most of us yearn for the simpler life that the tenets of deep ecology advocate. However, there are far too many people and far too little land for us all to embrace this lifestyle. Photographed in Humboldt County, California.

VOLUNTARY SIMPLICITY

You have just had a very brief introduction to a few aspects of environmental ethics. The question that now needs to be addressed is, "How should you live?" An increasing number of people in the United States and other highly developed nations have embraced the concept of **voluntary simplicity**, which recognizes that individual happiness and quality of life are not necessarily linked to the accumulation of material goods. People who embrace voluntary simplicity recognize that a person's values and character define that individual more than how many things he or she owns. This belief requires a change in behavior as people purchase and use fewer items than they might have formerly. It is a commitment at the individual level toward saving the planet for future generations.

Many individuals, businesses, and communities are developing innovative ways to make do with less. Some communities have organized "tool libraries" from which people check out power and hand tools for a minimal fee. Such sharing reduces the requirement for every household to own a complete set of garden and maintenance tools. Other items, such as luggage, household appliances, cookbooks, and recreational equipment, could also be shared or rented at low cost within a community.

Thousands of people in Europe and, to a lesser extent, the United States, have embraced the principle of voluntary simplicity by signing up with car sharing programs (**FIGURE 2.8**). Such programs, which are designed for people who use a car once every day or so, offer an economical alternative to individual ownership. Car sharing also may reduce the number of cars manufactured. Studies show that most car sharers drive significantly fewer miles than they did before they joined the program.

Car sharing FIGURE 2.8

The city of Philadelphia reduced the number of vehicles provided to city employees by partnering with PhillyCarShare, a local non-profit organization. The program provides city employees with access to a fleet of hybrid gas–electric vehicles.

"PhillyCarShare is a Philadelphia-based non-profit organization dedicated to reducing automobile dependency in the Philadelphia region through community-based car sharing—improving neighborhood livability; increasing transit ridership, walking, and biking; improving air quality; promoting economic development; mitigating development; mitigating parking demand; and increasing accessibility while alleviating the burdens of auto ownership."
—from the PhillyCarShare Mission Statement

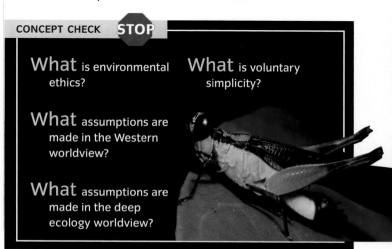

CONCEPT CHECK STOP

What is environmental ethics?

What is voluntary simplicity?

What assumptions are made in the Western worldview?

What assumptions are made in the deep ecology worldview?

Environmental Justice

I n the early 1970s, the Board of National Ministries of the American Baptist Churches coined the term *eco-justice* to link social ethics and environmental ethics. At the local level, eco-justice represents opposition to the environmental inequities faced by low-income minority communities in both rural and urban areas. Many studies indicate that low-income communities and/or communities of color are more likely to have chemical plants, hazardous waste facilities, sanitary landfills, sewage treatment plants, and incinerators in their neighborhoods (**FIGURE 2.9**). A 1990 study at Clark Atlanta University, for example, found that six of Houston's eight incinerators were located on less expensive land in predominantly black neighborhoods. Such communities often have a limited involvement in the political process and may not even be aware of their exposure to higher levels of pollutants.

Because people in low-income communities frequently lack access to sufficient health care, they may not be treated adequately for exposure to environmental contaminants. The high incidence of asthma in many minority communities may be caused or exacerbated by exposure to environmental pollutants. Few studies, except those documenting lead contamination, have examined how environmental pollutants interact with other socioeconomic factors to cause health problems. Although anecdotal evidence abounds, there is currently little scientific evidence that shows to what extent a polluted environment is responsible for the disproportionate health problems of poor and minority communities. For example, a 1997 health study of residents in San Francisco's highly polluted Bayview Hunters Point area found that hospitalizations for chronic illnesses were the highest rates in the state. Researchers failed, however, to isolate an increased exposure to toxic pollutants as the primary cause of these illnesses.

In addition to concerns about pollution in their neighborhoods, low-income communities may not receive equal benefits from federal cleanup programs. A

A children's playground overlooks a pulp mill FIGURE 2.9

Poor minority neighborhoods often have the most polluted and degraded environments. Photographed in Kingsport, Tennessee.

1992 paper published in the *National Law Journal* reported that toxic waste sites in white communities were cleaned up faster and better than those in communities of blacks, Hispanics, or other minorities.

ENVIRONMENTAL JUSTICE AND ETHICAL ISSUES

There is an increasing awareness that environmental decisions such as where to locate a hazardous waste landfill have important ethical dimensions. The most basic ethical dilemma centers on the rights of the poor and disenfranchised versus the rights of the rich and powerful. Whose rights should have priority in these decisions? The challenge is to find and adopt solutions that respect all social groups, including those yet to be born. **Environmental justice** is a funda-

> ▪ **environmental justice** The right of every citizen, regardless of age, race, gender, social class, or other factor, to adequate protection from environmental hazards.

mental human right in an ethical society. Although we may never completely eliminate past environmental injustices, we have a moral imperative to prevent them today so that the negative effects of pollution do not disproportionately affect any segment of society.

In response to these concerns, a growing environmental justice movement has emerged at the grassroots level as a strong motivator for change. Advocates of the environmental justice movement are calling for special efforts to clean up hazardous sites in low-income neighborhoods, from inner-city streets to Native American reservations. Many advocates cite the need for more research on human diseases that environmental pollutants may influence.

CONCEPT CHECK **STOP**

What is environmental justice?

Which communities are exposed to a disproportionate share of environmental hazards?

An Overall Plan for Sustainable Living

www.wiley.com/college/berg

LEARNING OBJECTIVES

Define carrying capacity and list two ways to stay within Earth's carrying capacity.

Explain why we should respect and care for Earth's biological diversity.

Describe environmental goals at local, national, and global levels.

Environmental problems are based in part on the large disparities in the standards of living experienced in different parts of the world. The United States and other highly developed countries enjoy an abundance of what the world has to offer. However, we are not managing the world's resources sustainably. If the world's resources were considered a bank account, we are living off the principal and not off the interest. Living off the interest alone, without touching the principal, would be sustainable. Our way of living, however, is clearly unsustainable because it is rapidly exhausting the quantity and quality of the natural resources that will be available to people in the future.

Let's now examine a plan for building a sustainable society. This plan is loosely based on the principles for sustainable living presented in *Caring for the Earth*, a report that the World Conservation Union, the United Nations Environment Programme, and the World Wide Fund for Nature jointly published in 1991.

CONTROL HUMAN POPULATION GROWTH AND IMPROVE THE QUALITY OF HUMAN LIFE

The ability of a given environment to absorb wastes and renew itself ultimately determines its **carrying capacity**. Great disparities in living standards exist for humans in different parts of the world. These living standards have a profound effect on the carrying capacity of an area: The higher the living standards, the lower the area's carrying capacity. The poorer people living in developing countries have far less than their share of the world's resources. However, population growth rates tend to be highest where poverty is most intense, draining the carrying capacity. Adopting acceptable levels of population and consumption that local environments can support will correct this situation.

Worldwide attention to family planning, making it available for everyone, can help stabilize the human population (**FIGURE 2.10**). To stay within Earth's car-

carrying capacity The maximum population that can be sustained by a given environment or by the world as a whole.

rying capacity, it will be necessary not only to reach a stable population but also to greatly reduce excessive consumption and waste. We must manage excessive consumption and waste in an integrated fashion coupled with educational programs, so that people understand the limits of Earth's carrying capacity.

The ultimate goal of development is to improve the quality of human life, making it possible for humans throughout the world to enjoy long, healthy, and fulfilling lives. A serious complication lies in the very unequal distribution of the world's resources. Those who live in highly developed countries, a rapidly shrinking 19 percent of the global population, control about 79 percent of the world's finances, as measured by summing gross domestic products. In contrast, more than half of the world's population currently lives on less than $2 per day. Among the world's poor people, roughly one in seven receives less than 80 percent of the minimum caloric intake that the United Nations recommends; as a result, their brains and bodies do not develop properly. For many of the world's women and children, life is an endless struggle for survival, centering on the daily requirements for firewood, clean water, and food (**FIGURE 2.11**). Such poverty, along with the enormous pressures of human population growth and consumption rates, are global problems that can't be solved without modifying the standard of living enjoyed in highly developed countries.

Not only does failing to confront the problem of poverty around the world continue to make it impossible to attain global sustainability, but it also is morally indefensible. Most people would find unacceptable the deaths of 30,000 infants and children each day, most of which could be prevented by access to adequate food or basic medical techniques and supplies. Everyone must have a reasonable share of Earth's productivity or our civilization will eventually come unraveled. As U.S. president Franklin Delano Roosevelt said in his second inaugural address in 1937, "The test of our progress is not whether we add more to the abundance of those who have much; it is whether we provide enough for those who have too little."

The universal education of children and the reduction of illiteracy are of critical importance in

Family planning in Egypt FIGURE 2.10

Women at a health clinic learn about family planning and birth control.

Children at work FIGURE 2.11

These children are gathering firewood near the Ethiopian village of Meshal. In many developing countries, young children gather firewood, carry water, and work in the fields with their parents. Older children often have wage-producing jobs that add to the family income.

gineers live in less developed countries. How are these countries to decide, on the basis of their limited knowledge, whether to join international agreements concerning the environment or how best to manage their own natural resources? The training and employment of professional scientists and engineers in developing countries is a matter of high priority.

RESPECT AND CARE FOR EARTH'S BIOLOGICAL DIVERSITY

At the most basic level, we are part of Earth's web of life, having evolved within it and as part of it. Humans are entirely dependent on that web, with all of its complex interactions, for our survival. Many people therefore question whether it is morally acceptable for us to continue destroying the web of life so rapidly. Can we justify, in ethical terms, the fact that we are driving thousands of species to extinction that are, as far as we know, our only living companions in the universe? Our individual rights as humans are ultimately affected by the level of respect we hold for one another and for the natural environment that supports us.

Pragmatically, we have a clear interest in protecting Earth's **biological diversity**. We obtain from living organisms all of our food, most of our medicines, building and clothing materials, and many other products. In addition, communities of organisms and the natural environment in which they live provide an enormous array of **ecosystem services** without which we would not survive. The ultimate reason for caring for the community of life on Earth is therefore a selfish one.

With respect to biological diversity, some 80 percent of the species of plants,

raising and maintaining appropriate standards of living in every country. Because responses to environmental problems depend on the public's awareness and understanding of the issues and the underlying scientific concepts involved, environmental education is critical to appropriate decision making. The emphasis on environmental education has grown dramatically. For example, three international treaties supporting environmental education went into effect between 1975 and 1990.

One of the greatest barriers to equalizing the gap between highly developed and developing nations is the lack of trained professionals in the latter group. Only about 10 percent of the world's scientists and en-

> **biological diversity** The number and variety of Earth's organisms.

> **ecosystem services** Important environmental benefits, such as clean air to breathe, clean water to drink, and fertile soil in which to grow crops, that the natural environment provides.

Yanomami children in Brazil enjoy a photographer's camera. Intrusion into isolated areas such as the Amazon Basin threatens both biological diversity and the cultures of indigenous people that have lived in harmony with nature for hundreds of generations.

Young trees in Scotland FIGURE 2.13

These evergreens are being cultivated as part of a reforestation project on land unsuitable for growing crops.

animals, fungi, and microorganisms are found in developing countries. How will these relatively poor countries sustainably manage and conserve these precious resources? Biological diversity is an intrinsically local problem: Each nation must address it for the sake of its own people's future, as well as for the world at large (**FIGURE 2.12**). The problem is more complicated because scientists have not yet identified and described most kinds of organisms. Biological diversity, like most problems of sustainable development, can be addressed adequately only if the number of scientists and engineers in developing countries is increased.

Farmland, grazing land, forest plantations, and similar modified ecosystems must be managed efficiently so that they produce as much as possible and thus reduce the pressure on natural lands everywhere (**FIGURE 2.13**). All of the food and other agricultural products that sustain human life are grown on an area about the size of South America. Very little additional land—land that is not now cultivated—is suitable for agriculture. Obviously, improving agricultural productivity is one of the highest priorities in achieving future global sustainability, although residents of a rich country like the United States, where food is inexpensive, find it difficult to appreciate this goal properly.

Each country should develop a comprehensive plan for the protection of its natural areas, with international assistance provided where needed. Biological diversity will not survive without our direct intervention, as 6.5 billion humans have placed so many of Earth's organisms at risk. For our own future good, we must reduce these risks.

WORK ON DEVELOPMENT AND CONSERVATION AT LOCAL, NATIONAL, AND GLOBAL LEVELS

Any long-term improvement in the world's condition must start with individuals—our values, attitudes, and practices. Each of us makes a difference, and ultimately our collective activities make the world what it is. In richer parts of the world, apathy, ignorance, and wasteful consumption have negative effects. In less developed countries, survival may be such an overwhelming concern that conservation in any sense of the word may seem irrelevant.

People are generally concerned about the environment, but their concerns do not naturally translate into action. Education helps people understand the reasons for changing lifestyle practices that may be highly ingrained or traditional. Many people believe that individual actions do not really make a difference, yet in fact, they are the only actions that do. If people have the opportunity to understand the way the natural world functions, they appreciate their own place in it and value actions that are sustainable. Accurate information must be made widely available, giving the media an important part to play.

Most communities care deeply about the sustainable management of their own environments, whether they express their views precisely in that way or not. Acting together makes people a strong and effective force, regardless of whether their community is wealthy or poor. In the United States and similar highly developed countries, action often takes place through the ballot box, but there is no substitute for direct action in any part of the world. Developing effective local governments that are responsive to the need for sustainable community development is an essential element in achieving overall success.

People often speak of throwing garbage away, but all that enters Earth, with minor exceptions, is sun-light; and all that leaves is some of the radiated heat originating from that sunlight. Thus there is truly no "away," no place where our garbage, our carbon dioxide, our pollution—anything that we produce or concentrate—can go (**FIGURE 2.14**). Yet we treat Earth as if it had no limits—as if there were plenty of room in which to dispose of things.

Imagine a similar situation: If we simply took into our homes all that we need or wish to consume, and then just left behind right there in the house any part that we did not use, there would soon be no room for us! Earth is precisely the same. That is basically why each community must learn to take care of itself: to make and use at least the great majority of what it consumes locally, and to dispose of it locally.

In general, actions that give people more control over their own lives enable them to move their communities toward sustainability. These actions include access to resources and the ability to play a role in managing them,

The solid waste that an average U.S. family produces annually FIGURE 2.14

The piles of cans, bottles, and newspapers on the left are what the average family now recycles. The trashcans and bags on the right are filled with the solid waste that remains after recycling. Photographed in San Diego, California.

A new degree of commitment to environmental issues has steadily grown within many religious faiths and organizations. Espousing convictions some see as "creation care," religious groups actively champion such issues as controlling global warming and protecting endangered species.

- In 1998 the National Council of Churches in the United States, representing 34 Christian denominations, began a major educational and advocacy program focused on the responsibility Christians share for the environment, specifically climate-warming issues. Jewish and Greek Orthodox groups have joined the effort.

- In 1999 an unusual alliance of environmentalists and Christians in southern California sued the U.S. Fish and Wildlife Service for refusing to designate critical habitat for seven endangered animal species.

- In 2000 more than 2,000 religious leaders met at the U.N. Millennium World Peace Summit of Religious and Spiritual Leaders. Protecting the environment was a major topic of discussion.

- In 2001 the U.N. Environment Programme and the Islamic Republic of Iran sponsored an International Seminar on Religion, Culture, and Environment. They considered ways to counter environmental degradation.

Environmental action at the community level
FIGURE 2.15

These women are part of the Green Belt Movement, which has planted millions of trees in Kenya.

access to education and training, and the right to participate in decision making (FIGURE 2.15). Local communities empowered by their nations will function well, and so, therefore, will the world as a whole.

Just as individuals and communities must commit themselves to new ethical standards if they are to approach the goal of sustainability in their own lives, so nations must establish fully their commitment to the principles of a sustainable society. No national plan is successful without a statement of its sustainability goals, accompanied by an educational program that explains why these goals are so important.

Although the necessity of creating a global alliance for sustainable development is obvious, it has proven very difficult in practice. Successfully dealing with global problems depends on international cooperation that contributes to an integrated global action. The strengthening of the United Nations as an effective force for global sustainability would contribute greatly to the creation of a sustainable, healthy, peaceful, and prosperous world.

WHAT KIND OF WORLD DO YOU WANT?

Perhaps the most important single lesson you will learn in this text is that those who live in highly developed countries are at the core of problems facing the global environment today. Highly developed countries consume a disproportionate share of resources and must reduce their levels of consumption if we are going to achieve global sustainability. We drain resources from the entire globe and thus contribute to a future in which neither our children nor our grandchildren will live in anything like the affluence that we experience now.

This book is about consequences. In the last 40 years or so, we have recognized the impact of human activities on the biosphere. We now understand this impact enough to know that we can't continue as we have and expect any sort of viable future for our species. If all

we do, as a result of this new knowledge, is make a few changes in consumer choices and write a few letters, it won't be nearly enough.

People hoping to preserve a livable world must become the next pioneers. A pioneer is one who ventures into unexplored territory, a process that is simultaneously terrifying and profoundly exciting. The unexplored territory in this case is the development of a truly different way for humans to exist in the world. No models exist for this kind of change. You must help create the political will for it with your numbers and your commitment. You must create the economic power with your thoughtful decisions as both consumers and leaders. You must create social change with your acceptance and respect of the differences among peoples.

Many people think that to accomplish this change will require reconnecting with the natural environment. It means, at a personal level, taking opportunities to be outside (even if it is a city park or a backyard garden), to listen to the wind, and to look at the exquisite variety of plants, insects, and other life-forms with which we coexist (FIGURE 2.16). Humans evolved in nature. Our immensely complex and multidimensional brains evolved precisely because we interacted with growing things, weather patterns, and other animals. The world we have created screens us from all that. The sophisticated devices we imagined and manufactured—such as televisions, computers, and automobiles—now define our world. One of your challenges will be to use technology as a tool but not to let it define your interaction with the world.

The new environmental revolution will require that you revalue yourself and your life according to a different set of ideals. Wealth and material possessions have come to mean success, at a tremendous cost to the planet. Such ideas are deeply imbedded and extremely compelling and will be difficult to change. You may need to throw off some of the myths of Western culture, such as the belief that "faster" and "more" inevitably mean "better." You will at least be called on to examine those myths very deeply and very thoughtfully and to decide what they mean to you.

The choices you make will have a greater impact on the future than those that any generation has had before. Even choosing to do nothing will have profound consequences for the future. At the same time, you face an incredible opportunity. Margaret Mead (1901–1978), the noted American anthropologist, once said, "Never doubt that a small group of thoughtful, committed people can change the world; indeed, it's the only thing that ever has!" This is a time in history when the best of human qualities—vision, courage, imagination, and concern—will play a critical role in establishing the nature of tomorrow's world.

CONCEPT CHECK STOP

What is Earth's carrying capacity? How is carrying capacity related to sustainability?

How are environmental goals best met at the local level? at the national level? at the global level?

What ecosystem services do living things provide?

The 2002 World Summit

In 1992 representatives from most of the world's countries met in Rio de Janeiro, Brazil, for the U.N. Conference on Environment and Development. Countries attending the conference examined environmental problems that are international in scope: pollution and deterioration of the planet's atmosphere and ocean, a decline in the number and kinds of organisms, and destruction of forests.

In addition, the 1992 participants adopted *Agenda 21,* a complex action plan of sustainable development for the 21st century in which future economic development, particularly in developing countries, will be reconciled with environmental protection. *Agenda 21* recommended more than 2,500 actions to deal with our most urgent environmental, health, and social problems.

In 2002 the United Nations sponsored the World Summit on Sustainable Development in Johannesburg, South Africa, to assess the progress and failures in the 10 years since the 1992 summit. The 2002 summit (pictured at the right) considered population growth, poverty, the depletion of fresh water, food security, the use of unsustainable energy sources, and land degradation and habitat loss.

Some of the world's problems have worsened since 1992. The human population has grown by 1 billion. The per capita incomes in more than 80 countries are lower than they were in 1992. Between 2 and 5 million people die each year from water-related diseases, a result of the unavailability of basic sanitation services for about 2.4 billion people. About one-third of Earth's soil is so degraded that it can't be used to grow food.

Thus, with few exceptions, we have made little progress in solving the world's most serious environmental problems. The real world is operating much as it always has. Also, many governments have shifted their attention to other challenges such as terrorism and worsening international tensions. Meanwhile, scientific warnings on important environmental problems such as global climate change have grown increasingly severe.

The 2002 summit reached conceptual agreements and produced a timetable for meeting goals to reduce poverty and environmental degradation. But these agreements work only if the world's nations enforce them. The 2002 summit, like the 1992 summit, has little power to accomplish anything.

Despite the lack of significant progress at the international level since 1992, national, state, and local levels have made important progress. Many countries have enacted more stringent air pollution laws, including the phasing out of leaded gasoline. More than 100 countries have created sustainable-development commissions. Also, the World Bank has invested $8.5 billion in sustainable development projects around the world.

U.S. delegates at a World Summit press briefing.

CHAPTER SUMMARY

1 Human Use of the Earth

1. **Sustainable development** is economic growth that meets the needs of the present without compromising the ability of future generations to meet their own economic needs. Environmentally sound decisions, economically viable decisions, and socially equitable decisions interact to promote sustainable development.

2. **Sustainable consumption** is the use of goods and services that satisfy basic human needs and improve the quality of life but that also minimize the use of resources so they are available for future use. Population growth, economic activities, technology choices, social values, and government policies add complex dimensions to the concept of sustainable consumption.

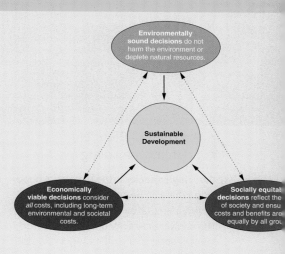

Environmentally **sound decisions** do not harm the environment or deplete natural resources.

Sustainable Development

Economically **viable decisions** consider *all* costs, including long-term environmental and societal costs.

Socially equitable **decisions** reflect the of society and ensu costs and benefits are equally by all grou

2 Human Values and Environmental Problems

1. **Environmental ethics** is a field of applied ethics that considers the moral basis of environmental responsibility and how far this responsibility extends. Environmental ethicists consider how humans should relate to the natural environment.

2. An **environmental worldview** is a worldview that helps us make sense of how the environment works, our place in the environment, and right and wrong environmental behaviors. The **Western worldview** is an understanding of our place in the world based on human superiority and dominance over nature, the unrestricted use of natural resources, and increased economic growth to manage an expanding industrial base. The **deep ecology worldview** is an understanding of our place in the world based on harmony with nature, a spiritual respect for life, and the belief that humans and all other species have an equal worth.

3. **Voluntary simplicity** recognizes that individual happiness and quality of life are not necessarily linked to the accumulation of material goods. People who embrace voluntary simplicity recognize that a person's values and character define that individual more than how many things he or she owns.

3 Environmental Justice

1. **Environmental justice** is the right of every citizen, regardless of age, race, gender, social class, or other factor, to adequate protection from environmental hazards.

CHAPTER SUMMARY

4 An Overall Plan for Sustainable Living

1. **Carrying capacity** is the maximum population that can be sustained by a given environment or by the world as a whole. The carrying capacity of a given environment is determined ultimately by its ability to absorb wastes and renew itself. Worldwide attention to family planning is one way to stay within Earth's carrying capacity; we can help the population stabilize if we make family planning available for everyone. Addressing world poverty and raising standards of living in every country through the education of children and the reduction of illiteracy are other ways to achieve sustainability.

2. We have an ethical obligation to respect and care for other species, which are, as far as we know, our only living companions in the universe. We have a clear interest in protecting Earth's **biological diversity**, the number and variety of Earth's organisms, because we obtain from them all of our food, most of our medicines, building and clothing materials, and many other products. **Ecosystem services** are important environmental benefits, such as clean air to breathe, clean water to drink, and fertile soil in which to grow crops, that the natural environment provides.

3. Any long-term improvement in the condition of the world must start at the local level with individuals; each of us makes a difference, and it is ultimately our collective activities that make the world what it is. At the national level, constitutional documents and other statements of national policy should lay out commitments to the principles of a sustainable society and make possible the generation of appropriate laws and suitable economic policies.

KEY TERMS

- **sustainable development** p. 26
- **sustainable consumption** p. 27
- **environmental ethics** p. 29
- **environmental worldview** p. 30
- **Western worldview** p. 30
- **deep ecology worldview** p. 31
- **environmental justice** p. 35
- **carrying capacity** p. 36
- **biological diversity** p. 37
- **ecosystem services** p. 37

CRITICAL AND CREATIVE THINKING QUESTIONS

1. State if each of the following statements reflects the Western worldview, the deep ecology worldview, or both. Explain your answers.
   a. Species exist to be used by humans.
   b. All organisms, humans included, are interconnected and interdependent.
   c. There is a unity between humans and nature.
   d. Humans are a superior species capable of dominating other organisms.
   e. Humans should protect the environment.
   f. Nature should be used, not preserved.
   g. Economic growth will help Earth manage an expanding human population.
   h. Humans have the right to modify the environment to benefit society.
   i. All forms of life are intrinsically valuable and therefore have the right to exist.

2. How are sustainable consumption and voluntary simplicity related?

3. What social groups generally suffer the most from environmental pollution and degradation? What social groups generally benefit from this situation?

4. In its broadest sense, how does environmental justice relate to highly developed countries and less developed countries?

5. What is the role of education in changing personal attitudes and practices that affect the environment?

6. Write a one-page essay describing what kind of world you want to leave for your children.

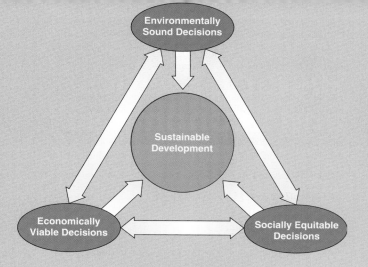

7. Are an improved standard of living and a reduction in the level of consumption mutually exclusive? Give an example that supports your answer.

8. Write a paragraph that briefly describes your environmental worldview.

9. Development is sometimes equated with economic growth. Explain the difference between sustainable development and development as an indicator of economic growth.

10. How do the three factors shown in the figure interact to promote sustainable development?

What is happening in this picture **?**

These fishermen have seen their catches decline in recent years. Why do you think fish shortages are occurring around the world?

How does Garrett Hardin's description of the tragedy of the commons in medieval Europe relate to the ocean's fisheries today?

Name an additional example of a global commons other than the one shown in this photograph.

Environmental History, Politics, and Economics

OLD-GROWTH FORESTS
OF THE PACIFIC NORTHWEST

During the late 1980s and 1990s, an environmental controversy began in western Oregon, Washington, and northern California that continues today. At stake were thousands of jobs and the future of large tracts of *old-growth coniferous forest*. The northern spotted owl *(see photograph)*, listed as a threatened species since 1990 under the *Endangered Species Act*, came to symbolize the confrontation.

Environmentalists consider old-growth forests—forests that have never been logged—a national treasure. In 1991 a court ordered the suspension of logging in about 1.2 million hectares (3 million acres) of federal forest in the Pacific Northwest, home to the northern spotted owl and 40 other endangered or threatened species. The timber industry stated that thousands of jobs would be lost.

In 1993, President Bill Clinton convened a summit in Portland, Oregon. The 1994 *Northwest Forest Plan* that arose from this summit represented a compromise between environmental and timber interests. The plan protected habitat of endangered species and allowed some logging to resume on federal forests in Washington, Oregon, and northern California. Thanks to a healthy infusion of state and federal aid to the area, many timber workers were retrained for other careers.

Further legislative wrangling led to changes on both sides of the issue. Loggers gained greater access to parts of the forest that the Northwest Forest Plan had declared off-limits, but the courts ordered the federal agencies overseeing logging on federal land to complete surveys of endangered and threatened species before granting the timber industry permission to log.

In this chapter, we consider environmental science in the context of history, politics, and economics.

CHAPTER OUT

Conservation and Preservation
of Resources p. 48

Environmental History p. 49

Environmental Legislation p. 57

Environmental Economics p. 60

Case Study: Environmental Prob
in Central and Eastern Europe p.

Conservation and Preservation of Resources

Resources are any part of the natural environment used to promote the welfare of people or other species. Examples of resources include air, water, soil, forests, minerals, and wildlife. **Conservation** is the sensible and careful management of natural resources. Humans have practiced conservation of natural resources for thousands of years. Three thousand years ago, the Phoenicians terraced hilly farmland to prevent soil erosion. More than two thousand years ago, the Greeks practiced crop rotation to maintain yields on farm-lands, and the Romans practiced irrigation. Other Europeans gradually adopted and further refined these and other conservation techniques (**FIGURE 3.1**).

Conservation is different from preservation. Conservation involves sustainability—that is, use of resources without inflicting excessive environmental damage, so that the resources are available not only for current needs but also for the needs of future generations. **Preservation** is concerned with setting aside undisturbed areas, maintaining them in a pristine state, and protecting them from human activities that might alter their "natural" state.

Conservation did not become a popular movement until the early 20th century. At that time, expanding industrialization, coupled with enormous growth in the human population, began to increase pressure on the world's supply of natural resources.

Soil conservation FIGURE 3.1

Plowing and planting fields in curves that conform to the natural contours of the land reduce soil erosion.

CONCEPT CHECK STOP

What is conservation?

How does conservation differ from preservation?

What is preservation?

Environmental History

From the establishment of the first permanent English colony at Jamestown, Virginia, in 1607, the first two centuries of U.S. history were a time of widespread environmental destruction. Land, timber, wildlife, rich soil, clean water, and other resources were cheap and seemingly inexhaustible. The European settlers did not dream that the bountiful natural resources of North America would one day become scarce. During the 1700s and most of the 1800s, many Americans had a **frontier attitude**, a desire to conquer nature and put its resources to use in the most lucrative manner possible.

PROTECTING FORESTS

The great forests of the Northeast were cut down within a few generations, and, shortly after the Civil War in the 1860s, loggers began deforesting the Midwest at an alarming rate. Within 40 years, they had deforested an area the size of Europe, stripping Minnesota, Michigan, and Wisconsin of virgin forest. By 1897 the sawmills of Michigan had processed 160 billion board feet of white pine, leaving less than 6 billion board feet standing in the whole state.

During the 19th century, many U.S. naturalists began to voice concerns about conserving natural resources. **John James Audubon** (1785–1851) painted lifelike portraits of birds and other animals in their natural surroundings that aroused widespread public interest in the wildlife of North America (**FIGURE 3.2**). **Henry David Thoreau** (1817–1862), a prominent U.S. writer, lived for 2 years on the shore of Walden Pond

Tanagers FIGURE 3.2

This portrayal is one of 500 engravings in Audubon's classic *The Birds of America,* completed in 1844. Shown are two male Louisiana tanagers (also called western tanagers, top) and male and female scarlet tanagers (bottom).

near Concord, Massachusetts. There he observed nature and contemplated how people could economize and simplify their lives to live in harmony with the natural world. **George Perkins Marsh** (1801–1882) was a farmer, linguist, and diplomat at various times during his life. Today he is most remembered for his book *Man and Nature*, published in 1864, which provided one of the first discussions of humans as agents of global environmental change.

In 1875 a group of public-minded citizens formed the American Forestry Association with the intent of influencing public opinion against the wholesale destruction of America's forests. Sixteen years later, in 1891, the *Forest Reserve Act* (which was part of the General Land Law Revision Act) gave the president the authority to establish forest reserves on public (federally owned) land. Benjamin Harrison (1833–1901), Grover Cleveland (1837–1908), William McKinley (1843–1901), and Theodore Roosevelt (1858–1919) used this law to put a total of 17.4 million hectares (43 million acres) of forest, primarily in the West, out of the reach of loggers.

In 1907 angry Northwest congressmen pushed through a bill stating that national forests could no longer be created by the president but would require an act of Congress. Roosevelt signed the bill into law, but not before designating 21 new national forests that totaled 6.5 million hectares (16 million acres).

Roosevelt appointed **Gifford Pinchot** (1865–1946) the first head of the U.S. Forest Service. Both Roosevelt and Pinchot were **utilitarian conservationists** who viewed forests in terms of their usefulness to people—such as in providing jobs and renewable re-

■ **utilitarian conservationist**
A person who values natural resources because of their usefulness to us but uses them sensibly and carefully.

■ **biocentric preservationist**
A person who believes in protecting nature from human interference because all forms of life deserve respect and consideration.

sources. Pinchot supported expanding the nation's forest reserves and managing them scientifically, for instance harvesting trees only at the rate at which they re-grow. Today, national forests are managed for multiple uses, from biological habitats to recreation to timber harvest to cattle grazing.

ESTABLISHING NATIONAL PARKS AND MONUMENTS

Congress established the world's first national park in 1872 after a party of Montana explorers reported on the natural beauty of the canyon and falls of the Yellowstone River. Yellowstone National Park now includes parts of Idaho, Montana, and Wyoming. In 1890 the *Yosemite National Park Bill* established the Yosemite and Sequoia national parks in California, largely in response to the efforts of a single man, naturalist and writer **John Muir** (1838–1914) (FIGURE 3.3). Muir was a **biocentric preservationist**. Muir also founded the *Sierra Club*, a national conservation organization that is still active on a range of environmental issues.

President Theodore Roosevelt *(left)* **and John Muir** FIGURE 3.3

Photo was taken on Glacier Point above Yosemite Valley, California.

In 1906 Congress passed the *Antiquities Act,* which authorized the president to set aside sites that had scientific, historic, or prehistoric importance. By 1916 there were 16 national parks and 21 national monuments, under the loose management of the U.S. Army. Today there are 58 national parks and 73 national monuments under the management of the National Park Service.

Controversy over battles such as the Hetch Hetchy Valley generated a strong sentiment that the nation should better protect its national parks (**Fig-ure 3.4**). In 1916 Congress created the National Park Service to manage the national parks and monuments for the enjoyment of the public, "without impairment." It was this clause that gave a different outcome to another battle, fought in the 1950s between conservationists and dam builders over the construction of a dam within Dinosaur National Monument. No one could deny that to fill the canyon with 400 feet of water would "impair" it. This victory for conservation established the "use without impairment" clause as the firm backbone of legal protection afforded our national parks and monuments.

Hetch Hetchy Valley in Yosemite FIGURE 3.4

Some environmental battles involving the protection of national parks were lost. John Muir's Sierra Club fought such a battle with the city of San Francisco over its efforts to dam a river and form a reservoir in the Hetch Hetchy Valley, which lay within Yosemite National Park and was as beautiful as Yosemite Valley. In 1913 Congress voted to approve the dam. A view in Hetch Hetchy Valley before (**A**) and after (**B**) Congress approved the dam.

CONSERVATION IN THE MID-20TH CENTURY

During the Great Depression, the federal government financed many conservation projects to provide jobs for the unemployed. During his administration, **Franklin Roosevelt** (1882–1945) established the Civilian Conservation Corps, which employed 500,000 young men to plant trees, make paths and roads in national parks and forests, build dams to control flooding, and perform other activities that protected natural resources.

During the droughts of the 1930s, windstorms carried away much of the topsoil in parts of the Great Plains, forcing many farmers to abandon their farms and search for work elsewhere. The *American Dust Bowl* alerted the United States to the need for soil conservation, and President Roosevelt formed the Soil Conservation Service in 1935.

Aldo Leopold (1886–1948) was a wildlife biologist and environmental visionary who was extremely influential in the conservation movement of the mid- to late-20th century (**FIGURE 3.5**). His textbook, *Game Management,* published in 1933, supported the passage of a 1937 act in which new taxes on sporting weapons and ammunition funded wildlife management and research. Leopold also wrote philosophically about humanity's relationship with nature and about the need to conserve wilderness areas in *A Sand County Almanac,* published in 1949. Leopold argued persuasively for a land ethic and the sacrifices that such an ethic requires.

Aldo Leopold had a profound influence on many American thinkers and writers, including **Wallace Stegner** (1909-1993), who penned his famous "Wilderness Essay" in 1962. Stegner's essay, written to a commission that was conducting a national inventory of wilderness lands, helped create support for the passage of the *Wilderness Act* of 1964. Stegner wrote:

> Something will have gone out of us as a people if we ever let the remaining wilderness be destroyed; if we permit the last virgin forests to be turned into comic books and plastic cigarette cases; if we drive the few remaining members of the wild species into zoos or to extinction; if we pollute the last clean air and dirty the last clean streams and push our paved roads through the last of the silence, so that never again will Americans be free in their own country from the noise, the exhausts, the stinks of human and automotive waste . . .
>
> We simply need that wild country available to us, even if we never do more than drive to its edge and look in. For it can be a means of reassuring ourselves of our sanity as creatures, a part of the geography of hope.

Aldo Leopold FIGURE 3.5

Leopold's *A Sand County Almanac* is widely considered an environmental classic.

During the 1960s, public concern about pollution and resource quality began to increase, in large part due to marine biologist **Rachel Carson** (1907–1964). Carson wrote about interrelationships among living organisms, including humans, and the natural environment (**FIGURE 3.6**). Her most famous work, *Silent Spring*, was published in 1962. In it Carson wrote against the indiscriminate use of pesticides:

> *Pesticide sprays, dusts, and aerosols are now applied almost universally to farms, gardens, forests, and homes—nonselective chemicals that have the power to kill every insect, the "good" and the "bad," to still the song of birds and the leaping of fish in the streams, to coat the leaves with a deadly film, and to linger on in soil—all this though the intended target may be only a few weeds or insects. Can anyone believe it is possible*

> *to lay down such a barrage of poisons on the surface of the earth without making it unfit for all life? They should not be called "insecticides," but "biocides."*

Silent Spring heightened public awareness and concern about the dangers of uncontrolled use of DDT and other pesticides, including poisoning birds and other wildlife and contaminating human food supplies. Ultimately, the book led to restrictions on the use of certain pesticides. Around this time, the media began to increase its coverage of environmental incidents, such as hundreds of deaths in New York City from air pollution (1963); closed beaches and fish kills in Lake Erie from water pollution (1965); and detergent foam in a creek in Pennsylvania (1966).

In 1968, when the population of Earth was "only" 3.5 billion people, ecologist **Paul Ehrlich** published *The Population Bomb*. In it he described the damage occurring to Earth's life support system because it was supporting such a huge population, including the depletion of essential resources such as fertile soil, groundwater, and other living organisms. Ehrlich's book raised the public's awareness of the dangers of overpopulation and triggered debates on how to deal effectively with population issues.

THE ENVIRONMENTAL MOVEMENT

Until 1970 the voice of **environmentalists**, people concerned about the environment, was heard in the United States primarily through societies such as the Sierra Club and the National Wildlife Federation. There was no generally perceived **environmental movement** until the spring of 1970, when **Gaylord Nelson**, former senator of Wisconsin, urged Harvard graduate student **Denis Hayes** to organize the first nationally celebrated Earth Day. This event awakened U.S. environmental consciousness to population growth, overuse of resources, and pollution and degradation of the environment. On Earth Day 1970, an estimated 20 million people in the United States planted trees, cleaned roadsides and riverbanks, and marched in parades to demonstrate their support of improvements in resource conservation and environmental quality.

Rachel Carson FIGURE 3.6

Carson's book, *Silent Spring*, heralded the beginning of the environmental movement.

Earth Day 1970 in New York City

FIGURE 3.7

In the years that followed the first Earth Day (FIGURE 3.7), environmental awareness and the belief that individual actions could repair the damage humans were doing to Earth became a pervasive popular movement. Musicians and other celebrities popularized environmental concerns. Many of the world's religions—such as Christianity, Judaism, Islam, Hinduism, Buddhism, Taoism, Shintoism, Confucianism, and Jainism—embraced environmental themes such as protecting endangered species and controlling global warming.

By Earth Day 1990, the movement had spread around the world, signaling the rapid growth in environmental consciousness. An estimated 200 million people in 141 nations demonstrated to increase public awareness of the importance of individual efforts ("Think globally, act locally"). The theme of Earth Day 2000, "Clean Energy Now," reflected the dangers of global climate change and what individuals and communities could do: replace fossil fuel energy sources, which produce greenhouse gases, with solar electricity, wind power, and the like. However, by 2000 many environmental activists had begun to think that the individual actions Earth Day espouses, while collectively important, are not as important as pressuring governments and large corporations to make environmentally friendly decisions. FIGURE 3.8 shows a timeline of selected environmental events since Earth Day 1970.

CONCEPT CHECK STOP

What role did each of the following have in U.S. environmental history: protecting forests; establishing and protecting national parks and monuments; conservation in the mid-20th century; the environmental movement of the late-20th century?

How did Aldo Leopold influence the conservation movement of the mid- to late-20th century?

What was the environmental contribution of Rachel Carson?

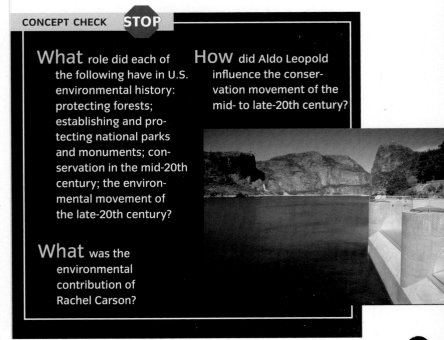

Timeline of selected environmental events, from 1970 to the present FIGURE 3.8

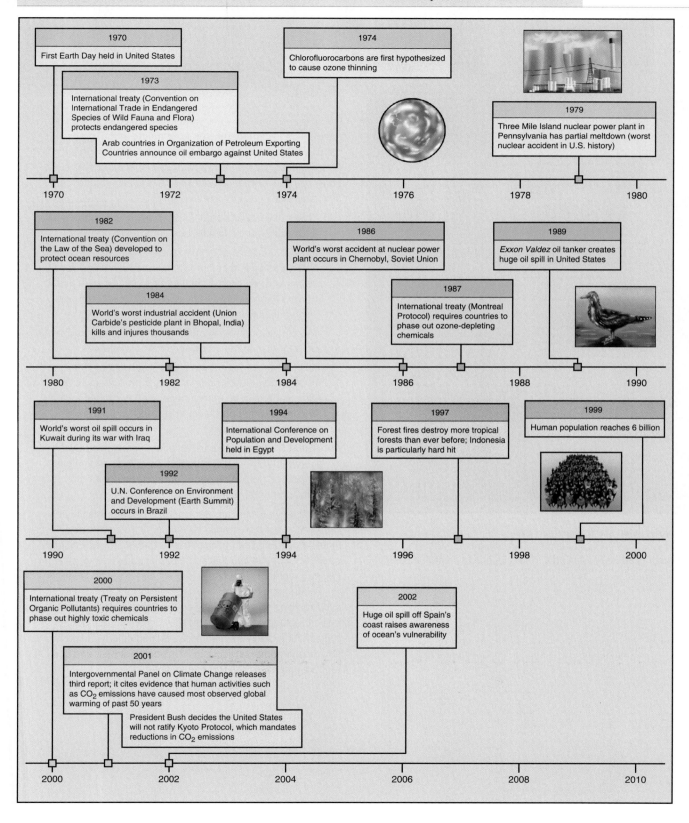

1970
First Earth Day held in United States

1973
International treaty (Convention on International Trade in Endangered Species of Wild Fauna and Flora) protects endangered species

Arab countries in Organization of Petroleum Exporting Countries announce oil embargo against United States

1974
Chlorofluorocarbons are first hypothesized to cause ozone thinning

1979
Three Mile Island nuclear power plant in Pennsylvania has partial meltdown (worst nuclear accident in U.S. history)

1970   1972   1974   1976   1978   1980

1982
International treaty (Convention on the Law of the Sea) developed to protect ocean resources

1984
World's worst industrial accident (Union Carbide's pesticide plant in Bhopal, India) kills and injures thousands

1986
World's worst accident at nuclear power plant occurs in Chernobyl, Soviet Union

1987
International treaty (Montreal Protocol) requires countries to phase out ozone-depleting chemicals

1989
Exxon Valdez oil tanker creates huge oil spill in United States

1980   1982   1984   1986   1988   1990

1991
World's worst oil spill occurs in Kuwait during its war with Iraq

1992
U.N. Conference on Environment and Development (Earth Summit) occurs in Brazil

1994
International Conference on Population and Development held in Egypt

1997
Forest fires destroy more tropical forests than ever before; Indonesia is particularly hard hit

1999
Human population reaches 6 billion

1990   1992   1994   1996   1998   2000

2000
International treaty (Treaty on Persistent Organic Pollutants) requires countries to phase out highly toxic chemicals

2002
Huge oil spill off Spain's coast raises awareness of ocean's vulnerability

2001
Intergovernmental Panel on Climate Change releases third report; it cites evidence that human activities such as CO_2 emissions have caused most observed global warming of past 50 years

President Bush decides the United States will not ratify Kyoto Protocol, which mandates reductions in CO_2 emissions

2000   2002   2004   2006   2008   2010

EnviroDiscovery   Environmental Literacy

Because responses to environmental problems depend on the public's awareness and understanding of the issues and the underlying scientific concepts involved, environmental education is critical to appropriate decision making. The emphasis on environmental education has grown dramatically over the years:

- Three international treaties supporting environmental education went into effect between 1975 and 1990.

- In 1990, 22 university presidents from 13 nations issued a declaration of their commitment to environmental education and research at their institutions. More than 300 university presidents from at least 40 countries have since followed suit.

- More than 30 states require some form of environmental education in primary and secondary schools.

- The National Environmental Education Act of 1990 requires the Environmental Protection Agency to provide leadership in promoting environmental education and increasing public awareness and knowledge of environmental issues.

- The U.N. Decade of Education for Sustainable Development (2005–2014) is dedicated to improving basic education, including public understanding about environmental sustainability.

The North American Association for Environmental Education has issued guidelines for educators to help them select materials such as textbooks and films that are based on sound scientific evidence and that present a balanced perspective on environmental problems.

However, a backlash against environmental education occurred beginning in the late 1990s. Some conservative research groups criticized what they perceived as a biased presentation of environmental issues, particularly the promotion of environmental activism, in schools. Fairness and accuracy are emphasized in these guidelines.

Environmental education

Fourth graders in Marietta, Georgia, study a small stream.

Environmental Legislation

Well-publicized ecological disasters, such as the 1969 oil spill off the coast of Santa Barbara, California, and overwhelming public support for the Earth Day movement, resulted in the January 1970 signing of the *National Environmental Policy Act (NEPA)*. The Environmental Protection Agency (EPA) was created in July of the same year. A key provision of NEPA states that the federal government must consider the environmental impact of a proposed federal action, such as financing highway or dam construction, when making decisions about that action. NEPA provides the basis for developing detailed **environmental impact statements (EISs)** to accompany every federal recommendation or proposal for legislation. An EIS is a document that describes the nature of the proposal, why it is needed, short- and long-term environmental impacts of the proposal, and possible alternatives to the proposed action that would create fewer adverse effects. NEPA also requires solicitation of public comments when preparing an EIS, which generally provides a broader perspective on the proposal and its likely effects.

NEPA established the **Council on Environmental Quality** to monitor the required EISs and report directly to the president. Because this council had no enforcement powers, NEPA was originally considered innocuous, more a statement of good intentions than a regulatory policy. During the next few years, however, environmental activists took people, corporations, and the federal government to court to challenge their EISs or use them to block proposed development. The courts decreed that EISs thoroughly analyze the environmental consequences of anticipated projects on soil, water, and organisms and that EISs be made available to the public (**FIGURE 3.9**). These rulings put very sharp teeth into NEPA—particularly the provision for public

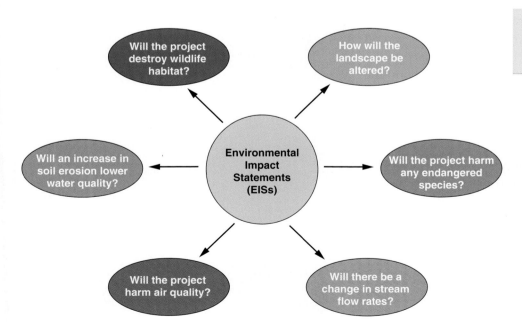

Environmental impact statements
FIGURE 3.9

These detailed statements help federal agencies consider the environmental impacts of proposed activities. When the environmental impacts are judged too severe, alternative actions are considered.

scrutiny, which places intense pressure on federal agencies to respect EIS findings.

NEPA revolutionized environmental protection in the United States. Federal agencies manage federal highway construction, flood and erosion control, military projects, and many other public works. They oversee nearly one-third of the land in the United States. Federal holdings include fossil fuel and mineral reserves, millions of hectares of public grazing land, and public forests. Since 1970 very little has been done to any of them without some sort of environmental review. NEPA has also influenced environmental legislation in at least 36 states and in other countries, including Canada, Australia, France, New Zealand, and Sweden.

Although almost everyone agrees that NEPA has successfully reduced adverse environmental impacts of federal activities and projects, it has its critics. Some environmentalists complain that EISs are sometimes incomplete or that reports are ignored when decisions are made. Other critics think the EISs delay important projects ("paralysis by analysis") because the documents are too involved, take too long to prepare, and are often the targets of lawsuits.

ENVIRONMENTAL REGULATIONS

Once an environmental problem becomes widely recognized, the process of environmental regulation begins with a U.S. congressperson drafting legislation. Ideally, before the legislation is drafted, the tradeoffs for several proposed alternative actions are evaluated. This process, known as **full cost accounting**, is a valuable economic tool in environmental decision making.

After the legislation is passed and the president signs it, it usually goes to the Environmental Protection Agency (EPA), which was created to translate the law's language into regulations that specify allowable levels of pollution. Before the regulations officially become law, several rounds of public comments allow affected parties to present their views; the EPA is required

full cost accounting The process of evaluating and presenting to decision makers the relative benefits and costs of various alternatives.

to respond to all of these comments. Then the Office of Management and Budget reviews the new regulations. Implementation and enforcement of the new law often fall to state governments, which must send the EPA details for achieving the goals of the new regulations.

ACCOMPLISHMENTS OF ENVIRONMENTAL LEGISLATION

During the period since Earth Day 1970, Congress has passed almost 40 major environmental laws that address a wide range of issues, such as endangered species, clean water, clean air, energy conservation, hazardous wastes, and pesticides. This tough interlocking mesh of laws greatly increased federal regulation of pollution to improve environmental quality.

Despite imperfections, environmental legislation has had overall positive effects. Since 1970,

- Eight national parks have been established (**FIGURE 3.10**), and the National Wilderness Preservation System now totals more than 43 million hectares (106 million acres).

Mesquite Flat Dunes in Death Valley National Park
FIGURE 3.10

This national park, formerly a national monument, was created in 1994.

- Millions of hectares of farmland particularly vulnerable to erosion have been withdrawn from production, reducing soil erosion by more than 60 percent.

- Many endangered species are recovering, and the American alligator, California gray whale, and bald eagle have recovered enough to be removed from the endangered species list. (However, dozens of other species, such as the manatee and Kemp's ridley sea turtle, have suffered further declines or extinction since 1970.)

Although we still have a long way to go, pollution control efforts through legislation have been particularly successful. According to the EPA's "Draft Report on the Environment 2003,"

- Emissions of 6 important air pollutants have dropped by more than 25 percent.

- Since 1990, levels of wet sulfate, a major component of acid rain, have dropped by 20 to 30 percent.

- Release of toxic chemicals into water and air from industrial sources declined by 48 percent since 1988.

- Fewer rivers and streams are in violation of water quality standards. However, fish-consumption advisories relating to specific toxins such as mercury or polychlorinated biphenyls (PCBs) have increased, possibly because more consistent monitoring has led to more accurate measurements of these pollutants.

- In 2002, 94 percent of the U.S. population received water from community water systems that met health-based drinking-water standards, up from 79 percent in 1993 (FIGURE 3.11).

- Of the 1,498 contaminated sites listed on the Superfund National Priorities List in 2002, 846 are now cleaned up. This figure is up from 149 in 1992.

In the 1960s and 1970s, pollution was often obvious—witness the Cuyahoga River in Cleveland, Ohio, which burst into flames from the oily pollutants on its

Water treatment plant FIGURE 3.11

The water supply for a town or city is treated before use so it is safe to drink. Photographed in Miami, FLorida.

surface several times, including a highly publicized burn in 1969. Legislators, the media, and the public typically perceive things like burning rivers as serious threats that require immediate attention without regard to the cost. Now that the most obvious pollution problems in the United States are largely addressed, more and more people look at environmental cleanup in terms of benefit versus cost. Thus, economics is increasingly important in environmental legislation and policymaking.

CONCEPT CHECK **STOP**

Why is the National Environmental Policy Act the cornerstone of U.S. environmental law?

What are environmental impact statements?

Environmental Economics

LEARNING OBJECTIVES

Explain how economics is related to natural capital. Make sure you include sources and sinks.

Give two reasons why the national income accounts are incomplete estimates of national economic performance.

Distinguish among the following economic terms: marginal cost of pollution, marginal cost of pollution abatement, and optimum amount of pollution.

Describe various approaches to pollution control, including command and control regulation and incentive-based regulation.

Economics is the study of how people use their limited resources to try to satisfy their unlimited wants. Seen through an economist's eyes, the world is one large marketplace, where resources are allocated to a variety of uses, and where goods—a car, a pair of shoes, a hog—and services—a haircut, a museum tour, an education—are consumed and paid for. In a free market, supply and demand determine the price of a good. If something in great demand is in short supply, its price will be high. High prices encourage suppliers to produce more of a good or service, as long as the selling price is higher than the cost of producing the good or service. This interaction of demand, supply, price, and cost underlies much of what happens in the U.S. economy, from the price of a hamburger to the cycles of economic expansion (increase in economic activity) and recession (slow-down in economic activity).

Economies depend on the natural environment as *sources* for raw materials and *sinks* for waste products (**FIGURE 3.12**). Both sources and sinks contribute to **natural capital**. According to economists, the environment provides natural capital for our production and consumption. Resource degradation and pollution represent the overuse of natural capital. *Resource degradation* is the overuse of sources, and *pollution* is the overuse of sinks; both threaten our long-term economic future.

natural capital Earth's resources and processes that sustain living organisms, including humans; includes minerals, forests, soils, groundwater, clean air, wildlife, and fisheries.

national income accounts A measure of the total income of a nation's goods and services for a given year.

NATIONAL INCOME ACCOUNTS AND THE ENVIRONMENT

Much of our economic well-being flows from natural, rather than human-made, assets like land, rivers, the ocean, oil, timber, and the air we breathe.

Ideally, for the purposes of economic and environmental planning, the **national income accounts** should measure the use and misuse of natural resources and the environment. Two measures used in national income accounting are *gross domestic product (GDP)* and *net domestic product (NDP)*. Both GDP and NDP provide estimates of national economic performance that are used to make important policy decisions.

Unfortunately, current national income accounting practices do not incorporate environmental factors. Two important conceptual problems exist with the way the national income accounts currently handle the economic use of natural resources and the environment: natural resource depletion and the cost and benefits of pollution control. Better accounting for environmental quality—both resource depletion and pollution—would help address whether for any given activity the benefits (both economic and environmental) exceed the costs.

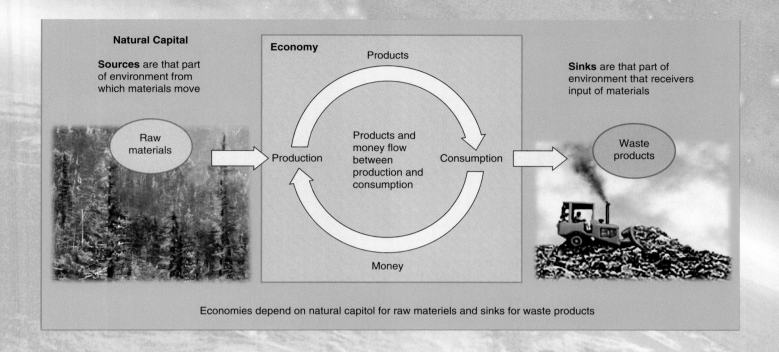

Natural Capital

Sources are that part of environment from which materials move

Raw materials

Economy

Products

Production

Products and money flow between production and consumption

Consumption

Money

Sinks are that part of environment that receivers input of materials

Waste products

Economies depend on natural capitol for raw materiels and sinks for waste products

Natural resource depletion

If a manufacturing firm produces some product (output) but in the process wears out a portion of its plant and equipment, the firm's output is counted as part of GDP, but the depreciation of capital is subtracted in the calculation of NDP. Thus NDP is a measure of the net production of the economy, after a deduction for used-up capital. In contrast, when an oil company drains oil from an underground field, the value of the oil produced is counted as part of the nation's GDP, but no offsetting deduction to NDP is made to account for the fact that nonrenewable resources were used up (FIGURE 3.13).

In principle, the draining of the oil field is a type of depreciation, and the oil company's net product should be accordingly reduced. The same point applies to any other natural resource that is depleted in the process of production. Natural capital is a very large part of a country's economic wealth, and we should treat it the same as human-made capital.

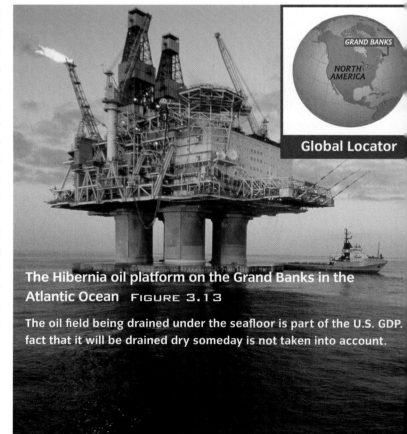

Global Locator

The Hibernia oil platform on the Grand Banks in the Atlantic Ocean FIGURE 3.13

The oil field being drained under the seafloor is part of the U.S. GDP. fact that it will be drained dry someday is not taken into account.

Pollution cleanup and GDP FIGURE 3.14

Workers in yellow protective clothing test toxic waste prior to safely disposing of it in a hazardous waste landfill. These cleanup costs should be added to the GDP accounts because the cleanup improves the environment.

The cost and benefits of pollution control

Imagine that a company has the following choices: It can produce $100 million worth of output and, at the same time, dump its wastes, polluting the local river. Alternatively, if the company uses 10 percent of its workers to properly dispose of its wastes, it avoids polluting but only gets $90 million of output. Under current national income accounting rules, if the firm chooses to pollute rather than not to pollute, it will make a larger contribution to GDP ($100 million rather than $90 million), because the national income accounts attach no explicit value to a clean river. In an ideal accounting system, the economic cost of environmental degradation is subtracted in the calculation of a firm's contribution to GDP, and activities that improve the environment—because they provide real economic benefits—are added to GDP (FIGURE 3.14). To summarize, we should subtract estimates of environmental damage from GDP.

Incorporating resource depletion and pollution into the national income accounting is important because GDP and related statistics are used continually in policy analyses. An increasing number of economists, government planners, and scientists support replacing GDP and NDP with a more comprehensive measure of national income accounting that includes estimates of both natural resource depletion and the environmental cost of economic activities.

AN ECONOMIST'S VIEW OF POLLUTION

An important aspect of the operation of a free-market system is that the person consuming a product should pay for all the cost of producing it. However, consumption or production of a product often has an **external cost**.

A product's market price does not usually reflect an external cost—that is, the buyer or seller doesn't pay for it. As a result, the market system generally does not operate in the most efficient way.

Consider the following example of an external cost. If an industry makes a product and, in so doing, also releases a pollutant into the environment, the product is bought at a price that reflects the cost of making it but not the cost of the pollutant's damage to the environment. This damage is the external cost of the product. Because this damage is not included in the product's price and because the consumer may not know that the pollution exists or that it harms the environment, the cost of the pollution has no impact on the consumer's decision to buy the product. As a result, consumers of the product may buy more of it than they would if its true cost, including the cost of pollution, were known or reflected in the selling price.

The failure to add the price of environmental damage to the cost of products generates a market force that increases pollution. From the perspective of economics, then, one of the causes of the world's pollution problem is the failure to consider negative external costs in the pricing of goods. We now examine industrial pollution from an economist's viewpoint, as a policy-making failure. Keep in mind, however, that lessons about the economics of industrial pollution also apply to other environmental issues (such as resource degradation) where harm to the environment is a consequence of economic activity.

■ **external cost**
A harmful environmental or social cost that is borne by people not directly involved in buying or selling a product.

■ **marginal cost of pollution** The added cost for all present and future members of society of an additional unit of pollution.

How much pollution is acceptable?

To assign a proper price to pollution, economists first try to answer the basic question "How much pollution should we allow in our environment?" Imagine two environmental extremes: a wilderness in which no pollution is produced but neither are goods, and a "sewer" that is completely polluted from excess production of goods. In our world, a move toward a better environment almost always entails a cost in terms of goods.

How do we, as individuals, as a country, and as part of the larger international community, decide where we want to be between the two extremes of a wilderness and a sewer? Economists analyze the marginal costs of environmental quality and of other goods to answer such questions. A **marginal cost** is the additional cost associated with one more unit of something. Two examples of marginal costs associated with pollution are the effects of pollution on human health and on organisms in the natural environment.

The trade-off between environmental quality versus more goods involves balancing marginal costs of two kinds: (1) the cost, in terms of environmental damage, of more pollution (the marginal cost of pollution) and (2) the cost, in terms of giving up goods, of eliminating pollution (the marginal cost of pollution abatement).

Determining the **marginal cost of pollution** involves assessing the risks associated with the pollution. Let's consider a simple example involving the marginal cost of sulfur dioxide, a type of air pollution produced during the combustion of fuels containing sulfur. Sulfur dioxide is removed from the atmosphere as acid rain, which causes damage to the environment, particulary aquatic ecosystems. Economists add up the harm of each additional unit of pollution—in this example, each ton of sulfur dioxide added to the atmosphere. As the total amount of pollution increases, the harm of each additional unit usually also increases, and

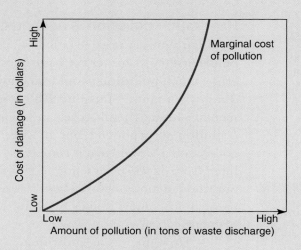

Marginal cost of pollution

FIGURE 3.15

At low pollution levels, the environment may absorb the damage, so that the marginal cost of one added unit of pollution is near zero. As the level of pollution rises, the cost in terms of human health and a damaged environment increases sharply. At very high levels of pollution, the cost soars.

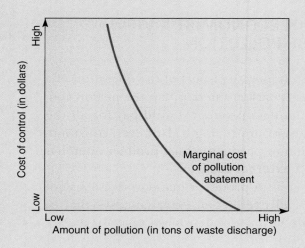

Marginal cost of pollution abatement

FIGURE 3.16

At high pollution levels, the marginal cost of eliminating one unit of pollution is low. As more and more pollution is eliminated from the environment, the cost of removing each additional (marginal) unit of pollution increases.

marginal cost of pollution abatement The added cost for all present and future members of society of reducing one unit of a given type of pollution.

cost-benefit diagram A diagram that helps policymakers make decisions about cost of a particular action and benefits that would occur if that action were implemented.

as a result, the curve showing the marginal cost of pollution slopes upward (FIGURE 3.15).

The **marginal cost of pollution abatement** tends to rise as the level of pollution declines, as shown in FIGURE 3.16. It is relatively inexpensive to reduce automobile exhaust emissions by half, but costly devices are required to reduce the remaining emissions by half again. For this reason, the curve showing the marginal cost of pollution abatement slopes downward.

In FIGURE 3.17, the two marginal-cost curves from Figures 3.15 and 3.16 are plotted together on one graph, called a **cost-benefit diagram**.

Economists use this diagram to identify the point at which the marginal cost of pollution equals the marginal cost of abatement—that is, the point where the two curves intersect. As far as economics is concerned, this point represents an **optimum amount of pollution**.

There are two major flaws in the economist's concept of optimum pollution. First, it is difficult to determine the true cost of environmental damage caused by pollution. Second, when economists add up pollution costs, they do not take into account the possible disruption or destruction of the environment. The web of relationships within the environment is extremely intricate and may be more vulnerable to pollution damage than is initially obvious, sometimes with disastrous results. This is truly a case where the whole is much greater than the sum of its parts, and it is inappropriate for economists to simply add up the cost of lost elements in a polluted environment.

optimum amount of pollution The amount of pollution that is economically most desirable.

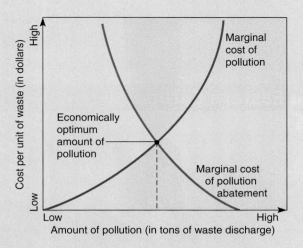

Cost-benefit diagram FIGURE 3.17

Economists identify the optimum amount of pollution as the amount at which the marginal cost of pollution equals the marginal cost of pollution abatement (the point at which the two curves intersect). If more pollution than the optimum is allowed, the social cost is unacceptably high. If less than the optimum amount of pollution is allowed, the pollution abatement cost is unacceptably high.

ECONOMIC STRATEGIES FOR POLLUTION CONTROL

Command and control regulations and incentive-based regulations are two ways that governments control pollution (FIGURE 3.18). To date, most pollution control efforts in the United States have involved **command and control regulation**. Sometimes com-

mand and control laws require use of a specific pollution control method, such as the use of catalytic converters in cars to decrease polluting exhaust emissions. In other cases, a quantitative goal is set. The Clean Air Act Amendments of 1990 established a goal of a 60 percent reduction in nitrogen oxide emissions in passenger cars by the year 2003. Usually, all polluters must comply with the same rules and regulations regardless of their particular circumstances.

Some economists criticize command and control regulation for being more costly than necessary. They think command and control regulation sets environmental pollution levels much lower than the economically optimum level of pollution. Most economists, whether progressive or conservative, prefer **incentive-based regulation** over command and control regulation. Incentive-based regulation is a market-oriented strategy. It seeks to use the economic forces of a free market to alleviate the pollution problem—that is, it depends on market incentives to reduce pollution and minimize the cost of control.

■ **command and control regulation** Pollution control laws that work by setting limits on levels of pollution.

■ **incentive-based regulation** Pollution control laws that work by establishing emission targets and providing industries with incentives to reduce emissions.

Economic Strategies for Pollution Control

Command and Control Regulations	Incentive-Based Regulations

Pollution control FIGURE 3.18

Command and control regulations restrict the emission of pollutants, often by requiring the use of a specific type of pollution abatement technology. Incentive-based regulations set emission targets and let the industry decide how to meet them.

CONCEPT CHECK STOP

What is natural capital? How is economics related to natural capital?

Why are national income accounts incomplete estimates of total national economic performance?

How do command and control regulation and incentive-based regulation differ regarding pollution control?

Environmental Problems in Central and Eastern Europe

The fall of the Soviet Union and communist governments in Central and Eastern Europe during the late 1980s revealed a grim legacy of environmental destruction. Chemicals leaking out of dump sites contaminated agricultural soil and water. Buildings and statues eroded, and entire forests died because of air pollution and acid rain. One of the most polluted areas in the world is the "Black Triangle," which consists of the bordering regions of eastern Germany, the northern Czech Republic, and southwest Poland.

Massive pollution affects human health in the region. Many people suffer from asthma, emphysema, chronic bronchitis, and other respiratory diseases. Levels of cancer, miscarriages, and birth defects are extremely high. Life expectancies are still lower there than in other industrialized nations. In 2005, the average Eastern European lived to age 69—10 years less than the average Western European.

The economic assumption behind communism was one of high production and economic self-sufficiency. Governments supported heavy industry—power plants, chemicals, metallurgy, and large machinery—at the expense of more environmentally benign service industries. As a result, Central and Eastern Europe became over-industrialized, and most of the region's plants lacked the pollution abatement equipment now required in factories in most industrialized countries.

Experts predict that eliminating the pollution legacy of communism will take decades. How much will it cost? The numbers are staggering. Improving the environment in eastern Germany alone will cost up to an estimated $300 billion.

While switching from communism to democracy with a free-market economy, current Central and Eastern European governments also face the responsibility of improving the environment. Hungary, Poland, and the Czech Republic have been relatively successful in moving toward a market economy, generating enough money to invest in environmental cleanup. From

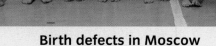

Birth defects in Moscow

Since 1973, at least 90 children with terminal-limb defects were born in two neighborhoods of Moscow. Medical researchers have not investigated limb defects thoroughly and do not know if their cause is environmental, genetic, or both.

1990 to 2000, sulfur dioxide emissions in the Czech Republic declined 86 percent, nitrogen oxide emissions from industries declined 69 percent, and particulate emissions declined 91 percent. Economic recovery is slow in other countries, such as Bulgaria, Romania, and Russia, and severe budgetary problems there have forced the environment to take a back seat to political and economic reform.

Pollution problems in former communist countries

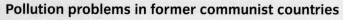

A former uranium processing plant produced radioactive waste that damaged this land in the Estonian town of Sillamae. Note the exceedingly high reading on the Geiger counter. Thousands of polluted sites exist in former communist countries, the result of rapid expansion of industrialization without regard for the environment. It will take decades for the contamination from these sites to be cleaned up.

CHAPTER SUMMARY

1 Conservation and Preservation of Resources

1. **Conservation** is the sensible and careful management of natural **resources**, such as air, water, soil, forests, minerals, and wildlife. **Preservation** is concerned with setting aside undisturbed areas, maintaining them in a pristine state, and protecting them from human activities.

2 Environmental History

1. The first two centuries of U.S. history were a time of widespread environmental destruction. During the 1700s and early 1800s, most Americans had a desire to conquer and exploit nature as quickly as possible. During the 19th century, many U.S. naturalists became concerned about conserving natural resources. The earliest conservation legislation revolved around protecting land—forests, parks, and monuments. By the late 20th century, environmental awareness had become a pervasive popular movement.

2. **John James Audubon**'s art aroused widespread interest in the wildlife of North America. **Henry David Thoreau** wrote about living in harmony with the natural world. **George Perkins Marsh** wrote about humans as agents of global environmental change. **Theodore Roosevelt** appointed **Gifford Pinchot** as the first head of the U.S. Forest Service. Pinchot supported expanding the nation's forest reserves and managing forests scientifically. The Yosemite and Sequoia national parks were established, largely in response to the efforts of naturalist **John Muir**. **Franklin Roosevelt** established the Civilian Conservation Corps and the Soil Conservation Service. In *A Sand County Almanac,* **Aldo Leopold** wrote about humanity's relationship with nature. **Wallace Stegner** helped create support for the passage of the Wilderness Act of 1964. **Rachel Carson** published *Silent Spring*, alerting the public about the dangers of uncontrolled pesticide use. **Paul Ehrlich** published *The Population Bomb,* which raised the public's awareness of the dangers of overpopulation.

3. A **utilitarian conservationist** is a person who values natural resources because of their usefulness to us but uses them sensibly and carefully. A **biocentric preservationist** is a person who believes in protecting nature because all forms of life deserve respect and consideration.

3 Environmental Legislation

1. Since 1970 the federal government has addressed many environmental problems. The National Environmental Policy Act (NEPA) was passed in 1970. NEPA established the **Council on Environmental Quality** to monitor required **environmental impact statements (EISs)** and report directly to the president.

2. By requiring EISs that are open to public scrutiny, NEPA initiated serious environmental protection in the United States. NEPA allows citizen suits, in which private citizens take violators, whether they are private industries or government-owned facilities, to court for noncompliance.

3. **Full cost accounting** is the process of evaluating and presenting to decision makers the relative benefits and costs of various alternatives.

CHAPTER SUMMARY

4 Environmental Economics

1. **Economics** is the study of how people use their limited resources to try to satisfy their unlimited wants. Economies depend on the natural environment as sources for raw materials and sinks for waste products. Both sources and sinks contribute to **natural capital**, which is Earth's resources and processes that sustain living organisms, including humans. Natural capital includes minerals, forests, soils, groundwater, clean air, wildlife, and fisheries.

2. **National income accounts** are a measure of the total income of a nation's goods and services for a given year. An **external cost** is a harmful environmental or social cost that is borne by people not directly involved in buying or selling a product. Many economists, government planners, and scientists support a comprehensive measure of national income accounting that includes estimates of both natural resource depletion and the environmental cost of economic activities.

3. From an economic point of view, the appropriate amount of pollution is a trade-off between harm to the environment and inhibition of development. The **marginal cost of pollution** is the added cost for all present and future members of society of an additional unit of pollution. The **marginal cost of pollution abatement** is the added cost for all present and future members of society of reducing one unit of a given type of pollution. Economists think the use of resources for pollution abatement should increase only until the cost of abatement equals the cost of the pollution damage. This results in the **optimum amount of pollution**—the amount of pollution that is economically most desirable.

4. To control pollution, government often uses **command and control regulations**, which are pollution control laws that work by setting limits on levels of pollution. **Incentive-based regulations** are pollution control laws that work by establishing emission targets and providing industries with incentives to reduce emissions.

KEY TERMS

- **utilitarian conservationist** p. 50
- **biocentric preservationist** p. 50
- **full cost accounting** p. 58
- **natural capital** p. 60
- **national income accounts** p. 60
- **external cost** p. 63
- **marginal cost of pollution** p. 63
- **marginal cost of pollution abatement** p. 64
- **cost-benefit diagram** p. 64
- **optimum amount of pollution** p. 64
- **command and control regulation** p. 65
- **incentive-based regulation** p. 65

CRITICAL AND CREATIVE THINKING QUESTIONS

1. The National Environmental Policy Act (NEPA) is sometimes called the "Magna Carta of environmental law." What is meant by such a comparison?

2. If you were a member of Congress, what legislation would you introduce to deal with each of the following problems?
   - Poisons from a major sanitary landfill are polluting your state's groundwater.
   - Acid rain from a coal-burning power plant in a nearby state is harming the trees in your state. Loggers and foresters are upset.
   - There is a high incidence of cancer in the area of your state where heavy industry is concentrated.

3. How would an economist approach each of the problems listed in question 2? How would an environmentalist?

4. Draw a diagram showing how the following are related: natural environment, sources, sinks, economy, production, consumption, raw materials, and waste products.

5. Draw a graph that shows the marginal cost of pollution, the marginal cost of pollution abatement, and the optimum amount of pollution.

6. Discuss the events that led to the Northwest Forest Plan of 1994.

7. Describe the extent of environmental destruction in formerly communist countries.

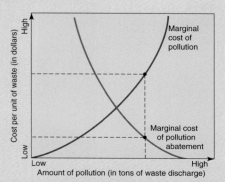

8. In the graph shown above, is the amount of pollution indicated by the vertical dashed line more or less than the economically optimum amount of pollution?

9. If you were an economist examining the previous graph, what would you recommend, increasing or decreasing pollution abatement measures? Why?

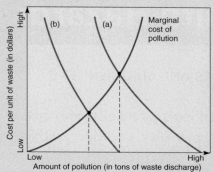

10. The graph above shows two curves, labeled a and b, that represent marginal cost of pollution abatement. In this hypothetical situation, technological innovations were developed between 2003 and 2006 that lowered the abatement cost. Which curve corresponds to 2003, and which to 2006? Explain your answer.

What is happening in this picture ?

- This photo was taken in Hong Kong in 1990. What event is taking place?

- Note the ages of the individuals in this photo. How do their ages contribute to growing environmental awareness?

- Do you think a major international celebration of Earth Day will be held in 2010? Why or why not?

Risk Analysis and Environmental Health Hazards

PESTICIDES AND CHILDREN

In recent years, attention to the health effects of household pesticides on children has increased. These pesticides appear to be a greater threat to children than to adults because children tend to play on floors and lawns, where they are exposed to greater concentrations of pesticide residues. Also, children are probably more sensitive to pesticides because their bodies are still developing. Several preliminary studies suggest that exposure to household pesticides may cause brain cancer and leukemia in children, but scientists must do more research before any firm conclusions are made.

Research supports the hypothesis that exposure to pesticides may affect the development of intelligence and motor skills of young children. One study, published in *Environmental Health Perspectives* in 1998, compared two groups of rural Yaqui Indian preschoolers in Mexico. These two nearly identical groups differed mainly in their exposure to pesticides: One group lived in a farming community where pesticides were used frequently *(see photograph)* and the other in an area where pesticides were rarely used. When asked to draw a person, most of the 17 children from the low-pesticide area drew recognizable stick figures *(see part a of inset)*, whereas most of the 34 children from the high-pesticide area drew meaningless lines and circles *(see part b)*. Additional tests of simple mental and physical skills revealed similar striking differences between the two groups of children.

CHAPTER OUTLINE

A Perspective on Risks p. 72

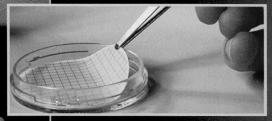

Environmental Health Hazards p. 75

Movement and Fate of Toxicants p. 79

How We Determine the Health Effects of Pollutants p. 83

The Precautionary Principle p. 88

Drawings of a person
(by 4-year-olds)

Foothills   Valley

54 mo.
(a)   female

54 mo.
(b)   female

A Perspective on Risks

LEARNING OBJECTIVES

Define risk and risk assessment.

Explain how the four steps of risk assessment help determine adverse health effects.

T hreats to our health, particularly from toxic chemicals in the environment, make big news. Many of these stories are more sensational than factual. If they were completely accurate, people would be dying everywhere—but in fact, human health is generally better today than at any time in our history, and our life expectancy continues to increase rather than decline. This does not mean that you should ignore chemicals that humans introduce into the environment. Nor does it mean you should discount the stories that the news media sometimes sensationalize. These stories serve an important role in getting the regulatory wheels of the government moving to protect us as much as possible from the dangers of our technological and industrialized world.

Risk is inherent in all our actions and in everything in our environment. All of us take risks every day of our lives. Walking on stairs involves a small risk, but a risk nonetheless, because sometimes people die from falls on stairs. Using household appliances is slightly risky, because sometimes people die from electrocution when they operate appliances with faulty wiring or use appliances in an unsafe manner. Driving or riding in a car, or flying in a jet, has risks that are easier for most of us to recognize. Yet few of us hesitate to get in a car or board a plane because of the associated risk. It is important to have an adequate understanding of the nature and size of risks before deciding what actions are appropriate to avoid them.

Although we sometimes speak of percentages, probabilities of risk are always calculated as fractions. If a risk is certain to occur, its probability is 1; if it is certain not to occur, its probability is 0. Most probabilities of risk are some number between 0 and 1. For example, according to the American Cancer Society, in 2002 about 170,000 Americans who smoked died of cancer. This translates into a probability of risk of 0.00059, or

risk The probability of harm (such as injury, disease, death, or environmental damage) occurring under certain circumstances.

Probability of risk of dying by selected causes, 1998 TABLE 4.1		
Cause of death	**U.S. deaths in 1998**	**Probability of risk**
Cardiovascular disease	940,600	3.5 of every 1,000 people
Cancer (all types)	541,500	2.0 of every 1,000 people
Accidents (including motor vehicle)	97,800	3.6 of every 10,000 people
Suicide	30,600	1.1 of every 10,000 people
Homicide	18,300	0.7 of every 10,000 people
Accidental falls	16,274	0.6 of every 10,000 people
Accidental poisonings by drugs	9,838	3.6 of every 100,000 people
Accidental drownings	3,964	1.5 of every 100,000 people
Fire	3,255	1.2 of every 100,000 people
Accidents by firearms	726	2.6 of every 1,000,000 people
Accidents (airplane)	692	2.5 of every 1,000,000 people

The four steps of risk assessment for adverse health effects FIGURE 4.1

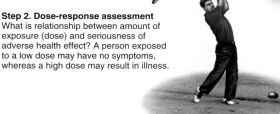

Step 1. Hazard identification
Does exposure to substance cause increased likelihood of adverse health effect such as cancer or birth defects?

Step 2. Dose-response assessment
What is relationship between amount of exposure (dose) and seriousness of adverse health effect? A person exposed to a low dose may have no symptoms, whereas a high dose may result in illness.

Low dose (no adverse effects)

Step 3. Exposure assessment
How much, how often, and how long are humans exposed to substance in question? Where humans live relative to emissions is also considered.

Step 4. Risk characterization
What is probability of individual or population having adverse health effect? Risk characterization evaluates data from dose-response assessment and exposure assessment (steps 2 and 3). Risk characterization indicates that Mexican-Americans, many of which are agricultural workers, are more vulnerable to pesticide exposure than other groups (see graph).

Agricultural workers have a greater exposure to chemicals such as pesticides

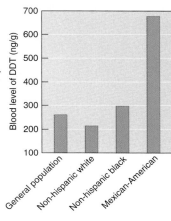

Blood level of DDT (ng/g) — General population, Non-hispanic white, Non-hispanic black, Mexican-American

---

about 6 of every 10,000 Americans. (TABLE 4.1 shows probabilities of risk of dying in a given year by selected causes.)

The four steps involved in **risk assessment** for adverse health effects are summarized in FIGURE 4.1. Once a risk assessment is performed, its results are evaluated with relevant political, social, and economic considerations to determine whether we should reduce or eliminate a particular risk and, if so, what we should do. This evaluation, which includes the development and implementation of laws to regulate hazardous substances, is known as risk management.

■ **risk assessment** The use of statistical methods to quantify the risks of an action so they can be compared and contrasted with other risks.

Risk assessment helps estimate the probability that an event will occur and lets us set priorities and manage risks in an appropriate way. As an example, consider a person who smokes a pack of cigarettes a day and drinks well water containing traces of the cancer-causing chemical trichloroethylene (in acceptable amounts, as established by the Environmental Protection Agency, or EPA). Without a knowledge of risk assessment, this person might buy bottled water in an attempt to reduce his or her chances of getting cancer. Based on risk assessment calculations, the annual risk from smoking is 0.00059, or 5.9×10^{-4}, whereas the annual risk from drinking

A Perspective on Risks 73

Despite known cancer risks—tobacco use causes at least 30 percent of all cancer deaths in the United States—many people continue to smoke.

Percentage of U.S. students (grades 9–12) who smoked cigarettes on 20 or more of the past 30 days TABLE 4.2			
Year	**Total**	**Females**	**Males**
2005	9.4	9.3	9.3
2003	9.7	9.7	9.6
2001	13.8	12.9	14.9
1999	16.8	15.6	17.9
1997	16.7	15.7	17.6
1995	16.1	15.9	16.3
1993	13.8	13.5	14.0
1991	12.7	12.4	13.0

water with EPA-accepted levels of trichloroethylene is 0.000000002, or 2.0×10^{-9}. This means that this person is almost *300,000 times* more likely to get cancer from smoking than to get it from ingesting such low levels of trichloroethylene. Knowing this, the person in our example would, we hope, stop smoking.

One of the most perplexing dilemmas of risk assessment is that people often ignore substantial risks but get extremely upset about minor risks. The average life expectancy of smokers is more than 8 years shorter than that of nonsmokers, and almost one-third of all smokers die from diseases that the habit causes or exacerbates. Yet many people get more upset over a one-in-a-million chance of getting cancer from pesticide residues on food than they do over the relationship between smoking and cancer. Perhaps part of the reason for this attitude is that behaviors such as diet, smoking, and exercise are parts of our lives that we can change if we choose to (**FIGURE 4.2** and **TABLE 4.2**). Risks over which most of us have no control, such as pesticide residues or nuclear wastes, tend to evoke more fearful responses.

CONCEPT CHECK **STOP**

What are risk and risk assessment?

What are the four steps of risk assessment?

Environmental Health Hazards

LEARNING OBJECTIVES

Define toxicology and distinguish between acute toxicity and chronic toxicity.

Explain why public water supplies are monitored for fecal coliform bacteria despite the fact that most strains of *E. coli* do not cause disease.

Describe the link between environmental changes and emerging diseases.

The human body is exposed to many kinds of chemicals in the environment. Both natural and synthetic chemicals are in the air we breathe, the water we drink, and the food we eat. All chemicals, even "safe" chemicals such as sodium chloride (table salt), are toxic if exposure is high enough. For example, a 1-year-old child will die from ingesting about 2 tablespoons of table salt; table salt is also harmful to people with heart or kidney disease. Chemicals with adverse effects are known as **toxicants**.

Toxicology involves (a) studying the effects of toxicants on living organisms, (b) studying the mechanisms that cause toxicity, and (c) developing ways to prevent or minimize adverse effects. (Creating exposure guidelines for specific toxicants is one of these ways.) The effects of toxicants following exposure can be immediate (*acute toxicity*) or prolonged (*chronic toxicity*). Symptoms of **acute toxicity** range from dizziness and nausea to death. In comparison, **chronic toxicity** generally produces damage to vital organs, such as the kidneys or liver. Human diseases that are the result of chronic toxicity are *noninfectious*—that is, they are not transmitted from one human host to another. Toxicologists know far less about chronic toxicity than they do about acute toxicity, partly because the symptoms of chronic toxicity often mimic those of chronic diseases associated with risky lifestyle patterns, poor nutrition, and aging.

toxicology The study of toxicants, chemicals with adverse effects on health.

acute toxicity Adverse effects that occur within a short period after high-level exposure to a toxicant.

chronic toxicity Adverse effects that occur after a long period of low-level exposure to a toxicant.

DISEASE-CAUSING AGENTS IN THE ENVIRONMENT

Disease-causing agents are *infectious* organisms, such as bacteria, viruses, protozoa, and parasitic worms that cause diseases. Typhoid, cholera, bacterial dysentery, polio, and infectious hepatitis are some of the more common bacterial or viral diseases that are transmissible through contaminated food and water. Diseases like these are considered environmental health hazards. Other human diseases, such as acquired immunodeficiency syndrome (AIDS), are not transmissible through the environment, and aren't discussed here.

The vulnerability of our public water supplies to waterborne disease-causing agents was dramatically demonstrated in 2000, when the first waterborne outbreak in North America of a deadly strain of *Escherichia coli* occurred in Ontario, Canada. Several people were killed, and several thousand became sick. Prior to this outbreak, this deadly *E. coli* strain was transmitted almost exclusively through contaminated food.

This and similar outbreaks raise concerns about the safety of our drinking water. Because sewage-contaminated water is an environmental threat to public health, periodic tests are made for the presence of sewage in our water supplies.

Some human diseases transmitted by polluted water TABLE 4.3

Disease	Type of organism	Symptoms
Cholera	Bacterium	Severe diarrhea, vomiting; fluid loss of as much as 20 quarts per day causes cramps and collapse
Dysentery	Bacterium	Infection of the colon causes painful diarrhea with mucus and blood in the stools; abdominal pain
Enteritis	Bacterium	Inflammation of the small intestine causes general discomfort, loss of appetite, abdominal cramps, and diarrhea
Typhoid	Bacterium	Early symptoms include headache, loss of energy, fever; later, a pink rash appears along with (sometimes) hemorrhaging in the intestines
Infectious hepatitis	Virus	Inflammation of liver causes jaundice, fever, headache, nausea, vomiting, severe loss of appetite, muscle aches, and general discomfort
Poliomyelitis	Virus	Early symptoms include sore throat, fever, diarrhea, and aching in limbs and back; when infection spreads to spinal cord, paralysis and atrophy of muscles
Cryptosporidiosis	Protozoon	Diarrhea and cramps that last up to 22 days
Amoebic dysentery	Protozoon	Infection of the colon causes painful diarrhea with mucus and blood in the stools; abdominal pain
Schistosomiasis	Fluke	Tropical disorder of the liver and bladder causes blood in urine, diarrhea, weakness, lack of energy, repeated attacks of abdominal pain
Ancylostomiasis	Hookworm	Severe anemia, sometimes symptoms of bronchitis

pathogen An agent (usually a microorganism) that causes disease.

The best indicator of sewage-contaminated water is the presence of the common intestinal bacterium *E. coli* because it doesn't appear in the environment except from human and animal feces. Tests for *E. coli* are used to indicate the possible presence of disease-causing agents (TABLE 4.3). Although most strains of coliform bacteria found in sewage do not cause disease, testing for these bacteria is a reliable way to indicate the likely presence of **pathogens** in water.

To test for the presence of *E. coli* in water, the **fecal coliform test** is performed (FIGURE 4.3). A small sample of water is passed through a filter to trap the bacteria, which are then transferred to a petri dish that contains nutrients. After an incubation period, the number of greenish colonies present indicates the number of *E. coli*. Safe drinking water should contain no more than 1 coliform bacterium per 100 mL of water (about ½ cup); safe swimming water should have no more than 200 per 100 mL of water; and general recreational water (for boating) should have no more than 2,000 per 100 mL. In contrast, raw sewage may contain several million coliform bacteria per 100 mL of water. Water pollution and purification are discussed further in Chapter 10.

ENVIRONMENTAL CHANGES AND EMERGING DISEASES

Human health has improved significantly over the past several decades, but environmental factors remain a significant cause of human disease in many areas of the world. **Epidemiologists**, scientists who investigate the outbreaks of both infectious and noninfectious diseases in a population, have established links between human health and human activities that alter the environment. The U.N. World Health Organization released a 1997

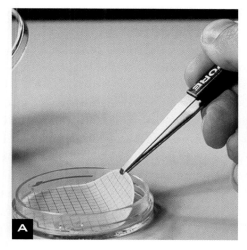

A

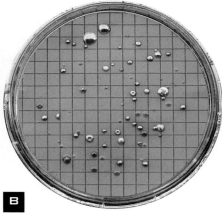

B

Fecal coliform test FIGURE 4.3

This test indicates the likely presence of disease-causing agents in water. A water sample is first passed through a filtering apparatus. **A** The filter disk is then placed on a medium that supports coliform bacteria for 24 hours. **B** After incubation, the number of bacterial colonies is counted. Each colony of *Escherichia coli* arose from a single coliform bacterium in the original water sample.

report that concluded that about 25 percent of disease and injury worldwide is related to human activities that cause environmental changes. The environmental component of human health is sometimes direct and obvious, as when people drink unsanitary water and contract dysentery, a waterborne disease that causes diarrhea. Diarrhea annually causes 4 million deaths worldwide, mostly in children.

The health effects of many human activities are complex and often indirect. The disruption of natural environments may give disease-causing agents an opportunity to thrive. Development activities such as cutting down forests, building dams, and expanding agriculture may bring more humans into contact with new or rare disease-causing agents. Such projects may increase the population and distribution of disease-carrying organisms such as mosquitoes, thereby increasing the spread of disease (**FIGURE 4.4**). Social factors may also contribute to disease epidemics. Highly concentrated urban populations promote the rapid spread

Road clearing in the Amazon rain forest
FIGURE 4.4

The drainage ditches on each side of the road provide standing water for mosquito larvae.

of infectious organisms among large numbers of people (FIGURE 4.5). Global travel also has the potential to contribute to the rapid spread of disease as infected individuals move easily from one place to another.

Consider malaria, a disease that mosquitoes transmit to humans. Each year, between 300 million and 500 million people worldwide contract malaria, and it causes more than one million deaths. About 60 different species of *Anopheles* mosquito transmit the parasites that cause malaria. Each mosquito species thrives in its own unique combination of environmental conditions (such as elevation, amount of precipitation, temperature, relative humidity, and availability of surface water).

In some regions of the world, such as Peru and Bangladesh, the incidence of malaria is increasing, partly because of environmental changes. Temporary pools of water in areas of recently cleared forest provide ideal mosquito breeding sites. The incidence of malaria is demonstrably higher in parts of the Amazon where forest has been cleared to make room for the increasing human population. In the Amazon, the construction of roads, which typically have drainage ditches on each side, has also benefited malaria-transmitting mosquitoes. Furthermore, human-induced changes in world climate may allow expansion of the malaria-transmitting mosquito into areas that are not currently part of the mosquito's range. During the present global warming trend, malaria has been observed at higher elevations in the tropics, which are warmer than they were previously.

One of the newest tools epidemiologists use is remote-sensing data gathered by low-flying aircraft or satellites. Epidemiologists studying malaria in the coastal areas of southern Chiapas, Mexico, used satellite images of landscapes to predict which localities

Pedestrians and traffic pack a street in Dhaka, Bangladesh
FIGURE 4.5

The development of cities, and the concentration of people living in them, permit the rapid spread of infectious disease-causing agents.

ASIA
BANGLADESH

Global Locator

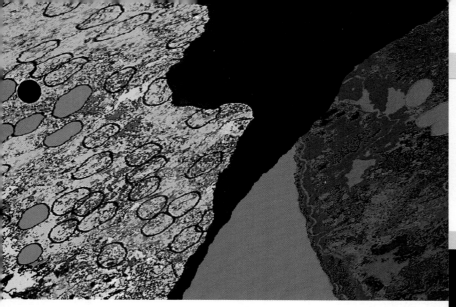

Satellite image of part of Mexico Figure 4.6

In this study of malaria risk, each village, surrounded by a 1-km buffer, is represented as a black ring; of these, the 10 villages predicted most at risk for malaria are shown as magenta circles. The remote-sensing model correctly identified 7 of the 10 villages (magenta circles with black dots) with the greatest abundance of mosquitoes and therefore the greatest risk of malaria transmission.

CONCEPT CHECK STOP

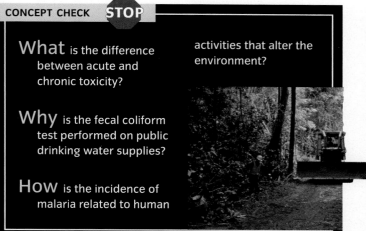

What is the difference between acute and chronic toxicity?

Why is the fecal coliform test performed on public drinking water supplies?

How is the incidence of malaria related to human activities that alter the environment?

would be most susceptible to malaria (Figure 4.6). They found that villages located near two kinds of land-scape—transitional wetlands and unmanaged pastures—tended to have a greater incidence of malaria. These villages were then targeted for interventions (such as spraying pesticides) to reduce the mosquito population.

Movement and Fate of Toxicants

LEARNING OBJECTIVES

Distinguish among persistence, bioaccumulation, and biological magnification of toxicants.

Discuss the mobility of persistent toxicants in the environment.

Describe the purpose of the Stockholm Convention on Persistent Organic Pollutants.

S ome chemically stable toxicants are particularly dangerous because they resist degradation and readily move around in the environment. These include certain pesticides, radioactive isotopes, heavy metals such as mercury, flame retardants such as PBDEs (polybrominated diphenyl ethers), and industrial chemicals such as PCBs (polychlorinated biphenyls).

The effects of the pesticide **DDT (dichloro-diphenyltrichloroethane)** on many bird species demonstrate the problem. Falcons, pelicans, bald eagles, ospreys, and many other birds are very sensitive to traces of DDT in their tissues. Substantial scientific evidence indicates that the DDT causes these birds to lay eggs with extremely thin, fragile shells that usually break during incubation, causing the chicks' deaths. After 1972, the year DDT was banned in the United States, the reproductive success of many birds improved.

The impact of DDT on birds is the result of (1) its persistence, (2) bioaccumulation, and (3) biological magnification. **Persistence** means that the substance is extremely stable and may take many years to break down into a less toxic form. When an organism can't metabolize (break down) or excrete a toxicant, it is simply stored, usually in fatty tissues. Over time, the organism may accumulate high concentrations of the

GEOGRAPHIC

Effect of DDT on birds FIGURE 4.7

A A bald eagle feeds its chick. B A comparison of the number of successful bald eagle offspring with the level of DDT residues in their eggs. Note that reproductive success improved after DDT levels decreased. (DDE is a dangerous chemical that certain microorganisms form from DDT.)

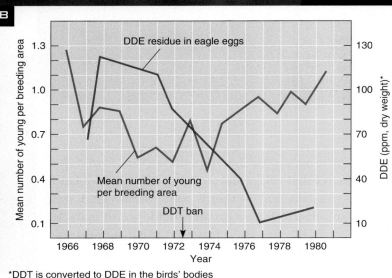

*DDT is converted to DDE in the birds' bodies

toxicant. The buildup of a persistent toxicant in an organism is **bioaccumulation** (FIGURE 4.7).

Organisms at the top of the food chain tend to store greater concentrations of bioaccumulated toxicants in their bodies than those lower on the food chain.

As an example of **biological magnification**, consider a food chain studied in a Long Island salt marsh that was sprayed with DDT over several years for mosquito control: algae and plankton → shrimp → American eel → Atlantic needlefish → ring-billed gull (FIGURE 4.8). All top carnivores, from fishes to humans, are at risk from biological magnification. Scientists therefore test pesticides to ensure they do not persist and accumulate in the environment.

> ■ **biological magnification**
> The increase in toxicant concentrations as the toxicant passes through successive levels of the food chain.

MOBILITY IN THE ENVIRONMENT

Persistent toxicants tend to move through the soil, water, and air, sometimes long distances. For example, pesticides applied to agricultural lands may wash into rivers and streams when it rains, harming fishes (FIGURE 4.9). If the pesticide level in their aquatic ecosystem is high enough, the fishes may die. At lower levels, the fishes may still suffer from symptoms of chronic toxicity like bone degeneration. These symptoms may decrease the fishes' competitiveness or increase their chances of being eaten by predators.

Mobility of persistent toxicants is also a risk for humans. In 1994 the Environmental Working Group, a private organization, analyzed five common herbicides (weed-killing chemicals) found in drinking water. It concluded that 3.5 million people in the Midwest face a slightly elevated cancer risk because of exposure to the herbicides. The EPA has since mandated a reduction in use of the five herbicides.

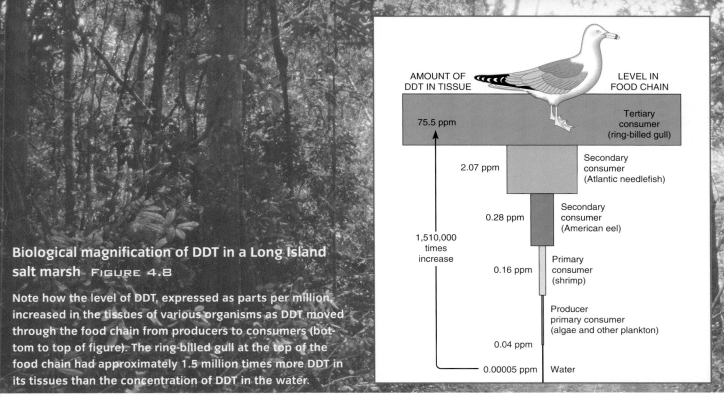

Biological magnification of DDT in a Long Island salt marsh FIGURE 4.8

Note how the level of DDT, expressed as parts per million, increased in the tissues of various organisms as DDT moved through the food chain from producers to consumers (bottom to top of figure). The ring-billed gull at the top of the food chain had approximately 1.5 million times more DDT in its tissues than the concentration of DDT in the water.

AMOUNT OF DDT IN TISSUE | LEVEL IN FOOD CHAIN

75.5 ppm — Tertiary consumer (ring-billed gull)

2.07 ppm — Secondary consumer (Atlantic needlefish)

0.28 ppm — Secondary consumer (American eel)

0.16 ppm — Primary consumer (shrimp)

Producer primary consumer (algae and other plankton)

0.04 ppm

1,510,000 times increase

0.00005 ppm — Water

Mobility of pesticides in the environment FIGURE 4.9

The intended pathway of pesticides in the environment is shown across the top of the figure, while the actual pathways are shown across the bottom.

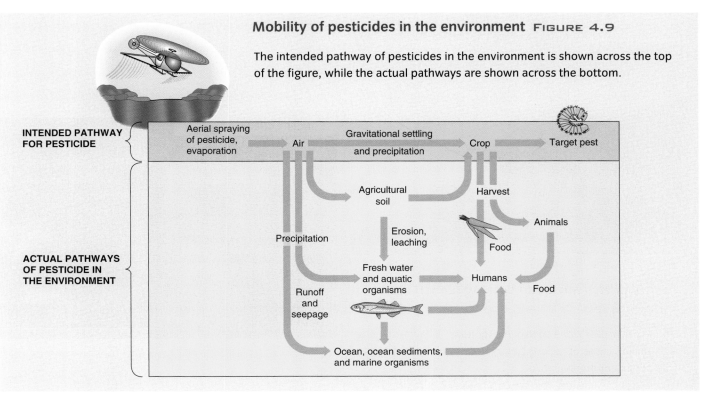

INTENDED PATHWAY FOR PESTICIDE

Aerial spraying of pesticide, evaporation → Air → Gravitational settling and precipitation → Crop → Target pest

ACTUAL PATHWAYS OF PESTICIDE IN THE ENVIRONMENT

Agricultural soil

Harvest

Animals

Precipitation

Erosion, leaching

Food

Fresh water and aquatic organisms

Humans

Food

Runoff and seepage

Ocean, ocean sediments, and marine organisms

THE GLOBAL BAN OF PERSISTENT ORGANIC POLLUTANTS

The *Stockholm Convention on Persistent Organic Pollutants*, which was adopted in 2001, is an important U.N. treaty that seeks to protect human health and the environment from the 12 most toxic **persistent organic pollutants**, or **POPs**, on Earth (TABLE 4.4). Some POPs disrupt the endocrine system (discussed later in this chapter), cause cancer, or adversely affect the developmental processes of organisms.

The Stockholm Convention requires countries to develop plans to eliminate the production and use of intentionally produced POPs. A notable exception to this requirement is that DDT is still produced and used to control malaria-carrying mosquitoes in countries where no affordable alternatives exist (FIGURE 4.10).

persistent organic pollutants (POPs) A group of persistent toxicants that bioaccumulate in organisms and travel thousands of kilometers through air to water to contaminate sites far removed from their source.

Persistent organic pollutants: The "Dirty Dozen" TABLE 4.4	
Persistent organic pollutant	**Use**
Aldrin	Insecticide
Chlordane	Insecticide
DDT (dichlorodiphenyl-trichloroethane)	Insecticide
Dieldrin	Insecticide
Endrin	Rodenticide and insecticide
Heptachlor	Fungicide
Hexachlorobenzene	Insecticide; fire retardant
Mirex™	Insecticide
Toxaphene™	Insecticide
PCBs (polychlorinated biphenyls)	Industrial chemical
Dioxins	By-product of certain manufacturing processes
Furans (dibenzofurans)	By-product of certain manufacturing processes

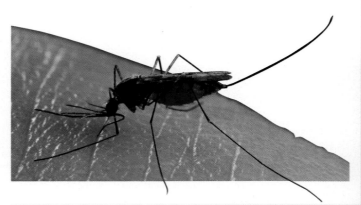

An *Anopheles* mosquito sucking human blood
FIGURE 4.10

DDT and other pesticides control mosquitoes, preventing more than 1 million cases of malaria per year. Unfortunately, DDT is also a persistent organic pollutant.

CONCEPT CHECK STOP

What is meant by a *persistent* toxicant?

What problems are associated with bioaccumulation and biological magnification?

What are POPs, and why has the international community banned them?

How We Determine the Health Effects of Pollutants

We measure toxicity by the dose at which adverse effects are produced. A **dose** of a toxicant is the amount that enters the body of an exposed organism. The **response** is the type and amount of damage that exposure to a particular dose causes. A dose may cause death *(lethal dose)* or cause harm but not death *(sub-lethal dose)*. Lethal doses, which are usually expressed in milligrams of toxicant per kilogram of body weight, vary depending on the organism's age, sex, health, and metabolism, and on how the dose was administered (all at once or over a period of time). The lethal doses, for humans, of many toxicants are known through records of homicides and accidental poisonings.

Because they weigh substantially less than adults, children are more susceptible to the effects of chemicals. Consider a toxicant with an LD_{50} of 100 mg/kg. A potentially lethal dose for a child who weighs 11.3 kg (25 lb) is $100 \times 11.3 = 1130$ mg, which is equal to a scant ¼ teaspoon if the chemical is a liquid. In comparison, the potentially lethal dose for an adult who weighs 68 kg (150 lb) is 6,800 mg, or about 2 teaspoons. This exercise demonstrates that we must protect children from exposure to environmental chemicals because lethal doses are smaller for children than for adults.

One way to determine acute toxicity is to administer different-sized doses to populations of laboratory animals, measure the responses, and use these data to predict the chemical effects on humans. The dose

LD_{50} values for selected chemicals TABLE 4.5	
Chemical	**LD_{50} (mg/kg)***
Aspirin	1,750.0
Ethanol	1,000.0
Morphine	500.0
Caffeine	200.0
Heroin	150.0
Lead	20.0
Cocaine	17.5
Sodium cyanide	10.0
Nicotine	2.0
Strychnine	0.8

*Administered orally to rats.

that is lethal to 50 percent of a population of test animals is called the **lethal dose–50 percent,** or **LD_{50}.** It is usually reported in milligrams of chemical toxicant per kilogram of body weight. An inverse relationship exists between the size of the LD_{50} and the acute toxicity of a chemical: The smaller the LD_{50}, the more toxic the chemical, and conversely, the greater the LD_{50}, the less toxic the chemical (**TABLE 4.5**). The LD_{50} is determined for all new synthetic chemicals—thousands are produced each year—as a way of estimating their toxic potential. It is generally assumed that a chemical with a low LD_{50} for several species of test animals is also very toxic in humans.

The **effective dose–50 percent,** or ED_{50}, measures a wide range of biological responses, such as stunted development in the offspring of a pregnant animal, reduced enzyme activity, or onset of hair loss. The ED_{50} is the dose that causes 50 percent of a population to exhibit whatever response is under study.

To develop a **dose–response curve,** scientists first test the effects of high doses and then work their way down to a **threshold** level, the maximum dose that has no measurable effect (or, alternatively, the minimum dose that produces a measurable effect) (**FIGURE 4.11**). Scientists assume that doses lower than the threshold level are therefore safe.

A growing body of evidence, however, suggests that for certain toxicants there is no safe dose. A threshold does not exist for these chemicals, and even the smallest amount causes a measurable response.

CANCER-CAUSING SUBSTANCES

Because cancer is so feared, for many years it was the only disease evaluated in chemical risk assessment. Environmental contaminants are linked to many serious health concerns, including other diseases, birth defects, damage to the immune system, reproductive problems, and damage to the nervous system or other body systems. We focus here on risk assessment as it relates to cancer, but non-cancer hazards are assessed in similar ways.

The most common method of determining whether a chemical causes cancer is to expose laboratory

> ■ **dose–response curve** In toxicology, a graph that shows the effect of different doses on a population of test organisms.

Dose–response curves FIGURE 4.11

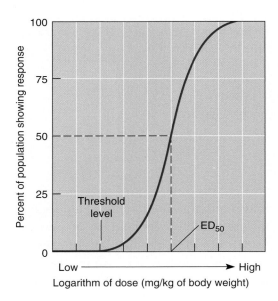

This hypothetical dose–response curve demonstrates two assumptions of classical toxicology: first, that the biological response increases as the dose is increased; second, that there is a safe dose—that is, a level of the toxicant at which no response occurs. Harmful responses occur only above a certain threshold level.

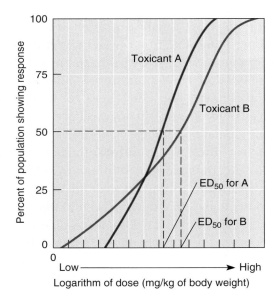

Dose–response curves for two hypothetical toxicants, A and B. As you can see, dose–response curves have a variety of shapes. In this example, A has a lower effective dose–50 percent (ED_{50}) than B. However, at lower doses, B is more toxic than A.

carcinogen Any substance (chemical, radiation, or virus) that causes cancer.

animals, such as rats, to extremely large doses of it and see whether they develop cancer. This method is indirect and uncertain, however. For one thing, although humans and rats are both mammals, they are different organisms and may respond differently to exposure to the same chemical. (Even rats and mice often respond differently to the same toxicant.)

Another problem is that the rats are exposed to massive doses of the suspected **carcinogen** relative to their body size, whereas humans are usually exposed to much lower amounts. Researchers must use large doses to cause cancer in a small group of laboratory animals within a reasonable amount of time. Otherwise, such tests would take years, require thousands of test animals, and be prohibitively expensive to produce enough data to have statistically significant results.

Risk assessment assumes that you can extrapolate (work backward) from the huge doses of chemicals and the high rates of cancer they cause in rats to determine the expected rates of cancer in humans exposed to lower amounts of the same chemicals. However, little evidence exists to indicate that extrapolating backward is scientifically sound. Even if you are reasonably sure that exposure to high doses of a chemical causes the same effects for the same reasons in both rats and humans, you cannot assume that these same mechanisms work at low doses in humans. The body metabolizes small and large doses of a chemical in different ways. For example, enzymes in the liver may break down carcinogens in small quantities, but an excessive amount of carcinogen might overwhelm the liver enzymes.

In short, extrapolating from one species to another and from high doses to low doses is filled with uncertainty and may overestimate or underestimate the toxicant's danger. However, animal carcinogen studies provide valuable information: A toxicant that does not cause cancer in laboratory animals at high doses is not likely to cause cancer in humans at levels found in the environment or in occupational settings.

Scientists do not currently have a reliable way to determine if exposure to small amounts of a substance causes cancer in humans. However, the

Measuring low doses of a toxicant (dioxin) in human blood serum FIGURE 4.12

Scientists are developing increasingly sophisticated methods of biomonitoring to analyze human tissues and fluids. Photographed at the Centers for Disease Control's Environmental Health Laboratory.

EPA is planning to change how toxic chemicals are evaluated and regulated. Toxicologists are developing methods to provide direct evidence of the risk involved in exposure to low doses of cancer-causing chemicals (**FIGURE 4.12**).

Epidemiological evidence, including studies of human groups accidentally exposed to high levels of suspected carcinogens, is also used to determine whether chemicals are carcinogenic. For example, in 1989 epidemiologists in Germany established a direct link between cancer and a group of chemicals called *dioxins* (see Table 4.4 on page 82). They observed the incidence of cancer in workers exposed to high concentrations of dioxins during an accident at a chemical plant in 1953 and found unexpectedly high levels of cancers in the digestive and respiratory tracts.

RISK ASSESSMENT OF CHEMICAL MIXTURES

Humans are frequently exposed to various combinations of chemical compounds, in the air we breathe, the food we eat, and the water we drink. For example, cigarette smoke contains a mixture of chemicals, as does automobile exhaust. Cigarette smoke is a mixture of air pollutants that includes hydrocarbons, carbon dioxide, carbon monoxide, particulate matter, cyanide, and a small amount of radioactive materials that come from the fertilizer used to grow the tobacco plants.

The vast majority of toxicology studies have been performed on single chemicals rather than chemical mixtures, and for good reason. Mixtures of chemicals interact in a variety of ways, increasing the level of complexity in risk assessment, a field already complicated by many uncertainties. Moreover, there are simply too many chemical mixtures to evaluate.

Chemical mixtures interact by **additivity**, **synergy**, or **antagonism**. When a chemical mixture is *additive*, the effect is exactly what you would expect, given the individual effects of each component of the mixture. If a chemical with a toxicity level of 1 is mixed with a different chemical, also with a toxicity level of 1, the combined effect of exposure to the mixture is 2. A chemical mixture that is *synergistic* has a greater combined effect than expected; two chemicals, each with a toxicity level of 1, might have a combined toxicity of 3. An *antagonistic* interaction in a chemical mixture results in a smaller combined effect than expected; for example, the combined effect of two chemicals, each with toxicity levels of 1, might be 1.3.

If toxicological studies of chemical mixtures are lacking, how is risk assessment for chemical mixtures assigned? Toxicologists use *additivity* to assign risk to mixtures—that is, they add the known effects of each compound in the mixture. Such an approach sometimes overestimates or underestimates the actual risk involved, but it is the best method currently available. The alternative—that of waiting for years or decades until numerous studies have been designed, funded, and completed—is unreasonable.

CHILDREN AND CHEMICAL EXPOSURE

The EPA estimates that 84 percent of U.S. homes use pesticide products, such as pest strips, bait boxes, bug bombs, flea collars, pesticide pet shampoos, aerosols, liquids, and dusts. These several thousand different household pesticides contain over 300 active ingredients and more than 2,500 inert ingredients.

Poison control centers in the United States annually receive more than 130,000 reports of exposure and possible poisoning from household pesticides. More than half of these incidents involve children.

There is also concern about children ingesting pesticide residues on food. The National Research Council published a 3-year study in 1993 called *Pesticides in the Diets of Infants and Children*. It recommended additional research on how pesticide residues on food affect the young.

Children and pollution   Consider the toxicants in air pollution. Air pollution is a greater health threat to children than it is to adults (FIGURE 4.13). The

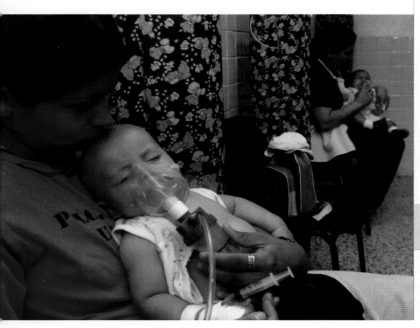

Air pollution and respiratory disease in children FIGURE 4.13

A Honduran mother gives oxygen to her baby, who suffers from environmentally linked respiratory disease. Farmers nearby burn land to prepare for the planting season; the resulting smoke triggers breathing problems, mostly in children and the elderly.

Smoking causes serious diseases such as lung cancer, emphysema, and heart disease and is responsible for the premature deaths of nearly half a million people in the United States each year. Cigarette smoking annually causes about 120,000 of the 140,000 deaths from lung cancer in the United States. Smoking also contributes to heart attacks and strokes and to cancer of the bladder, mouth, throat, pancreas, kidney, stomach, voice box, and esophagus.

Passive smoking, which is nonsmokers' chronic breathing of smoke from cigarette smokers, also increases the risk of cancer. Passive smokers suffer more respiratory infections, allergies, and other chronic respiratory diseases than other nonsmokers. Passive smoking is particularly harmful to infants and young children, pregnant women, the elderly, and people with chronic lung disease.

There is good news and bad news about smoking. The good news is that fewer people in highly developed nations such as the United States are smoking. The bad news is that more and more people are taking up the habit in China, Brazil, Pakistan, and other developing nations. Tobacco companies in the United States promote smoking abroad, and a substantial portion of our tobacco crop is exported. The World Health Organization (WHO) estimates that worldwide, 3 million people die each year of smoking-related causes. In an attempt to establish a global ban on tobacco advertising, WHO developed the Framework Convention on Tobacco Control. The treaty went into effect for signatory nations in February 2005 but had not been ratified by the United States as of June 2006.

A sugarcane worker smokes during a break in his long, hard day

U.S. tobacco companies export the cigarette habit abroad to compensate for a lower consumption in the United States. Photographed in the Philippines.

lungs continue to develop throughout childhood, and air pollution restricts lung development. In addition, a child has a higher metabolic rate than an adult and therefore needs more oxygen. To obtain this oxygen, a child breathes more air—about two times as much air per pound of body weight as an adult. This means that a child also breathes more air pollutants into the lungs. A 1990 study in which autopsies were performed on 100 Los Angeles children who died for reasons unrelated to respiratory problems found that more than 80 percent had subclinical lung damage, which is lung disease in its early stages, before clinical symptoms appear. (Los Angeles has some of the worst air quality in the world.)

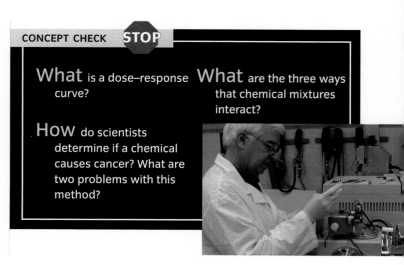

CONCEPT CHECK STOP

What is a dose–response curve?

What are the three ways that chemical mixtures interact?

How do scientists determine if a chemical causes cancer? What are two problems with this method?

The Precautionary Principle

LEARNING OBJECTIVE

Discuss the precautionary principle as it relates to the introduction of new technologies or products.

Y ou've probably heard the expression "An ounce of prevention is worth a pound of cure." This statement is the heart of the **precautionary principle** that many politicians and environmental activists advocate. According to the precautionary principle, we should not introduce new technology or chemicals until it is demonstrated that (a) the risks are small and (b) the benefits outweigh the risks. The precautionary principle puts the burden of proof on the developers of the new technology or substance. It asks them to prove the product is safe beyond a reasonable doubt, instead of society discovering its harmful effects after it has been introduced.

The precautionary principle is also applied to existing technologies when new evidence suggests they are more dangerous than originally thought. For example, when observations and experiments suggested that lead added to gasoline as an anti-knock ingredient was contaminating soil, particularly in inner cities near major highways, the precautionary principle led to the phase-out of leaded gasoline (**FIGURE 4.14**).

Children in the South Bronx of New York City
FIGURE 4.14

In addition to air pollution, these children are probably exposed to lead in the soil (from leaded gasoline) and in paint from old buildings. Children with low levels of lead in their blood may suffer from partial hearing loss, hyperactivity, attention deficit, lowered IQ, and learning disabilities.

precautionary principle Making decisions about adopting a new technology or chemical product by assigning the burden of proof of its safety to the developers of that technology or product.

Is Europe's ban of U.S. beef the result of its concern over the safety of hormone-treated beef or an excuse to support the European beef industry? Photographed on a range in Wyoming.

Precautionary principle or economic protectionism? FIGURE 4.15

To many people the precautionary principle is just common sense, given that science and risk assessment often cannot provide definitive answers to policy makers dealing with environmental and public health problems. The developers of the new technology or substance must prove it is safe beyond a reasonable doubt instead of society proving it is harmful after it has already been introduced. However, the precautionary principle does not require that developers provide absolute proof that their product is safe; such proof would be impossible to provide.

Certain laws and decisions in many European Union nations have incorporated the precautionary principle, and some laws in the United States have a precautionary tone. In 2000, Christine Todd Whitman, then governor of New Jersey, said in a speech to the National Academy of Sciences,

> Policy makers need to take a precautionary approach to environmental protection . . . We must acknowledge that uncertainty is inherent in managing natural resources, recognize it is usually easier to prevent environmental damage than to repair it later, and shift the burden of proof away from those advocating protection toward those proposing an action that may be harmful.

The precautionary principle has generated much controversy. Some scientists fear that the precautionary principle challenges the role of science and endorses making decisions without the input of science. Some critics contend that its imprecise definition reduces trade and limits technological innovations. For example, several European countries made precautionary decisions to ban beef from the United States and Canada because these countries use growth hormones to make the cattle grow faster. Europeans contend that the growth hormone might harm humans eating the beef, but the ban, in effect since 1989, is widely viewed as protecting the European beef industry (FIGURE 4.15). Another international controversy in which the precautionary principle is involved is the cultivation of genetically modified foods (discussed further in Chapter 14).

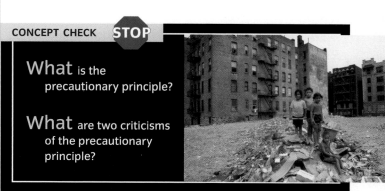

CONCEPT CHECK **STOP**

What is the precautionary principle?

What are two criticisms of the precautionary principle?

Endocrine Disrupters

Mounting evidence suggests that dozens of widely used industrial and agricultural chemicals are **endocrine disrupters**, which interfere with the normal actions of the endocrine system (the body's hormones) in humans and animals. These chemicals include chlorine-containing industrial compounds known as PCBs and dioxins, the heavy metals lead and mercury, pesticides like DDT, and certain plastics and plastic additives.

Hormones are chemical messengers that organisms produce to regulate their growth, reproduction, and other important biological functions. Some endocrine disrupters mimic the *estrogens,* a class of female sex hormones. Other endocrine disrupters mimic *androgens* (male hormones such as testosterone) or *thyroid hormones.* Like hormones,

endocrine disrupters are active at very low concentrations and therefore may cause significant health effects at relatively low doses.

Many endocrine disrupters appear to alter the reproductive development of various animal species. A chemical spill in 1980 contaminated Lake Apopka, Florida's third largest lake, with DDT and other agricultural chemicals that have known estrogenic properties. In the years following the spill, male alligators had low levels of testosterone and elevated levels of estrogen. The mortality rate for eggs in this lake is still extremely high—only 40 percent hatch, and half of these die within 10 days *(see photo at left).*

Humans may also be at risk from endocrine disrupters. The number of reproductive disorders, infertility cases, and hormonally related cancers (such as testicular cancer and breast cancer) appears to be increasing. However, we cannot make definite connections between environmental endocrine disrupters and human health problems at this time because of the limited number of human studies. Complicating such assessments is the fact that humans are also exposed to *natural,* hormone-mimicking substances in the plants we eat. For example, soy-based foods such as bean curd and soymilk contain natural estrogens.

Congress amended the *Food Quality Protection Act* and the *Safe Drinking Water Act* in 1996 to require the Environmental Protection Agency to develop a plan and establish priorities to test thousands of chemicals for their potential to disrupt endocrine systems. Chemicals testing positive are tested further to determine what specific damage, if any, they cause to reproduction and other biological functions. These tests, which may take decades to complete, should reveal the level of human and animal exposure to endocrine disrupters and the effects of this exposure.

Lake Apopka alligators

A young American alligator hatches from eggs that University of Florida researchers took from Lake Apopka, Florida. Few Lake Apopka eggs actually hatch, and many of the young alligators that do hatch have abnormalities in their reproductive systems. This young alligator may not leave any offspring.

1 A Perspective on Risks

1. A **risk** is the probability of harm (such as injury, disease, death, or environmental damage) occurring under certain circumstances. **Risk assessment** is the use of statistical methods to quantify the risks of an action so they can be compared and contrasted with other risks.

2. The four steps of risk assessment are: hazard identification (Does exposure cause an increased risk of an adverse health effect?); dose-response assessment (What is the relationship between the dose and the extent of the adverse health effect?); exposure assessment (How often and how long are humans exposed to the substance?); and risk characterization (What is the probability of having an adverse effect?).

2 Environmental Health Hazards

1. **Toxicology** is the study of **toxicants**, chemicals with adverse effects on health. **Acute toxicity** is adverse effects that occur within a short period after high-level exposure to a toxicant. **Chronic toxicity** is adverse effects that occur after a long period of low-level exposure to a toxicant.

2. A **pathogen** is an agent (usually a microorganism) that causes disease. Although most strains of coliform bacteria do not cause disease, the **fecal coliform test** is a reliable way to indicate the likely presence of pathogens in water.

3. About 25 percent of disease and injury worldwide is related to human-caused environmental changes. The environmental component of human health is sometimes direct, as when people drink unsanitary water and contract a waterborne disease. The health effects of other human activities are complex and indirect, as when the disruption of natural environments gives disease-causing agents an opportunity to break out of isolation.

3 Movement and Fate of Toxicants

1. Some toxicants exhibit **persistence**—they are extremely stable in the environment and may take many years to break down into less toxic forms. **Bioaccumulation** is the buildup of a persistent toxicant in an organism's body. **Biological magnification** is the increase in toxicant concentrations as the toxicant passes through successive levels of the food chain.

2. Persistent toxicants do not stay where they are applied but tend to move through the soil, water, and air, sometimes long distances. For example, pesticides applied to agricultural lands may wash into rivers and streams, harming fishes.

3. The Stockholm Convention on Persistent Organic Pollutants requires countries to eliminate the production and use of the 12 worst **persistent organic pollutants (POPs)**. POPs are a group of persistent toxicants that bioaccumulate in organisms and travel thousands of kilometers through air and water to contaminate sites far removed from their source.

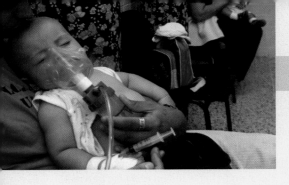

4 How We Determine the Health Effects of Pollutants

1. A **dose–response curve** is a graph that shows the effect of different doses on a population of test organisms. Scientists test the effects of high doses and work their way down to a **threshold** level, the maximum dose that has no measurable effect. It is assumed that doses lower than the threshold level will not have an effect on the organism and are therefore safe.

2. A **carcinogen** is any substance (chemical, radiation, or virus) that causes cancer. The most common method of determining whether a chemical is carcinogenic is to expose laboratory animals such as rats to extremely large doses of that chemical and see if they develop cancer. It is assumed you can extrapolate from the huge doses of chemicals and the high rates of cancer they cause in rats to determine the rates of cancer expected in humans exposed to lower amounts of the same chemicals.

3. When a chemical mixture is **additive**, the effect is exactly what you would expect, given the individual effects of each component of the mixture. A chemical mixture that is **synergistic** has a greater combined effect than expected. An **antagonistic** interaction in a chemical mixture results in a smaller combined effect than expected.

4. Because they weigh substantially less than adults, children are more susceptible to chemicals; the potentially lethal dose for a child is considerably less than the potentially lethal dose for an adult. Air pollution is a greater health threat to children than adults because pollution restricts lung development. Also, a child has a higher metabolic rate than an adult and therefore breathes more air and more air pollutants into the lungs.

5 The Precautionary Principle

1. The **precautionary principle** is making decisions about adopting a new technology or chemical product by assigning the burden of proof of its safety to the developers of that technology or product.

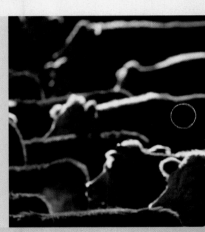

KEY TERMS

- **risk** p. 72
- **risk assessment** p. 73
- **toxicology** p. 75
- **acute toxicity** p. 75
- **chronic toxicity** p. 75
- **pathogen** p. 76
- **biological magnification** p. 80
- **persistant organic pollutants (POPs)** p. 82
- **dose–response curve** p. 84
- **carcinogen** p. 85
- **precautionary principle** p. 88

CRITICAL AND CREATIVE THINKING QUESTIONS

1. Which risk—an extremely small amount of a cancer-causing chemical in drinking water or smoking cigarettes—tends to generate the greatest public concern? Explain why this view is counterproductive.

2. Should public policymakers be more concerned with public risk perception or with actual risks? Explain your answer.

3. Describe the four steps of risk assessment for adverse health effects.

4. Is risk assessment important in toxicology? Explain.

5. Describe how a persistent pesticide might move around in the environment.

6. Why is air pollution a greater threat to children than it is to adults?

7. What is the Stockholm Convention on Persistent Organic Pollutants?

8. Select one of the two choices to complete the following sentence: "The absence of scientific certainty about the health effects of an environmental pollutant *is/is not* synonymous with the absence of risk." Explain your answer.

9. Would you support the United States adopting the precautionary policy in all of its legislation? Why or why not?

The figure to the right shows the organisms sampled in the Long Island salt marsh study of DDT (also see Figure 4.8 on page 81).

10. If DDT is sprayed on land to control insects, how does it get into the bodies of aquatic species?

11. Why does the Atlantic needlefish (5) contain more DDT in its body than an American eel (4)?

12. How does high concentration of DDT cause reproductive failure in birds at the top of the food chain?

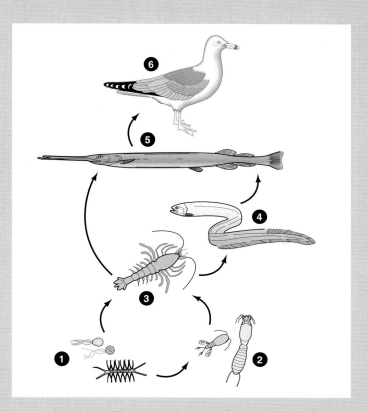

What is happening in this picture ?

- Why do we use animal testing to determine if a new pesticide causes cancer?

- This mouse developed cancer after exposure to high levels of a toxicant. What uncertainties are associated with extrapolating this result to low levels of exposure of that toxicant in humans?

- Chemical manufacturers have sometimes paid human subjects to be exposed to low doses of new chemicals, but the EPA currently has a moratorium on such testing. Suggest a possible reason for the moratorium. Do you consider such testing ethical? Why or why not?

How Ecosystems Work

LAKE VICTORIA'S ECOLOGICAL IMBALANCE

U ntil relatively recently, the world's second largest fresh water lake, Africa's Lake Victoria, was home to about 400 species of small, colorful fishes known as cichlids (pronounced sik´lids). The cichlids in Lake Victoria had remarkably different eating habits. Some grazed on algae; others consumed dead organic material at the lake bottom; and still others ate insects, shrimp, or other cichlids. These fishes thrived throughout the lake ecosystem and provided much needed protein to the diets of 30 million humans living near the lake.

Today, the aquatic community is different in Lake Victoria. More than half of the cichlids and other native fish species are now extinct. As a result of the disappearance of most of the algae-eating cichlids, the algal population has increased explosively. When these algae die, their decomposition uses up the dissolved oxygen in the water. The bottom zone of the lake, once filled with cichlids, is empty because it contains too little dissolved oxygen. Any fishes venturing into the oxygen-free zone suffocate. Local fishermen, who once caught and ate hundreds of different types of fishes, now catch only a few types.

When, in the early 1960s, the Nile perch *(inset)* was introduced into the lake, proponents thought its successful establishment would stimulate the local economy and help the fishermen. For about 20 years, as its population slowly increased, the Nile perch didn't have an appreciable effect on the lake. But in 1980, fishermen noticed they were harvesting increasing quantities of Nile perch and decreasing amounts of native fishes. By 1985, most of the annual catch was Nile perch, which was increasing in number because it had an abundant food supply—the cichlids.

In 1990 the water hyacinth, a South American plant, invaded Lake Victoria, adding to the ecological havoc *(see large photo)*. The World Bank has established the Lake Victoria Environmental Management Program to control the spread of the water hyacinth, but so far little progress has occurred.

CHAPTER OUT

What Is Ecology? p. 96

The Flow of Energy Through Ecosystems p. 99

The Cycling of Matter in Ecosystems p. 104

Ecological Niches p. 110

Interactions Among Organisms

CASE STUDY: Global Warming p.

What Is Ecology?

LEARNING OBJECTIVES

Define ecology.

Distinguish among the following ecological levels: population, community, ecosystem, landscape, and biosphere.

In the nineteenth century the German biologist Ernst Haeckel first developed the concept of ecology. He devised its name—*eco* from the Greek word for "house" and *logy* from the Greek word for "study." Thus, **ecology** literally means "the study of one's house." The environment—one's house—consists of two parts, the *biotic* (living) environment, which includes all organisms, and the *abiotic* (nonliving, or physical) environment, which includes such physical factors as living space, temperature, sunlight, soil, wind, and precipitation.

Ecology is the broadest field within the biological sciences, and it is linked to every other biological discipline. The universality of ecology also links it to other fields. Geology and earth science are extremely important to ecology, especially when ecologists examine the physical environment of planet Earth. Chemistry and physics are also important. Humans are biological organisms, and all of our activities have a bearing on ecology. Even economics and politics have profound ecological implications, as was presented in Chapter 3.

Ecologists are most interested in the levels of biological organization including or above the level of the individual organism. Individuals of the same species occur in **populations**. (A *species* is a group of similar organisms whose members freely interbreed with one another in the wild to produce fertile offspring; members of one species don't interbreed with other species of organisms.) A population ecologist might study a population of walruses or a population of marsh grass.

Populations are organized into **communities**. The number and kinds of species that live within a community, along with their relationships with one another, characterize communities. A community ecologist might study how organisms interact with one another—including feeding relationships (who eats whom)—in a coral reef community or in an alpine meadow community (**FIGURE 5.1**).

Ecosystem is a more inclusive term than community. An ecosystem includes all the biotic interactions of a community as well as the interactions between organisms and their abiotic environment. In an ecosystem, all of the biological, physical, and chemical components of an area form an extremely complicated interacting network of energy flow and materials cycling. An ecosystem ecologist might examine how energy, nutrient composition, or water affects the organisms living in a desert community or a coastal bay ecosystem.

The ultimate goal of ecosystem ecologists is to understand how ecosystems function. This isn't a simple task, but it is important because ecosystem processes collectively regulate global cycles of water, carbon, nitrogen, and oxygen essential to the survival of humans and all other organisms. As humans increasingly alter ecosystems for their own uses, the natural

ecology The study of the interactions among organisms and between organisms and their abiotic environment.

population A group of organisms of the same species that live together in the same area at the same time.

community A natural association that consists of all the populations of different species that live and interact together within an area at the same time.

ecosystem A community and its physical environment.

landscape A region that includes several interacting ecosystems.

An alpine meadow community FIGURE 5.1

Alpine flowers put on a colorful show during their brief blooming period. Photographed in National Park Berchtesgaden, Germany. *(Inset)* The golden-mantled ground squirrel, larger than a chipmunk, is a burrowing mammal common in alpine meadows in western North America.

functioning of ecosystems is changed, and we must learn if these changes will affect the *sustainability* of our life-support system.

Landscape ecology is a sub-discipline in ecology that studies the connections among ecosystems. Consider a **landscape** consisting of a forest ecosystem located adjacent to a pond ecosystem. One possible connection between these two ecosystems is the great blue heron, which eats fish, frogs, insects, crustaceans, and snakes along the shallow water of the pond but often builds nests and raises its young in the secluded treetops of the nearby forest (**FIGURE 5.2** on the following page). Landscapes, then, are based on larger land areas that include several ecosystems.

The organisms of the **biosphere**—Earth's communities, ecosystems, and landscapes—depend on one another and on the other realms of Earth's physical environment: the atmosphere, hydrosphere,

A connection between two ecosystems within a landscape FIGURE 5.2

A A great blue heron has caught a frog in a pond. B Herons usually nest in trees adjacent to the pond ecosystem.

and lithosphere. The *atmosphere* is the gaseous envelope surrounding Earth; the *hydrosphere* is Earth's supply of water—liquid and frozen, fresh and salty; and the *lithosphere* is the soil and rock of Earth's crust. Ecologists who study the biosphere examine the global interrelationships among Earth's atmosphere, land, water, and organisms.

The biosphere teems with life. Where do these organisms get the energy to live? And how do they harness this energy? Let's now examine the importance of energy to organisms, which survive only as long as the environment con-

| **biosphere** The layer of Earth containing all living organisms. |

tinuously supplies them with energy. You will revisit energy as it relates to human endeavors in many chapters throughout this text.

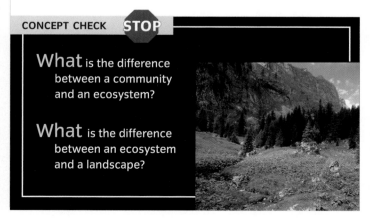

CONCEPT CHECK STOP

What is the difference between a community and an ecosystem?

What is the difference between an ecosystem and a landscape?

The Flow of Energy Through Ecosystems

Energy is the capacity or ability to do work. Organisms require energy to grow, move, reproduce, and maintain and repair damaged tissues. Energy exists as stored energy—called **potential energy**—or as **kinetic energy**, the energy of motion. Think of potential energy as an arrow on a drawn bow (**FIGURE 5.3**). When the string is released, this potential energy is converted to kinetic energy as the motion of the bow propels the arrow. Similarly, the grass a bison eats has chemical potential energy, some of which is converted to kinetic energy and heat as the bison runs across the prairie. Thus, energy changes from one form to another.

THE FIRST AND SECOND LAWS OF THERMODYNAMICS

Thermodynamics is the study of energy and its transformations. Two laws about energy apply to all things in the universe: the first and second laws of thermodynamics. According to the **first law of thermodynamics**, an organism may absorb energy from its surroundings, or it may

> ■ **first law of thermodynamics**
> Energy cannot be created or destroyed, although it can change from one form to another.

Potential and kinetic energy FIGURE 5.3

Potential energy is stored in the drawn bow (**A**) and is converted to kinetic energy (**B**) as the arrow speeds toward its target. Photographed in Athens, Greece, during the 2004 Summer Olympics.

Energy flow in the environment FIGURE 5.4

The sun powers photosynthesis, producing chemical energy stored in the leaves and seeds of this umbrella tree. Photographed in Hanging Rock State Park, North Carolina.

give up some energy into its surroundings, but the total energy content of the organism and its surroundings is always the same. An organism can't create the energy it requires to live. Instead, it must capture energy from the environment to use for biological work, a process involving the transformation of energy from one form to another. In **photosynthesis**, plants absorb the radiant energy of the sun and convert it into the chemical energy contained in the bonds of sugar molecules (**FIGURE 5.4**). Later, an animal that eats the plant may transform some of the chemical energy into the mechanical energy of muscle contraction, enabling the animal to walk, run, slither, fly, or swim.

As each energy transformation occurs, some of the energy is changed to heat energy that is released into the cooler surroundings. No organism can ever use this energy again for biological work; it is "lost"

■ **second law of thermodynamics** When energy is converted from one form to another, some of it is degraded into heat, a less usable form that disperses into the environment.

from the biological point of view. However, it isn't gone from a thermodynamic point of view because it still exists in the surrounding physical environment. The use of food to enable us to walk or run doesn't destroy the chemical energy once present in the food molecules. After you have performed the task of walking or running, the energy still exists in the surroundings as heat energy.

According to the **second law of thermodynamics**, the amount of usable energy available to do work in the universe decreases over time. The second law of thermodynamics is consistent with the first law—that is, the total amount of energy in the universe isn't decreasing with time. However, the total amount of energy in the universe available to do biological work is decreasing over time.

Less usable energy is more diffuse, or disorganized. *Entropy* is a measure of this disorder or randomness. Organized, usable energy has low entropy, whereas disorganized energy such as heat has high entropy. Another way to explain the second law of thermodynamics is that entropy, or disorder, in a system tends to increase over time. As a result of the second law of thermodynamics, no process requiring an energy conversion is ever 100 percent efficient because much of the energy is dispersed as heat, resulting in an increase in entropy. For example, an automobile engine, which converts the chemical energy of gasoline to mechanical energy, is between 20 and 30 percent efficient: Only 20 to 30 percent of the original energy stored in the chemical bonds of the gasoline molecules is actually transformed into mechanical energy, or work.

PRODUCERS, CONSUMERS, AND DECOMPOSERS

The organisms of an ecosystem are divided into three categories based on how they obtain nourishment: producers, consumers, and decomposers. Virtually all ecosystems contain representatives of all three groups, which interact extensively with one another, both directly and indirectly.

Many potential animal prey, such as woodchucks, run to their underground burrows to escape predators. Others have mechanical defenses, such as the barbed quills (in the porcupine) and a shell (in the pond turtle). Some animals live in groups—a herd of antelope, colony of honeybees, school of anchovies, or flock of pigeons. This social behavior decreases the likelihood of a predator catching one of them unaware because the group has so many eyes, ears, and noses watching, listening, and smelling for predators (**Figure 5.19A**).

Chemical defenses are common among animal prey. The South American poison arrow frog has poison glands in its skin and bright warning colors that experienced predators avoid. Some animals blend into their surroundings and so hide from predators. Certain caterpillars resemble twigs so closely you would never guess they are animals until they move (**Figure 5.19B**).

Plants possess adaptations that protect them from being eaten. The presence of spines, thorns, tough leathery leaves, or even thick wax on leaves discourages foraging herbivores from grazing. Other plants produce an array of protective chemicals that are unpalatable or even toxic to herbivores. The nicotine found in tobacco is so effective at killing insects that it is an ingredient in many commercial insecticides.

A

B

Avoiding predators FIGURE 5.19

A Two adult meerkats stand watch over a young meerkat pup. If one of the sentries spies a predator such as an eagle or a hawk, they will alert the other meerkats, and all will scramble into their burrows. Photographed in the Kalahari Desert, South Africa.

B Predators often overlook caterpillars that closely resemble small branches. Can you find the caterpillar? (*Hint:* The caterpillar is mimicking an adjacent twig.)

COMPETITION

Competition occurs when two or more individuals attempt to use an essential common resource such as food, water, shelter, living space, or sunlight.

Resources are often in limited supply in the environment, and their use by one individual decreases the amount available to others. If a tree in a dense forest grows taller than surrounding trees, it absorbs more of the incoming sunlight (**FIGURE 5.20**). Less sunlight is available for nearby trees that the taller tree shades. Competition occurs among individuals within a population *(intraspecific competition)* and between different species *(interspecific competition)*.

Competition isn't always a straightforward, direct interaction. Consider a variety of flowering plants that live in a young

competition The interaction among organisms that vie for the same resources in an ecosystem (such as food or living space).

pine forest and compete with conifers for such resources as soil moisture and soil nutrient minerals. Their relationship is more involved than simple competition. The flowers produce nectar that some insect species consume; these insects also prey on needle-eating insects, reducing the number of insects feeding on pines. It is therefore difficult to assess the overall effect of flowering plants on pines. If the flowering plants were removed from the community, would the pines grow faster because they were no longer competing for necessary resources? Or would the increased presence of needle-eating insects (caused by fewer omnivorous insects) inhibit pine growth?

Short-term experiments in which one competing plant species is removed from a forest community often have demonstrated improved growth for the remaining

Competition FIGURE 5.20

Plants compete for light and growing space in a forest. Taller trees reduce the amount of sunlight available for shorter ones. Photographed in Kerala State, India.

species. However, few studies have tested the long-term effects on forest species of removing one competing species. These long-term effects may be subtle, indirect, and difficult to assess. They may lower or negate the negative effects of competition for resources.

KEYSTONE SPECIES

Certain species are more crucial to the maintenance of their ecosystem than others. Such **keystone species** are vital in determining an ecosystem's species composition and how the ecosystem functions. Keystone species are usually not the most abundant species in the ecosystem. Although present in relatively small numbers, keystone species exert a profound influence on the entire ecosystem because they often affect the available amount of food, water, or some other resource.

Identifying and protecting keystone species are crucial goals of conservation biologists because if a keystone species disappears from an ecosystem, other organisms may become more common or more rare, or even disappear. One example of a keystone species is a top predator such as the gray wolf (FIGURE 5.21). Where wolves were hunted to extinction, the populations of deer and other herbivores increased explosively. As these herbivores overgrazed the vegetation, plant species that couldn't tolerate such grazing pressure disappeared. Smaller animals such as insects were lost from the ecosystem because the plants they depended on for food were now less abundant. Thus, the disappearance of the wolf resulted in an ecosystem with considerably less biological diversity.

Some scientists think we should abandon the concept of keystone species because it is problematic. For one thing, most of the information about keystone species is anecdotal. Scientists have performed few long-term studies to identify keystone species and to determine the nature and magnitude of their effects on the ecosystems they inhabit.

Gray wolf FIGURE 5.21

The wolf is considered a keystone species in its ecosystem.

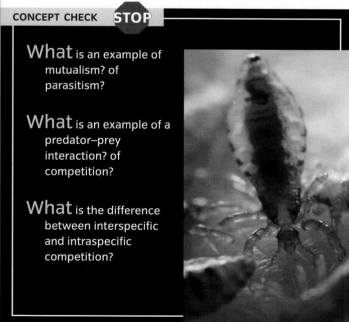

CONCEPT CHECK STOP

What is an example of mutualism? of parasitism?

What is an example of a predator–prey interaction? of competition?

What is the difference between interspecific and intraspecific competition?

Global Warming: Is There an Imbalance in the Carbon Cycle?

This vehicle averages perhaps 10 miles per gallon, which translates into a great deal of CO_2 emissions.

During the past two centuries, the level of carbon dioxide (CO_2) in the atmosphere has increased dramatically, caused by the burning of fossil fuels such as coal, oil, and natural gas, and the clearing and burning of forests. Environmental scientists are increasingly concerned that the rising levels of CO_2 may change Earth's climate. CO_2 levels rose from 315 parts per million (ppm) in 1958 to 377 ppm in 2004—a 20 percent increase. CO_2 in the atmosphere allows solar radiation to pass through but does not allow heat to radiate into space. Instead, the heat is radiated back to Earth's surface. As the CO_2 accumulates, it may trap enough heat to warm the planet.

Observations confirm that our climate is changing rapidly, particularly in the past few decades. The 1980s, 1990s, and early 2000s saw some of the warmest years since scientists started keeping records in 1820. The 10 warmest years have occurred since 1990. Earth's temperature in 2005 was the warmest in the weather records; the second warmest temperature on record occurred in 1998, and the third warmest in 2002.

Environmental scientists estimate that if trends don't change, Earth's mean temperature could rise 1.4° to 5.8°C (2.5° to 10.4°F) by the end of the 21st century. This temperature increase would make the atmosphere warmer than tree ring and glacial ice analyses estimate for the past 1,000 years. This warming could produce major shifts in patterns of rainfall and initiate melting of the West Antarctic and Southern Greenland ice sheets, as did the last warm period 120,000 years ago. Such melting would cause the ocean level to rise, an alarming scenario, as it might put many major cities at least partly under water.

An increasing number of environmental scientists are concerned that human-induced global climate warming may have started. The U.S. National Academy of Science has stated that global climate warming may be the most pressing international issue of the 21st century. In 2001 the U.N. Intergovernmental Panel on Climate Change (IPCC) released a report based on the consensus of hundreds of climate scientists worldwide. The IPCC declared that the human-produced increase in greenhouse gas concentrations are likely responsible for most of the observed warming in the last 50 years.

An international climate warming conference was held in Kyoto, Japan, in December 1997. Although many of the participating countries had conflicting positions, an initial treaty known as the Kyoto Protocol was formulated. This stipulated that highly developed countries must cut their emissions of CO_2 and other gases that cause warming by an average of 5.2 percent by 2012. Unless many countries ratify and observe the treaty, CO_2 emissions are expected to continue to increase. (The United States signed the protocol in 1998 at a follow-up meeting in Buenos Aires, but the Bush administration withdrew the United States from that commitment.)

Energy experts at the U.S. Department of Energy say that energy conservation and efficiency measures could accomplish much of the U.S. emissions reductions required if we were to observe the Kyoto Protocol. We could save the equivalent of 60 million metric tons of CO_2 emissions each year if automotive vehicles were designed to get better gas mileage *(see photo)*. Installing high-efficiency wind turbines to generate electricity in states such as Texas and California could save as much as 20 million metric tons of CO_2 emissions per year. We say more about global warming in Chapter 9.

CHAPTER SUMMARY

1 What is Ecology?

1. **Ecology** is the study of the interaction among organisms and between organisms and their abiotic environment.

2. A **population** is a group of organisms of the same species that live together in the same area at the same time. A **community** is a natural association that consists of all the populations of different species that live and interact together within an area at the same time. An **ecosystem** is a community and its physical environment. A **landscape** is a region that includes several interacting ecosystems. The **biosphere** is the layer of Earth containing all living organisms.

2 The Flow of Energy Through Ecosystems

1. **Energy** is the capacity or ability to do work. According to the **first law of thermodynamics**, energy can't be created or destroyed, although it can change from one form to another. As a result of the **second law of thermodynamics**, when energy is converted from one form to another, some of it is degraded into heat, a less usable form that disperses into the environment.

2. A **producer** manufactures large organic molecules from simple inorganic substances. A **consumer** can't make its own food and uses the bodies of other organisms as a source of energy and body-building materials. **Decomposers** are microorganisms that break down dead organic material and use the decomposition products to supply themselves with energy.

3. **Energy flow** is the passage of energy in a one-way direction through an ecosystem.

3 The Cycling of Matter in Ecosystems

1. **Biogeochemical cycles** are the processes by which matter cycles from the living world to the nonliving, physical environment and back again. Carbon dioxide is the important gas of the carbon cycle; carbon enters the living world by photosynthesis and returns to the abiotic environment when organisms respire. The hydrologic cycle continuously renews the supply of water and involves an exchange of water among the land, the atmosphere, and organisms. There are five steps in the nitrogen cycle: nitrogen fixation, nitrification, ammonification, assimilation, and dentrification. The phosphorus cycle has no biologically important gaseous compounds; phosphorus erodes from rock and is absorbed by plant roots.

See Figures 5.8, 5.9, 5.10, and 5.12 for additional details.

4 Ecological Niches

1. An **ecological niche** is the totality of an organism's adaptations, its use of resources, and the lifestyle to which it is fitted. An organism's ecological niche includes its **habitat**, its distinctive lifestyle, and its role in the community.

2. **Resource partitioning** is the reduction in competition for environmental resources, such as food, that occurs among coexisting species as a result of each species' niche differing from the others in one or more ways.

5 Interactions Among Organisms

1. **Symbiosis** is any intimate relationship or association between members of two or more species. **Mutualism** is a symbiotic relationship in which both species benefit. **Commensalism** is a symbiotic relationship in which one species benefits and the other species is neither harmed nor helped. **Parasitism** is a symbiotic relationship in which one species (the *parasite*) benefits at the expense of the other (the *host*).

2. **Predation** is the consumption of one species (the prey) by another (the predator). During **coevolution** between predator and prey, the predator evolves more efficient ways to catch prey (such as pursuit and ambush), and the prey evolves better ways to escape the predator (such as flight, association in groups, and camouflage).

3. **Competition** is the interaction among organisms that vie for the same resources in an ecosystem (such as food or living space). Competition occurs among individuals within a population (intraspecific competition) or between species (interspecific competition).

4. A **keystone species** is crucial in determining the nature and structure of the entire ecosystem in which it lives. Though present in relatively small numbers, keystone species have a disproportionate effect on the ecosystem.

KEY TERMS

- **ecology** p. 96
- **population** p. 96
- **community** p. 96
- **ecosystem** p. 96
- **landscape** p. 96

- **biosphere** p. 98
- **first law of thermodynamics** p. 99
- **second law of thermodynamics** p. 100
- **energy flow** p. 102
- **ecological niche** p. 110

- **symbiosis** p. 113
- **predation** p. 116
- **competition** p. 118

CRITICAL AND CREATIVE THINKING QUESTIONS

1. To function, ecosystems require an input of energy. Where does this energy come from?

2. After an organism uses energy, what happens to it?

3. In ecosystems, matter cycles, whereas energy flows in a linear fashion. Explain the distinction.

4. What is a biogeochemical cycle? Why is the cycling of matter essential to the continuance of life?

5. Describe how organisms participate in each of these biogeochemical cycles: carbon, nitrogen, and phosphorus.

6. How are food chains important in biogeochemical cycles?

7. How are the many cichlid species in Lake Victoria an example of resource partitioning?

8. In both parasitism and predation, one organism benefits at the expense of another. What is the difference between the two processes?

9. Some biologists think protecting keystone species would help preserve biological diversity in an ecosystem. Do you agree? Explain your answer.

10. How do human effects on climate warming make it more difficult to reach the goal of environmental sustainability?

This figure shows the components of a simple food chain. Use it to answer Questions 11–13.

11. Identify the producers, consumers, and decomposers in the food chain. How many trophic levels are represented?

12. Describe or indicate the flow of food and energy within this system.

13. Which forms of energy are present within this chain?

What is happening in this picture ?

- This caterpillar has inflated its thorax to make its body look like the head of a snake. Suggest a possible reason the caterpillar does this.

- Note the two large spots. What do they resemble? Why would this animal have such conspicuous spots?

- If a hungry bird saw this caterpillar, do you think it would have second thoughts before eating it? Why or why not?

Ecosystems and Evolution

THE FLORIDA EVERGLADES

The Everglades, in the southernmost part of Florida, is a vast expanse of predominantly saw-grass wetlands dotted with small islands of trees. It is a haven for wildlife, including alligators *(see photograph)*, snakes, panthers, otters, raccoons, and thousands of birds—great blue herons, snowy egrets, great white herons, roseate spoonbills, and osprey, to name just a few. At one time, this "river of grass" drifted south in a slow-moving sheet of water from Lake Okeechobee to Florida Bay.

The Everglades today is about half its original size of 1.6 million hectares (4 million acres) and suffers from many serious environmental problems. Wading bird populations have dropped 93 percent since 1930, and the area is now home to 50 endangered or threatened species. Rapid changes in the Everglades illustrate how misguided human activities can cause more harm than good.

More than 70 years of engineering projects—deemed necessary to protect the human population from storm-related flooding—have reduced the quantity of water flowing into the Everglades, restricting the natural recharging process there. The water that does enter is polluted by fertiliz-ers and pesticides used in agriculture. The flood-control measures diverted water from the wetlands, creating dry spaces that were then converted to agricultural or residential use. These developments caused habitat fragmentation, contributing to the Everglades' declining wildlife populations.

Florida's unique river of grass will never return completely to its original condition because there are too many sugar plantations and too many cities in the region. However, state and federal governments are working on a massive restoration project to undo some of the environmental damage. The plan incorporates controls on agricultural runoff, the conversion of some agricultural land to marshes, and a massive project to reengineer the area's entire system of canals, levees, and pumps so that a more natural flow of water will be restored to the Everglades. Restoration efforts will take more than 20 years and cost $8 billion. As the Everglades is restored to a scaled-down version of what it originally was, all who appreciate its uniqueness hope that the wildlife will rebound.

CHAPTER OUT

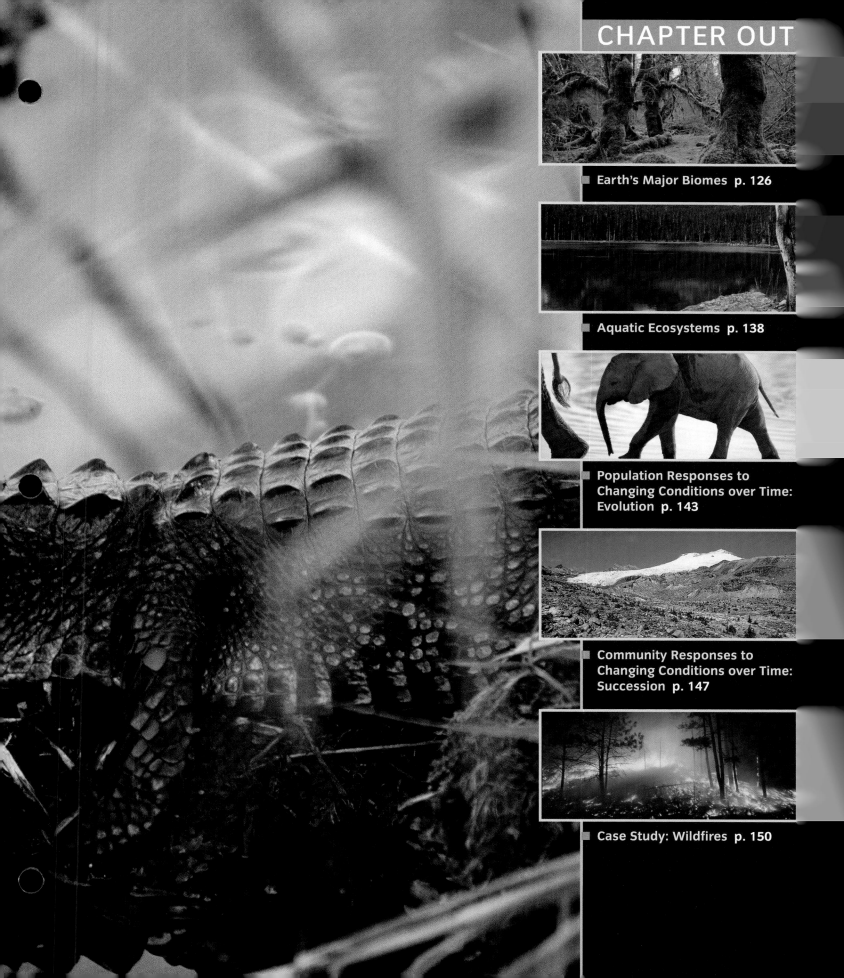

Earth's Major Biomes p. 126

Aquatic Ecosystems p. 138

Population Responses to
Changing Conditions over Time:
Evolution p. 143

Community Responses to
Changing Conditions over Time:
Succession p. 147

Case Study: Wildfires p. 150

Earth's Major Biomes

LEARNING OBJECTIVES

Define biome and discuss how biomes are related to climate.

Briefly describe the nine major terrestrial biomes, giving attention to the climate, soil, and characteristic organisms of each.

Earth has many different *climates*, which are based primarily on temperature and precipitation differences. Characteristic organisms have adapted to each climate. Because it is so large in area, a **biome** encompasses many interacting ecosystems (**FIGURE 6.1**). In terrestrial ecology, a biome is considered the next level of ecological organization above community, ecosystem, and landscape.

Near the poles, temperature is generally the overriding climate factor defining a biome, whereas in temperate and tropical regions, precipitation becomes more significant than temperature, as shown in **FIGURE 6.2**. Other *abiotic* factors to which certain biomes are sensitive include extreme temperatures as well as rapid temperature changes, fires, floods, droughts, and strong winds. Elevation also affects biomes: Changes in vegetation with increasing elevation resemble the changes in vegetation observed in going from warmer to colder climates.

biome A large, relatively distinct terrestrial region with similar climate, soil, plants, and animals regardless of where it occurs in the world.

The world's terrestrial biomes FIGURE 6.1

Although sharp boundaries are shown in this highly simplified map, biomes actually grade together at their boundaries. Use the legend below to identify the locations of the different biomes.

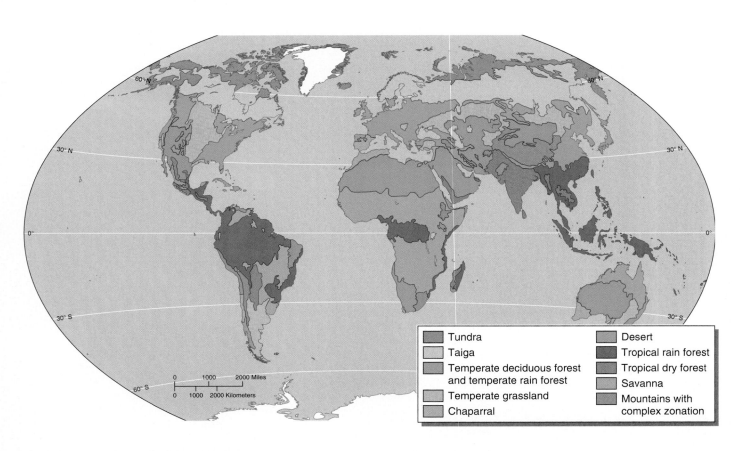

Tundra	Desert
Taiga	Tropical rain forest
Temperate deciduous forest and temperate rain forest	Tropical dry forest
Temperate grassland	Savanna
Chaparral	Mountains with complex zonation

www.wiley.com/
college/berg

Cold   Arctic

Tundra

Taiga

DECREASING TEMPERATURE

INCREASING LATITUDE

Temperate
deciduous forest   Temperate
grassland   Chaparral   Temperate
desert

Tropical
rain forest   Dry tropical
forest   Savanna   Moist tropical
desert   Dry tropical
desert

Hot

Tropics

Wet   DECREASING PRECIPITATION   Dry

Two climate factors, temperature and precipitation, have a predominant effect on biome distribution. In higher latitudes, temperature is the more important of the two, as mean annual temperatures decline poleward. In temperate and tropical zones, precipitation is more significant. Biomes differ in the relative amounts of precipitation that they receive and in the seasonal distribution of precipitation.

TUNDRA

Tundra (or *arctic tundra*) occurs in the extreme northern latitudes wherever the snow melts seasonally (**FIGURE 6.3**).

The Southern Hemisphere has no equivalent of the arctic tundra because it has no land in the corresponding latitudes. A similar ecosystem located in the higher elevations of mountains, above the tree line, is called *alpine tundra* to distinguish it from arctic tundra. Although the arctic tundra's growing season is short, the days are long. Above the Arctic Circle, the sun does not set at all for many days in midsummer, although the amount of light at midnight is one tenth that at noon. There is little precipitation, and most of the yearly 10 to 25 cm (4 to 10 in) of rain or snow falls during the summer months.

Most tundra soils formed when glaciers began retreating after the last Ice Age, about 17,000 years ago. These soils are usually nutrient-poor and have little *detritus* such as dead leaves and stems, animal droppings, or remains of organisms. Although the tundra's surface soil thaws during the summer, beneath it lies a layer of *permafrost,* permanently frozen ground that varies in depth and thickness. Permafrost interferes with drainage, so the thawed upper zone of soil is usually waterlogged during the summer. Limited precipitation, in combination with low temperatures, flat topography (or surface features), and the layer of permafrost, produces a landscape of broad, shallow lakes and ponds, sluggish streams, and bogs.

Tundra supports relatively few species compared to other biomes, but the species that do occur there often exist in great numbers. Mosses, lichens, grasses, and grasslike sedges are the dominant plants. Stunted trees and shrubs grow only in sheltered locations. As a rule, tundra plants seldom grow taller than 30 cm (12 in).

The year-round animal life of the tundra includes lemmings, voles, weasels, arctic foxes, snowshoe hares, ptarmigan, snowy owls, and musk oxen. In the summer, caribou migrate north to the tundra to graze on sedges, grasses, and dwarf willow. Dozens of bird species also migrate north in summer to nest and feed on abundant insects. Mosquitoes, blackflies, and deerflies survive the winter as eggs or pupae and appear in great numbers during summer weeks.

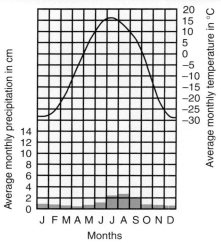

Arctic tundra FIGURE 6.3

A caribou buck forages for food in the tundra. Because of the tundra's short growing season and permafrost, only small, hardy plants grow in the northernmost biome that encircles the Arctic Ocean. Photographed in Denali National Park, Alaska. Climate graph shows monthly temperatures and precipitation for Fort Yukon, Alaska.

tundra The treeless biome in the far north that consists of boggy plains covered by lichens and small plants such as mosses; it has harsh, very cold winters and extremely short summers.

BOREAL FOREST

Just south of the tundra is the **boreal forest**, or northern coniferous forest, which stretches across North America and Eurasia (**FIGURE 6.4**). There is no biome comparable to the boreal forest in the Southern Hemisphere. Winters are extremely cold and severe, although not as harsh as those in the tundra. Boreal forest receives little precipitation, perhaps 50 cm (20 in) per year, and its soil is typically acidic and mineral-poor, with a deep layer of partly decomposed pine and spruce needles at the surface. Where permafrost occurs, it is found deep under the surface. Boreal forest has numerous ponds and lakes dug by grinding ice sheets during the last Ice Age.

Black and white spruces, balsam fir, eastern larch, and other conifers (cone-bearing evergreens) dominate the boreal forest. Conifers have many drought-resistant adaptations, such as needle-like leaves whose minimal surface area prevents water loss by evaporation. Such an adaptation helps conifers withstand the drought of the northern winter, when roots cannot absorb water through the frozen ground. Being evergreen, conifers resume photosynthesis as soon as warmer temperatures return.

The animal life of the boreal forest consists of some larger species such as caribou, which migrate from the tundra for winter; wolves; brown and black bears; and moose. However, most boreal mammals are medium-sized to small, including rodents, rabbits, and smaller predators such as lynx, sable, and mink. Birds are abundant in the summer but migrate to warmer climates for winter. There are few amphibians and reptiles except in the southern areas of the boreal forest. Insects are plentiful.

> **boreal forest**
> A region of coniferous forest (such as pine, spruce, and fir) in the Northern Hemisphere; located just south of the tundra.

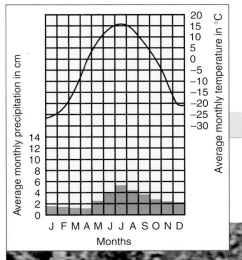

Boreal forest FIGURE 6.4

A herd of caribou runs past a boreal forest. These coniferous forests occur in cold regions of the Northern Hemisphere adjacent to the tundra. Photographed in Northwest Territories, Canada. Climate graph shows monthly temperatures and precipitation for Fort Smith, Northwest Territories, Canada.

Temperate rain forest FIGURE 6.5

Temperate rain forest FIGURE 6.5

This temperate biome has large amounts of precipitation. Photographed in Olympic National Park in Washington State. Climate graph shows monthly temperatures and precipitation for Estacada, Oregon.

TEMPERATE RAIN FOREST

A coniferous **temperate rain forest** occurs on the northwest coast of North America. Similar vegetation exists in southeastern Australia and in southern South America. Annual precipitation in this biome is high, from 200 to 380 cm (80 to 152 in), and condensation of water from dense coastal fogs augments the precipitation. The proximity of temperate rain forest to the coastline moderates its temperature so that the seasonal fluctuation is narrow; winters are mild and summers are cool. Temperate rain forest has relatively nutrient-poor soil, though its organic content may be high. Cool temperatures slow the activity of bacterial and fungal decomposers. Thus, needles and large fallen branches and trunks accumulate on the ground as litter that takes many years to decay and release nutrient minerals to the soil.

temperate rain forest A coniferous biome with cool weather, dense fog, and high precipitation.

The dominant vegetation in the North American temperate rain forest is large evergreen trees such as western hemlock, Douglas fir, western red cedar, Sitka spruce, and western arborvitae (FIGURE 6.5). Temperate rain forests are rich in epiphytes, which are smaller plants that grow on the trunks and branches of large trees. Epiphytes in this biome are mainly mosses, club mosses, lichens, and ferns, all of which also carpet the ground. Squirrels, wood rats, mule deer, elk, numerous bird species, and several species of amphibians and reptiles are common temperate rainforest animals. (Recall that you explored the issues surrounding old-growth temperate rain forests of the Pacific Northwest in Chapter 3.)

Zonation in a large lake: Bear Lake in Rocky Mountain National Park, Colorado.

(Inset, below) A lake is a standing-water ecosystem surrounded by land. The littoral zone is the shallow-water area around the lake's edge. The limnetic zone is the open, sunlit water away from the shore. The profundal zone, under the limnetic zone, is below where light penetrates.

Littoral zone

Limnetic zone

Profundal zone

Flowing-water ecosystems are highly variable. The surrounding environment changes greatly between a river's source and its mouth (FIGURE 6.12). Forest shades certain parts of the stream's course, while other parts are exposed to direct sunlight. Groundwater may well up through sediments on the bottom in one particular area, making the water temperature cooler in summer or warmer in winter than adjacent parts of the stream or river. The kinds of organisms found in flowing water vary greatly from one stream to another, depending primarily on the strength of the current. In streams with fast currents, some inhabitants have adaptations like suckers, with which they attach themselves to rocks to prevent being swept away. Some stream inhabitants have flattened bodies to slip under or between rocks. Other inhabitants such as fish are streamlined and muscular enough to swim in the current.

Features of a typical river FIGURE 6.12

A The river begins at a source, often high in mountains and fed by melting snows or glaciers. Headwater streams flow downstream rapidly, often over rocks (as rapids) or bluffs (as waterfalls). Along the way, tributaries feed into the river, adding to its flow. As the river's course levels out, the river flows more slowly and winds from side to side, forming bends called meanders. The flood plain is the relatively flat area on either side of the river that is subject to flooding. The delta is a fertile, low-lying plain at the river's mouth that forms from sediments the slow-moving river deposits as it empties into the ocean. B Aerial view of meanders in the Tambopata River, Peru.

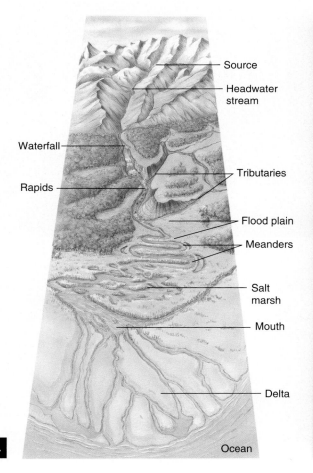

FRESHWATER WETLANDS

Freshwater wetlands include **marshes**, dominated by grasslike plants, and **swamps**, dominated by woody trees or shrubs (FIGURE 6.13). Wetland soils are water-logged for variable periods and are therefore anaerobic (without oxygen). They are rich in accumulated organic materials, partly because anaerobic conditions discourage decomposition.

> **freshwater wetlands** Lands that shallow fresh water covers for at least part of the year; wetlands have a characteristic soil and water-tolerant vegetation.

Wetlands provide excellent wildlife habitat for migratory waterfowl and other bird species, as well as for beaver, otters, muskrats, and game fish.

In addition to providing unique wildlife habitat, wetlands serve other important environmental functions, known as **ecosystem services**. When rivers flood their banks, wetlands are capable of holding or even absorbing the excess water, thereby helping to control flooding. The floodwater then drains slowly back into the rivers, providing a steady flow of water throughout the year. Wetlands also serve as groundwater recharging areas. One of their most important roles is to trap and hold pollutants in the flooded soil, thereby cleansing and purifying the water.

> **ecosystem services** Important environmental benefits, such as clean air to breathe, clean water to drink, and fertile soil in which to grow crops, that the natural environment provides.

Freshwater swamp FIGURE 6.13

Freshwater swamps are inland areas covered by water and dominated by trees, such as bald cypress. Photographed in Big Cypress Swamp, Florida.

THE DOMAINS AND KINGDOMS OF LIFE

Biologists arrange organisms into logical groups to try to make sense of the remarkable diversity of life that has evolved on Earth (**FIGURE 6.17**). For hundreds of years, biologists regarded organisms as falling into two broad categories—plants and animals. With the development of microscopes, however, it became increasingly obvious that many organisms did not fit very well into either the plant kingdom or the animal kingdom. For example, bacteria have a *prokaryotic* cell structure: They lack organelles enclosed by membranes, including a nucleus. This feature, which separates bacteria from all other or-

ganisms, is far more fundamental than the differences between plants and animals, which have similar cell structures. Hence, it became clear that bacteria were neither plants nor animals. Furthermore, certain microorganisms such as *Euglena,* which is both motile (has a hairlike flagellum for movement) and photosynthetic, seem to possess characteristics of both plants and animals.

These and other considerations eventually led to the three domain/six kingdom system of classification that many biologists use today. The prokaryotes fall into two groups that are sufficiently distinct from one another to be classified into two domains, **Archaea** and **Eubacteria**. The eukaryotes, organisms with eu-

The three domains and six kingdoms of life FIGURE 6.17

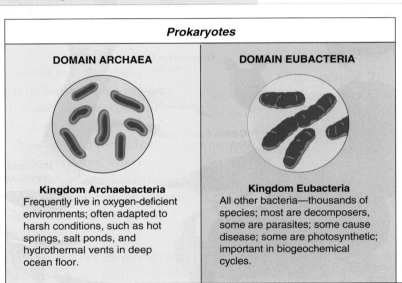

Prokaryotes

DOMAIN ARCHAEA

Kingdom Archaebacteria
Frequently live in oxygen-deficient environments; often adapted to harsh conditions, such as hot springs, salt ponds, and hydrothermal vents in deep ocean floor.

DOMAIN EUBACTERIA

Kingdom Eubacteria
All other bacteria—thousands of species; most are decomposers, some are parasites; some cause disease; some are photosynthetic; important in biogeochemical cycles.

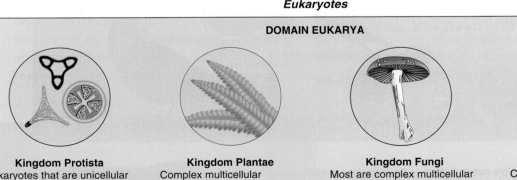

Eukaryotes

DOMAIN EUKARYA

Kingdom Protista
Eukaryotes that are unicellular or relatively simple multicellular organisms, such as algae, protozoa, slime molds, and water molds; important in aquatic food chains; algae are important producers.

Kingdom Plantae
Complex multicellular eukaryotes; most use radiant energy to manufacture food molecules by photosynthesis; play important role as producers and as source of atmospheric oxygen.

Kingdom Fungi
Most are complex multicellular eukaryotes that secrete digestive enzymes into their food and then absorb the predigested nutrients; decomposers; some are parasites; some cause disease.

Kingdom Animalia
Complex multicellular eukaryotes that ingest their food and then digest it inside their bodies; consumers— herbivores, carnivores, omnivores, and detritivores.

karyotic cells, are classified in domain **Eukarya**. Eukaryotic cells have a high degree of internal organization, containing nuclei, chloroplasts (in photosynthetic cells), and mitochondria.

Each of the six kingdoms is assigned to one of the three domains. Kingdom Archaebacteria corresponds to the domain Archaea, and kingdom Eubacteria corresponds to the domain Eubacteria. The remaining four kingdoms are classified in domain Eukarya.

Although the three domain/six kingdom system is a definite improvement over the two-kingdom system, it is not perfect. Most of its problems concern the kingdom Protista, which includes some organisms that may be more closely related to members of other kingdoms than to other protists. For example, green algae are protists that are clearly similar to plants but do not appear to be closely related to other protists, such as slime molds and brown algae.

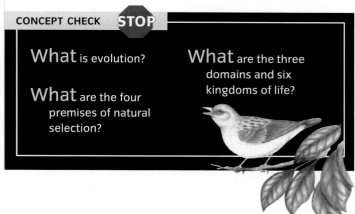

CONCEPT CHECK STOP

What is evolution?

What are the four premises of natural selection?

What are the three domains and six kingdoms of life?

Community Responses to Changing Conditions over Time: Succession

LEARNING OBJECTIVES

Define ecological succession.

Distinguish between primary and secondary succession.

A community of organisms does not spring into existence full-blown. By means of **ecological succession**, a given community develops gradually through a sequence of species. Certain organisms colonize an area; over time others replace them, and eventually the replacements are themselves replaced by still other species.

The actual mechanisms that underlie succession are not clear. In some cases, it may be that an earlier resident species modifies the environment in some way, thereby making it more suitable for a later species to colonize. It is also possible that prior residents live there in the first place because there is little competition from other species. Later, as more invasive species arrive, the original species are displaced.

Ecologists initially thought that succession inevitably led to a stable and persistent community, known as a *climax community,* such as a forest. But more recently, this traditional view has fallen out of favor. The apparent stability of a "climax" forest is probably the result of how long trees live relative to the human life span. It is now recognized that mature climax communities are not in a state of stable equilibrium, but rather in a state of continual disturbance. Over time, a mature community changes in species composition and in the relative abundance of each species, despite the fact that it retains an overall uniform appearance.

Succession is usually described in terms of the changes in the plant species growing in a given area, although each stage of the succession may also have its own kinds of animals and other organisms. Ecological succession is measured on the scale of tens, hundreds, or thousands of years, not the millions of years involved in the evolutionary time scale.

ecological succession The process of community development over time, which involves species in one stage being replaced by different species.

Primary succession on glacial moraine FIGURE 6.18

A After the glacier's retreat, lichens initially colonize the barren landscape, followed by mosses and small shrubs.

During the past 200 years, glaciers have retreated in Glacier Bay, Alaska. Although these photos were not taken in the same area, they show some of the stages of primary succession on glacial moraine (rocks, gravel, and sand that a glacier deposits).

B At a later time, dwarf trees and shrubs colonize the area.

C Still later, spruces dominate the community.

Primary succession is the change in species composition over time in a previously uninhabited environment (FIGURE 6.18). No soil exists when primary succession begins. Bare rock surfaces, such as recently formed volcanic lava and rock scraped clean by glaciers, are examples of sites where primary succession may take place. Details vary from one site to another, but on bare rock, lichens are often the most important element in the *pioneer community,* which is the initial community that develops during primary succession.

Lichens secrete acids that help break the rock apart, beginning the process of soil formation. Over time, mosses and drought-resistant ferns may replace the lichen community, followed in turn by tough grasses and herbs. Once soil accumulates, low shrubs may replace the grasses and herbs; over time, forest trees in several distinct stages would replace the shrubs. Primary succession on bare rock from a pioneer community to a forest community often occurs in this sequence: lichens → mosses → grasses → shrubs → trees.

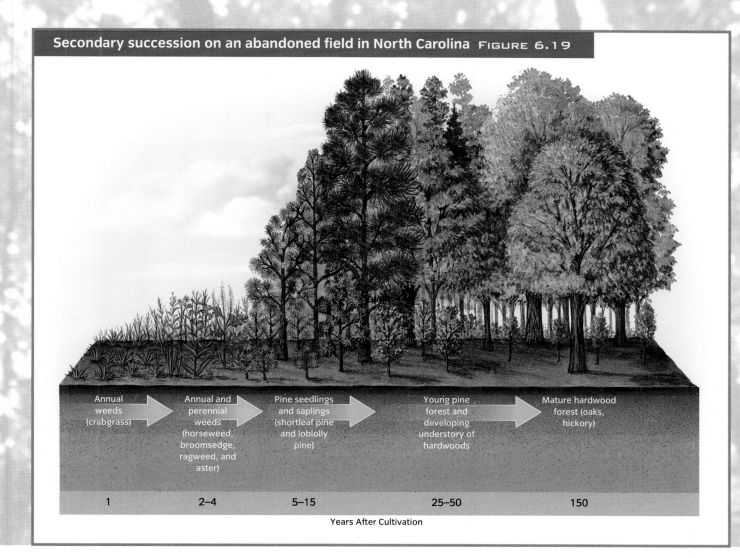

Secondary succession on an abandoned field in North Carolina FIGURE 6.19

| Annual weeds (crabgrass) | → | Annual and perennial weeds (horseweed, broomsedge, ragweed, and aster) | → | Pine seedlings and saplings (shortleaf pine and loblolly pine) | → | Young pine forest and developing understory of hardwoods | → | Mature hardwood forest (oaks, hickory) |

| 1 | 2–4 | 5–15 | 25–50 | 150 |

Years After Cultivation

Secondary succession is the change in species composition that takes place after some disturbance destroys the existing vegetation; soil is already present (FIGURE 6.19). Abandoned farmland or an open area caused by a forest fire are common examples of sites where secondary succession occurs. Biologists have studied secondary succession on abandoned farmland extensively. Although it takes more than 100 years for secondary succession to occur at a single site, a single researcher can study old-field succession in its entirety by observing different sites undergoing succession in the same general area. (The biologist may examine county tax records to determine when each field was abandoned.) Secondary succession on abandoned farmland in the southeastern United States proceeds in this sequence: crabgrass → horseweed, broomsedge, and other weeds → pine trees → hardwood trees.

CONCEPT CHECK **STOP**

What is ecological succession?

How does primary succession differ from secondary succession?

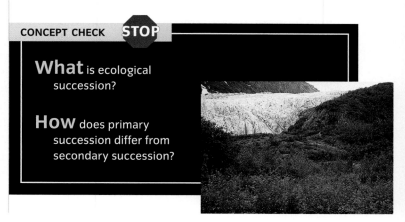

Wildfires

Wildfires—fires started by lightning—are an important environmental force in many geographic areas, especially places with wet seasons followed by dry seasons. Vegetation that grows during the wet season dries to tinder during the dry season. When lightning strikes, it ignites the dry organic material, and wind spreads the fire through the area. At the peak of the wildfire season in the American West, an area prone to wildfires, hundreds of new wildfires can break out each day.

Fires have several effects on the environment. First, combustion frees the minerals locked in dry organic matter. The ashes left by fire are rich in potassium, phosphorus, calcium, and other nutrient minerals essential for plant growth. Thus, vegetation flourishes after a fire. Second, fire removes plant cover and exposes the soil, which stimulates the germination of seeds that require bare soil and encourages the growth of shade-intolerant plants. Third, fire causes increased soil erosion because it removes plant cover, leaving the soil more vulnerable to wind and water.

Fires were a part of the natural environment long before humans appeared, and many terrestrial ecosystems have adapted to fire. Grasses adapted to wildfire have underground stems and buds. After fire kills the aerial parts, the

Wildfire in the western United States.

untouched underground parts send up new sprouts. Fire-adapted trees such as bur oak and ponderosa pine have thick, fire-resistant bark; others, like jack pines, depend on fire for successful reproduction, because the fire's heat opens the cones and releases the seeds.

Human interference also has an effect on the frequency and intensity of wildfires, even when the actual goal is fire prevention. When fire is excluded from a fire-adapted ecosystem, organic litter

accumulates. As a result, when a fire does occur, it burns hotter and is much more destructive than ecologically helpful. Decades of fire suppression in the West are partly responsible for the massively destructive fires that occurred in recent years in Arizona, Colorado, New Mexico, and Utah. *Prescribed burning* is an ecological management tool that allows for controlled burning to reduce organic litter and suppress fire-sensitive trees in fire-adapted areas.

1 Earth's Major Biomes

1. A **biome** is a large, relatively distinct terrestrial region with characteristic climate, soil, plants, and animals, regardless of where it occurs; a biome encompasses many interacting ecosystems. Near the poles, temperature is generally the overriding climate factor in determining biome distribution, whereas in temperate and tropical regions, precipitation is more significant.

2. **Tundra** is the treeless biome in the far north that consists of boggy plains covered by lichens and small plants such as mosses; it has harsh, very cold winters and extremely short summers. **Boreal forest** is a region of coniferous forest in the Northern Hemisphere, located just south of the tundra. **Temperate rain forest** is a coniferous biome with cool weather, dense fog, and high precipitation. **Temperate deciduous forest** is a forest biome that occurs in temperate areas where annual precipitation ranges from about 75 cm to 126 cm. **Tropical rain forest** is a lush, species-rich forest biome that occurs where the climate is very moist throughout the year. **Chaparral** is a biome with mild, moist winters and hot, dry summers; vegetation is typically small-leafed evergreen shrubs and small trees. **Temperate grassland** is a grassland with hot summers, cold winters, and less rainfall than is found in the temperate deciduous forest biome. **Savanna** is a tropical grassland with widely scattered trees or clumps of trees. **Desert** is a biome in which the lack of precipitation limits plant growth; deserts are found in both temperate and tropical regions.

2 Aquatic Ecosystems

1. In aquatic ecosystems, important environmental factors include salinity, amount of dissolved oxygen, and availability of light for photosynthesis.

2. Freshwater ecosystems include standing-water, flowing-water, and freshwater wetlands (marshes and swamps). A **flowing-water ecosystem** is a freshwater ecosystem such as a river or stream in which the water flows in a current. A **standing-water ecosystem** is a body of fresh water surrounded by land and whose water does not flow, such as a lake or pond. **Freshwater wetlands** are lands that shallow fresh water covers for at least part of the year; wetlands have a characteristic soil and water-tolerant vegetation. An **estuary** is a coastal body of water, partly surrounded by land, with access to the open ocean and a large supply of fresh water from a river. Water in an estuary is brackish rather than truly fresh. Temperate estuaries usually contain **salt marshes**, whereas tropical estuaries are lined with **mangrove forests**.

3 Population Responses to Changing Conditions over Time: Evolution

1. **Evolution** is the cumulative genetic changes in populations that occur during successive generations.

2. **Natural selection** is the tendency of better-adapted individuals—those with a combination of genetic traits better suited to environmental conditions—to survive and reproduce, increasing their proportion in the population. Natural selection is based on four premises established by Charles Darwin: (1) Each species produces more offspring than will survive to maturity. (2) The individuals in a population exhibit inheritable variation in their traits. (3) Organisms compete with one another for the resources needed to survive. (4) Those individuals with the most favorable combination of traits are most likely to survive and reproduce, passing their genetic traits to the next generation.

3. Earth's organisms are classified within a system of three domains and six kingdoms. The prokaryotes occupy two distinct domains, each containing one kingdom: Kingdom Archaebacteria corresponds to the domain **Archaea**, and kingdom Eubacteria corresponds to the domain **Eubacteria**. The Eukaryotes are clasified in the domain **Eukarya**, which contains four kingdoms: Protista, Plantae, Fungi, and Animalia.

4 **Community Responses to Changing Conditions over Time: Succession**

1. **Ecological succession** is the process of community development over time, which involves species in one stage being replaced by different species.

2. **Primary succession** is the change in species composition over time in an environment that was not previously inhabited by organisms; examples include bare rock surfaces, such as recently formed volcanic lava and rock scraped clean by glaciers. **Secondary succession** is the change in species composition that takes place after some disturbance destroys the existing vegetation; soil is already present. Examples include abandoned farmland and open areas caused by forest fires.

KEY TERMS

- ■ **biome** p. 126
- ■ **tundra** p. 128
- ■ **boreal forest** p. 129
- ■ **temperate rain forest** p. 130
- ■ **temperate deciduous forest** p. 131
- ■ **tropical rain forest** p. 132
- ■ **chaparral** p. 133
- ■ **temperate grassland** p. 135
- ■ **savanna** p. 136
- ■ **desert** p. 137
- ■ **standing-water ecosystem** p. 138
- ■ **flowing-water ecosystem** p. 140
- ■ **freshwater wetlands** p. 141
- ■ **ecosystem services** p. 141
- ■ **estuary** p. 142
- ■ **evolution** p. 143
- ■ **natural selection** p. 144
- ■ **ecological succession** p. 147

CRITICAL AND CREATIVE THINKING QUESTIONS

1. What two climate factors are most important in determining an area's characteristic biome?

2. What climate and soil factors produce each of the major terrestrial biomes?

3. In which biome do you live? If your biome does not match the description given in this book, how do you explain the discrepancy?

4. Which biomes are best suited for agriculture? Explain why each of the biomes you did not specify is less suitable for agriculture.

5. What environmental factors are most important in determining the kinds of organisms found in aquatic environments?

6. Distinguish between freshwater wetlands and estuaries, and between flowing-water and standing-water ecosystems.

7. During the mating season, male giraffes slam their necks together in fighting bouts to determine which male is stronger and can therefore mate with females. Explain how the long necks of giraffes may have evolved, using Darwin's theory of evolution by natural selection.

8. Describe the stages in old-field succession.

9. Although most salamanders have four legs, the aquatic salamander shown below resembles an eel. It lacks hind limbs and has very tiny forelimbs. Propose a hypothesis to explain how limbless salamanders evolved according to Darwin's theory of natural selection.

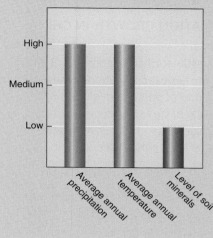

10. How could you test the hypothesis you proposed in question 9?

11. Which biome discussed in this chapter is depicted by the information in the graph? Explain your answer.

What is happening in this picture ?

This picture shows expensive homes built in the chaparral of the Santa Monica Mountains. Based on what you have learned in this chapter, what environmental problem might threaten these homes?

Sometimes people have removed the chaparral vegetation to prevent fires from damaging their homes. Where that has occurred, the roots no longer hold the soil in place. What could happen when the winter rains come?

Human Population Change and the Environment

7

SLOWING POPULATION GROWTH IN CHINA

China, with an estimated population of 1.3 billion people, has the largest population in the world. Recognizing that its rate of population growth had to decrease or the quality of life for everyone in China would be compromised, in 1971 the Chinese government began to pursue birth control seriously. It urged couples to marry later, increase spacing between children, and have fewer children.

In 1979 China instigated a more aggressive plan to reduce the birth rate, announcing incentives to promote later marriages and one-child families *(see inset)*. A couple who signed a pledge to limit themselves to a single child might be eligible for incentives like medical care and schooling for the child, cash bonuses, preferential housing, and retirement funds. Penalties were instituted, including fines and the surrender of all of these privileges, if a second child was born.

China's aggressive plan brought about the most rapid and drastic reduction in fertility in the world, from 5.8 births per woman in 1970 to 1.7 births per woman today. However, the plan was controversial and unpopular because it compromises individual freedom of choice. In some instances, social pressures caused women who were pregnant with a second child to get an abortion.

In China, sons are valued more highly than daughters because sons carry on the family name and traditionally provide old-age security for their parents. A disproportionate number of male babies have been born in recent years, suggesting that some expectant parents determine the sex of their fetus and abort it if it is female. In the past, parents required to conform to the one-baby policy abandoned or killed thousands of newborn baby girls because they wanted a boy instead.

In 1984 the one-child family policy was relaxed in rural China, where 70 percent of all Chinese live. China's recent population control program has relied on education, publicity campaigns *(see larger photograph)*, and fewer penalties to achieve its goals.

CHAPTER OUT

Population Ecology p. 156

Human Population Patterns p. 1

Demographics of Countries p. 1

Stabilizing World Population p.

Population and Urbanization p.

CASE STUDY: Urban Planning in Curitiba, Brazil p. 180

Population Ecology

NATIONAL GEOGRAPHIC

LEARNING OBJECTIVES

Define population ecology.

Explain the four factors that produce changes in population size.

Define biotic potential and carrying capacity.

Individuals of a given species are part of a larger organization called a *population*. Populations exhibit characteristics that are distinct from those of the individuals that compose them. Some of the features characteristic of populations but not of individuals are birth and death rates, growth rates, and age structure. Studying populations of other species provides insight into some of the processes that affect the growth of human populations. Understanding human population change is important because the size of the human population is central to most environmental problems and their solutions.

Scientists who study **population ecology** try to determine the processes common to all populations (**FIGURE 7.1**). Population ecologists study how a population responds to its environment—such as how individuals in a given population compete for food or other resources, and how predation, disease, and other environmental pressures affect that population. Environmental pressures like these prevent populations—whether of bacteria or maple trees or giraffes—from increasing indefinitely.

population ecology
The branch of biology that deals with the number of individuals of a particular species found in an area and how and why those numbers increase or decrease over time.

What do these populations have in common? FIGURE 7.1

A A school of schoolmaster fish swims off the coast of Florida.

B A flock of mute swans crowds a lake in Great Britain. What we learn about one population helps us make predictions about other populations.

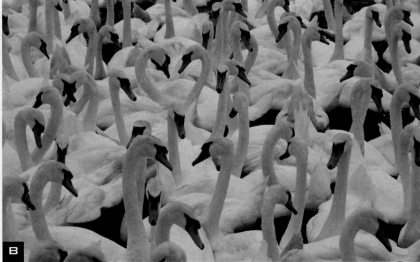

HOW DO POPULATIONS CHANGE IN SIZE?

Populations of organisms, whether sunflowers, eagles, or humans, change over time. On a global scale, this change is due to two factors: the rate at which individual organisms produce offspring (the birth rate), and the rate at which individual organisms die (the death rate) (FIGURE 7.2A). In humans, the birth rate (b) is usually expressed as the number of births per 1000 people per year, and the death rate (d) as the number of deaths per 1000 people per year. The **growth rate (r)** of a population is the birth rate (b) minus the death rate (d), or $r = b - d$. Growth rate is also referred to as *natural increase* in human populations.

If organisms in the population are born faster than they die, the growth rate is more than zero, and population size increases. If organisms in the population die faster than they are born, the growth rate is less than zero, and population size decreases. If the growth rate is equal to zero, births and deaths match, and population size is stationary despite continued reproduction and death.

In addition to birth and death rates, **dispersal**—movement from one region or country to another—affects local populations. There are two types of dispersal: **immigration (i)**, in which individuals enter a population and increase its size, and **emigration (e)**, in which individuals leave a population and decrease its size. The growth rate (r) of a local population must take into account birth rate (b), death rate (d), immigration (i), and emigration (e) (FIGURE 7.2B).

Factors that interact to change population size FIGURE 7.2

A On a global scale, the change in a population is due to the number of births and deaths.

B In local populations, such as the population of the United States, the number of births, deaths, immigrants, and emigrants affect population size.

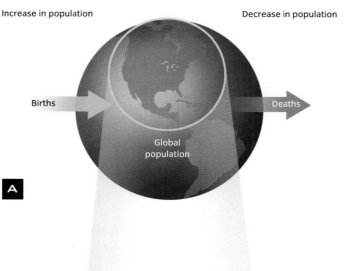

Increase in population · Decrease in population

Births → Global population → Deaths

A

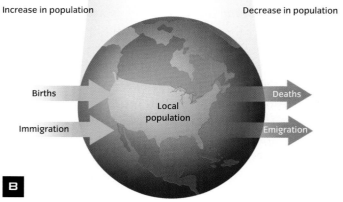

Increase in population · Decrease in population

Births → Local population → Deaths
Immigration → → Emigration

B

MAXIMUM POPULATION GROWTH

Different species have different **biotic potentials**. Several factors influence a species' biotic potential: the age at which reproduction begins, the fraction of the life span during which an individual can reproduce, the number of reproductive periods per lifetime, and the number of offspring produced during each period of reproduction. These factors, called *life history characteristics*, determine whether a particular species has a large or a small biotic potential.

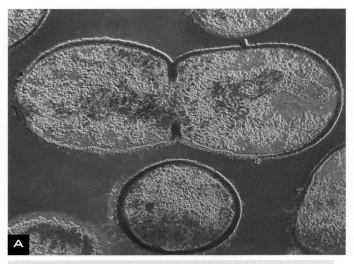

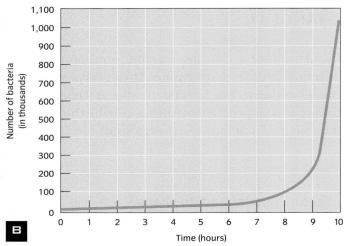

Exponential population growth FIGURE 7.3

A A micrograph of *Streptococcus* bacteria. The elongated cell is in the process of dividing.

B When bacteria divide at a constant rate and their numbers are graphed, the

curve of exponential population growth has a characteristic J shape.

Generally, larger organisms like blue whales have the smallest biotic potentials, whereas microorganisms have the greatest biotic potentials. Under ideal conditions (that is, in an environment with unlimited resources), certain bacteria reproduce by dividing in half every 30 minutes (FIGURE 7.3A). At this rate of growth, a single bacterium increases to a population of more than 1 million in just 10 hours, and exceeds 1 billion in 15 hours. If you plot population number versus time, the graph takes on the characteristic J shape of **exponential population growth** (FIGURE 7.3B). When a population grows exponentially, the larger the population gets, the faster it grows. It doubles, then doubles again, then again, but each time the doubling occurs within a shorter period. Regardless of species, whenever a population grows at its biotic potential, population size plotted versus time gives the same J-shaped curve. The only variable is time.

> **exponential population growth** The accelerating population growth that occurs when optimal conditions allow a constant reproductive rate.

ENVIRONMENTAL RESISTANCE AND CARRYING CAPACITY

Certain populations may exhibit exponential population growth for a short period. However, organisms don't reproduce indefinitely at their biotic potentials because the environment sets limits, which are collectively called **environmental resistance**. Examples of environmental resistance include limited food, water, shelter, and other essential resources (resulting in increased competition for these things) as well as increased disease and predation.

Using the earlier example, we find that bacteria never reproduce unchecked for an indefinite period because they run out of food and living space, and poisonous body wastes accumulate in their vicinity. With crowding, bacteria become more susceptible to parasites (high population densities facilitate the spread of infectious organisms such as viruses among individuals) and predators (high population densities increase the likelihood of a predator catching an individual). As the environment deteriorates, bacterial birth rate declines and death rate increases. The environmental conditions might worsen to a point where the death rate exceeds the birth rate, and as a result, the population decreases. Thus, the environment controls population size: As the population increases, so does environmental resistance, which limits population growth.

Over longer periods, the rate of population growth may decrease to nearly zero. This leveling out occurs at or near the environment's **carrying capacity (K)**. In nature, carrying capacity is dynamic and changes in response to environmental changes. An ex-

tended drought, for example, might decrease the amount of vegetation growing in an area, and this change, in turn, would lower the carrying capacity for deer and other herbivores in that environment.

When a population influenced by environmental resistance is graphed over a long period (**FIGURE 7.4**), the curve has an **S** shape. The curve shows the population's initial exponential increase (note the curve's **J** shape at the start, when environmental resistance is low). Then, the population number levels out as it approaches the carrying capacity of the environment. Although the **S** curve is an oversimplification of how most populations change over time, it fits some populations studied in the laboratory, as well as a few studied in nature.

Population growth as the carrying capacity is approached FIGURE 7.4

A *Paramecium* is a unicellular protist, a kind of microorganism.

B G.F. Gause, a Russian ecologist who conducted experiments during the 1930s, grew a population of *Paramecium* in a test tube. He supplied a limited amount of food daily and replenished the media occasionally to eliminate the buildup of metabolic wastes. Under these conditions, the population of *Paramecium* increased exponentially at first, but then the growth rate declined to zero, and the population size leveled off.

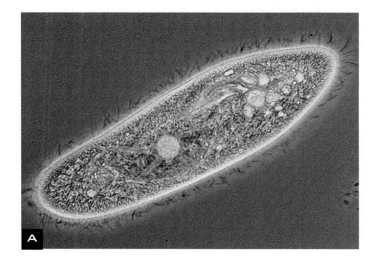

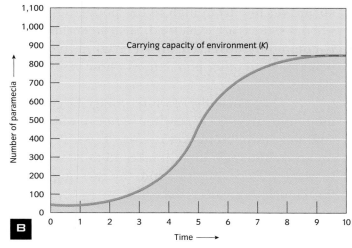

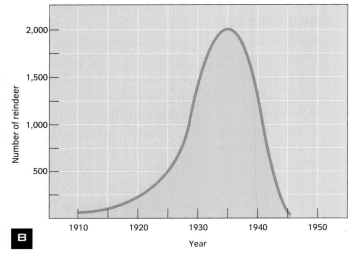

A population crash FIGURE 7.5

A A herd of reindeer on one of the Pribilof Islands in the Bering Sea, off the coast of Alaska.

B In 1910, a small herd of 26 reindeer was introduced on one of these islands. The herd's population increased exponentially for about 25 years until there were many more reindeer than the island could support, particularly in winter. The reindeer overgrazed the vegetation until the plant life was almost wiped out. Then, in slightly over a decade, as reindeer died from starvation, the number of reindeer plunged to less than 1 percent of the population at its peak.

A population rarely stabilizes at *K* (carrying capacity), as shown in Figure 7.4, but may temporarily rise higher than *K*. It will then drop back to, or below, the carrying capacity. Sometimes a population that overshoots *K* will experience a *population crash,* an abrupt decline from high to low population density when resources are exhausted. Such an abrupt change is commonly observed in bacterial cultures, zooplankton, and other populations whose resources are exhausted.

The availability of winter forage largely determines the carrying capacity for reindeer, which live in cold northern habitats (**FIGURE 7.5**). If reindeer overgraze the vegetation, it takes 15 to 20 years for it to recover. During that period, the carrying capacity for reindeer is greatly reduced.

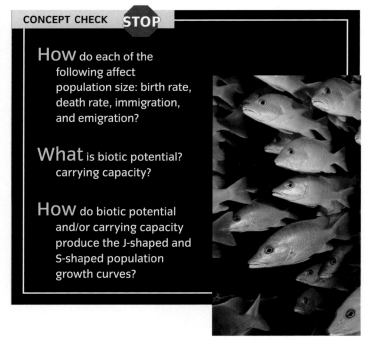

CONCEPT CHECK STOP

How do each of the following affect population size: birth rate, death rate, immigration, and emigration?

What is biotic potential? carrying capacity?

How do biotic potential and/or carrying capacity produce the J-shaped and S-shaped population growth curves?

Human Population Patterns

LEARNING OBJECTIVES

Summarize the history of human population growth.

Identify Thomas Malthus, relate his ideas on human population growth, and explain why he may or may not be wrong.

Explain why it is impossible to precisely determine how many people Earth can support—that is, Earth's carrying capacity for humans.

N ow that you have examined some of the basic concepts of population ecology, let's apply those concepts to the human population. **FIGURE 7.6** shows the increase in human population since 1800. The characteristic J curve of exponential population growth reflects the decreasing amount of time it has taken to add each additional billion people to our numbers. The United Nations projects the human population will reach 7 billion by 2013. Population experts predict that the population will level out during the 21st century, possibly forming the S curve observed in other species.

www.wiley.com/college/berg

One of the first people to recognize that the human population can't increase indefinitely was Thomas Malthus (1766–1834), a British economist. He pointed out that human population growth is not always desirable—a view contrary to the beliefs of his day and to those of many people even today. Noting that human population can increase faster than its food supply, he warned that the inevitable consequences of population growth would be famine, disease, and war. Since Malthus's time, the human population has increased from about 1 billion to more than 6 billion. On the surface, it seems that Malthus was wrong. Our population has grown dramatically because scientific advances have allowed food production to keep pace with population growth. Malthus's ideas may ultimately be proved correct, however, because we don't know whether this increased food production is sustainable. Have we achieved this increase in food production at the environmental cost of reducing the planet's ability to meet the needs of future populations?

Our world population was 6.5 billion in 2005, an increase of about 81 million from 2004. This increase isn't due to a rise in the birth rate (*b*). In fact,

the world birth rate has declined slightly during the past 200 years. The population growth is due instead to a dramatic *decrease* in the death rate (*d*) because greater food production, better medical care, and

Human population numbers, 1800 to present
FIGURE 7.6

Until recently, the human population has been increasing exponentially. There are now indications that the human population is beginning to level out, forming the S curve. It took thousands of years for the human population to reach 1 billion, a milestone that took place around 1800. It took 130 years to reach 2 billion (in 1930), 30 years to reach 3 billion (in 1960), 15 years to reach 4 billion (in 1975), 12 years to reach 5 billion (in 1987), and 12 years to reach 6 billion (in 1999).

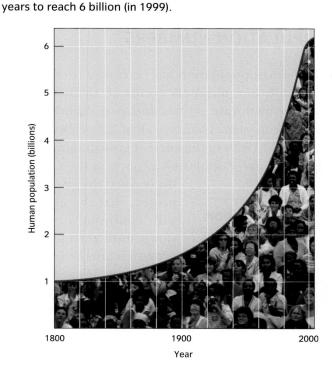

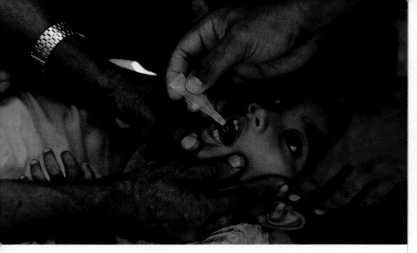

Advances in global health FIGURE 7.7

A child in Bangladesh receives a dose of oral polio vaccine. The Global Polio Eradication Initiative has almost reached its goal of eliminating polio, which at one time killed or crippled millions of children each year. Polio is still endemic in Nigeria, India, Afghanistan, and Pakistan, and it sometimes spreads from these countries to other countries.

improvements in water quality and sanitation practices have increased life expectancy for a great majority of the global population (FIGURE 7.7).

PROJECTING FUTURE POPULATION NUMBERS

The human population has reached a turning point. Although our numbers continue to increase, the world growth rate (*r*) has declined slightly over the past several years, from a peak of 2.2 percent per year in the mid-1960s to the current growth rate of 1.3 percent per year. Population experts at the United Nations and the World Bank project that the growth rate will continue to decrease slowly until **zero population growth** is attained toward the end of the twenty-first century. Exponential growth of the human population will end, and the **S** curve may replace the **J** curve.

zero population growth The state in which population remains the same size because the birth rate equals the death rate.

The United Nations periodically publishes population projections for the twenty-first century. The latest U.N. figures forecast that the human population will total between 7.9 billion (their "low" projection) and 10.9 billion (their "high" projection) in the year 2050, with 9.3 billion thought to be "most likely" (FIGURE 7.8). The estimates vary depending on fertility changes, particularly in less developed countries, because that is where almost all of the growth will take place.

Population projections must be interpreted with care because they vary depending on what assumptions are made. In projecting the world population will

Population projections to 2050 FIGURE 7.8

In 2000 the United Nations made three projections, each based on different fertility rates.

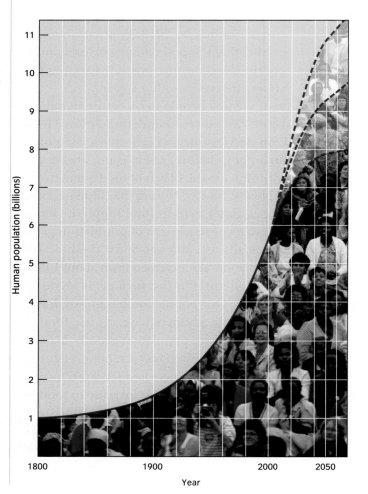

9. Tanzania, Argentina, and Poland have about the same population sizes (36 to 38 million each). The current total fertility rates of these countries are: Tanzania 5.7, Argentina 2.4, and Poland 1.2. Assuming that fertility rates decline to 2.0 by 2050 in Tanzania and Argentina, will there be a difference in population among the three countries? If so, which country will have the highest population in 2050? Explain your answer.

10. If you were to draw an age structure diagram for Poland, with a total fertility rate of 1.2, which of the following overall shapes would the diagram have? Explain why a country like Poland faces a population decline even if its fertility rate were to start increasing today.

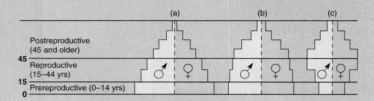

The following graph shows the ten countries with the largest populations in 2005. Use it to answer questions 11 and 12.

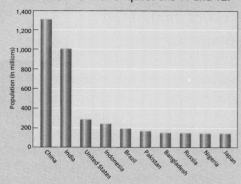

11. In 2050, population experts project that the Democratic Republic of Congo and Ethiopia will replace Russia and Japan in the top ten. Do Russia and Japan have high, medium, or low fertility rates? Do the Democratic Republic of Congo and Ethiopia have high, medium, or low fertility rates? Explain your answers.

12. The United States is projected to have 420 million people in 2050. Given that the U.S. fertility rate is relatively low and stable (2.0), why do you suppose the United States has one of the highest population growth rates in the developed world?

What is happening in this picture ?

The photo shows people—mainly displaced rural workers—picking through trash at the Smoky Mountain Dump in Manila, Philippines. Most of these people moved to Manila looking for work. Why didn't they find better employment?

They are looking mainly for scraps of plastic and metal, which they can sell. What valuable environmental service does such scavenging provide?

Air and Air Pollution 8

LONG-DISTANCE TRANSPORT OF AIR POLLUTION

Winds distribute certain hazardous air pollutants globally. *Persistent compounds* move through the air from warmer developing countries (where they are still used) to colder, highly developed nations, where they condense and form deposits on land and surface water. This process, in which volatile chemicals enter the atmosphere in warm regions and move to areas at higher, cooler latitudes, is known as the *global distillation effect*.

Dangerous levels of certain persistent toxic compounds are found in the Yukon (in northwestern Canada) and in other pristine arctic regions. These chemicals enter the food chain and become concentrated in the body fat of animals at the top (see *biological magnification* in Chapter 4). Fishes, seals, polar bears, and arctic people such as the Inuit are particularly vulnerable. When an Inuit woman consumes a single bite of raw whale skin, she ingests more PCBs than scientists think should be consumed in a week *(see photograph)*. The level of PCBs in the breast milk of that Inuit woman is five times higher than in the milk of women who live in southern Canada.

Examples of air pollution traveling from one continent to another were not well documented until recently. Now we know that atmospheric conditions cause pollutants from Asia to move east across the Pacific Ocean. In 1997 scientists detected pollutants in the air over the western United States that had been produced in Asia 6 days earlier. In 1998 a major dust storm in China produced a visible cloud of particulate matter that satellites tracked across the Pacific Ocean. When the polluted air reached the United States a few days later, an analysis found it to contain arsenic, copper, lead, and zinc from ore smelters in Manchuria.

No matter where you live, the air you breathe is often dirty and contaminated with pollutants. Air pollution also extends indoors to the air you breathe at home, in school, at work, and in your car. Because air pollution causes many health and environmental problems, most highly developed nations and many developing nations have established air quality standards for numerous air pollutants.

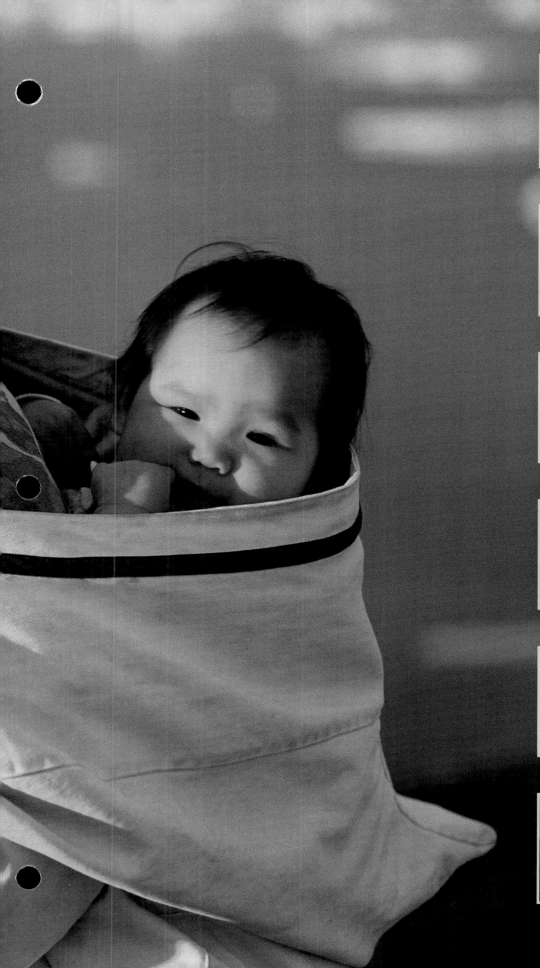

CHAPTER OUT

The Atmosphere p. 186

Types and Sources of Air Polluti

Effects of Air Pollution p. 194

Controlling Air Pollutants p. 200

Indoor Air Pollution p. 203

Case Study: Curbing Air Polluti
in Chattanooga p. 206

The Atmosphere

O xygen and nitrogen are the predominant gases in the **atmosphere**, accounting for about 99 percent of dry air (**TABLE 8.1**). Other gases make up the remaining 1 percent. In addition, water vapor and trace amounts of air pollutants are present in the air. The atmosphere becomes less dense as it extends outward into space.

Ulf Merbold, a German space shuttle astronaut, felt differently about the atmosphere after viewing it in space (**FIGURE 8.1**). "For the first time in my life, I saw the horizon as a curved line. It was accentuated by a thin seam of dark blue light—our atmosphere. Obviously, this wasn't the 'ocean' of air I had been told it was so many times in my life. I was terrified

atmosphere
The gaseous envelope surrounding Earth.

The atmosphere FIGURE 8.1

The "ocean of air" is a thin blue layer that separates the planet from the blackness of space.

Composition of the atmosphere* TABLE 8.1	
Nitrogen	78%
Oxygen	21%
Argon	0.93%
Carbon dioxide	0.04%
Other gases	0.03%

*Air also contains water vapor and various pollutants (methane, ozone, dust particles, microorganisms, and chlorofluorocarbons, or CFCs).

by its fragile appearance." The atmosphere is composed of five concentric layers—the troposphere, stratosphere, mesosphere, thermosphere, and exosphere (**FIGURE 8.2**). These layers vary in altitude and temperature, depending on the latitude and season.

The atmosphere performs several valuable **ecosystem services**. First, it protects Earth's surface from most of the sun's ultraviolet (UV) radiation and x-rays, and from lethal amounts of cosmic rays from space. Life as we know it would cease to exist without this shielding. Second, the atmosphere allows visible light and some infrared radiation to penetrate, both of which warm Earth's surface and the lower atmosphere. This interaction between solar energy and atmosphere is responsible for our weather and climate.

Organisms depend on the atmosphere for existence, but they also maintain and, in certain instances, modify its composition. Atmospheric oxygen is thought to have increased to its present level as a result of billions of years of photosynthesis. A balance between oxygen-producing photosynthesis and oxygen-using cellular respiration maintains the current level of oxygen.

A Layers of the atmosphere.

The outermost layer of the atmosphere, the **EXOSPHERE**, begins about 500 km above Earth's surface. The exosphere continues to thin until it converges with interplanetary space.

The **THERMOSPHERE** has steadily rising temperatures. Gases in the extremely thin air absorb x-rays and short-wave UV radiation, raising the temperature to 1,000°C or more. The thermosphere is important in long-distance communication because it reflects outgoing radio waves back toward Earth without the aid of satellites.

The **MESOSPHERE** is directly above the stratosphere. Temperatures drop steadily in the mesosphere to the lowest in the atmosphere—as low as −138°C.

The **STRATOSPHERE** has a steady wind but no turbulence; commercial jets fly here. The stratosphere contains a layer of ozone critical to life because it absorbs much of the sun's damaging ultraviolet (UV) radiation. Temperature increases with increasing altitude because the absorption of UV radiation by the ozone layer heats the air.

The **TROPOSPHERE** is the layer of atmosphere closest to Earth's surface. The temperature decreases with increasing altitude. Weather, including turbulent wind, storms, and most clouds, occurs here.

A

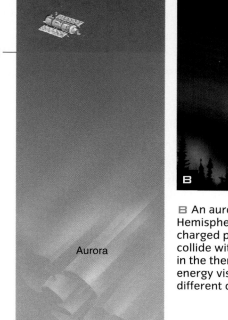

EXOSPHERE
500 km (310 mi)

THERMOSPHERE

Aurora

Meteors

80 km (50 mi)
MESOSPHERE
45 km (28 mi)
STRATOSPHERE
OZONE LAYER
10 km (6 mi)
Sea level
TROPOSPHERE

B An aurora in the Northern Hemisphere. Electrically charged particles from the sun collide with the gas molecules in the thermosphere, releasing energy visible as light of different colors.

C A thunderstorm in New Mexico. Thunderstorms develop from cumulus clouds that grow as deep as the troposphere (about 10 km). During a lightning flash, a negative charge moves from the bottom of the cloud to the ground, followed by an upward-moving charge along the same channel. The expansion of air around the lightning stroke produces sound waves, or thunder.

ATMOSPHERIC CIRCULATION

Variations in the amount of solar energy that reaches different areas on Earth cause differences in temperature, which then drive the circulation of the atmosphere. The very warm regions near the equator heat the air, causing it to expand and rise (FIGURE 8.3). As this warm air rises, it cools, and then it sinks again. Much of it recirculates almost immediately to the same areas it has left, but the remainder of the heated air splits and flows in two directions, toward the poles. The air chills enough to sink to the surface at about 30 degrees north and south latitudes. This descending air splits and flows over the surface in two directions.

Similar upward movements of warm air and its subsequent flow toward the poles also occur at higher latitudes farther from the equator. At the poles, the air cools, sinks, and flows back toward the equator, generally beneath the currents of warm air that simultaneously flow toward the poles. These constantly moving currents move heat from the equator toward the poles and cool the land over which they pass on their return. This continuous circulation moderates temperatures over Earth's surface.

In addition to these global circulation patterns, the atmosphere features smaller-scale horizontal movements, or **winds**. The motion of wind, with its eddies, lulls, and turbulent gusts, is difficult to predict. It results partly from fluctuations in atmospheric pressure and partly from the planet's rotation.

The gases that constitute the atmosphere have weight and exert a pressure—about 1013 millibars (14.7 lb per in^2) at sea level. Air pressure is variable, depending on altitude, temperature, and humidity. Winds tend to blow from areas of high atmospheric pressure to areas of low pressure, and the greater the difference between the high- and low-pressure areas, the stronger the wind.

Atmospheric circulation and heat exchange FIGURE 8.3

Atmospheric circulation transports heat from the equator to the poles *(left side of figure)*.

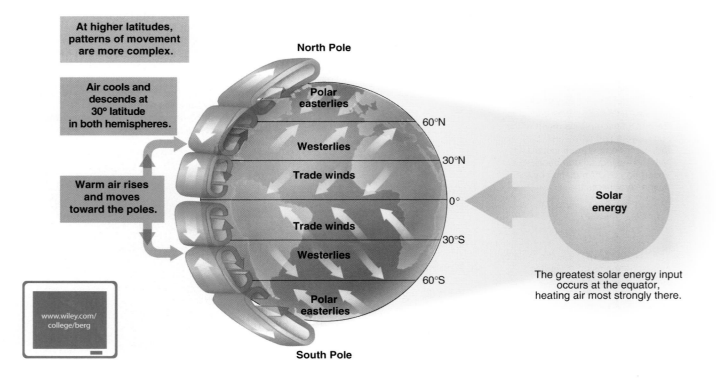

At higher latitudes, patterns of movement are more complex.

Air cools and descends at 30° latitude in both hemispheres.

Warm air rises and moves toward the poles.

North Pole

Polar easterlies

Westerlies

Trade winds

Trade winds

Westerlies

Polar easterlies

60°N

30°N

0°

30°S

60°S

South Pole

Solar energy

The greatest solar energy input occurs at the equator, heating air most strongly there.

www.wiley.com/college/berg

to 70 percent of the air pollutants in urban areas of Central America, and 50 to 60 percent in urban areas of India. During the 1990s the most rapid proliferation of motor vehicles worldwide occurred in Latin America, Asia, and Eastern Europe.

Lead pollution from heavily leaded gasoline is an especially serious problem in developing nations. The gasoline refineries in these countries are generally not equipped to remove lead from gasoline. (The United States was in the same situation until federal law mandated that U.S. refineries upgrade their equipment.) In Cairo, Egypt, for example, children's blood lead levels are more than two times higher than the level considered at-risk in the United States. Lead can retard children's growth and cause brain damage.

According to the World Health Organization, the five worst cities in the world in terms of exposing children to air pollution are Mexico City, Mexico; Beijing, China; Shanghai, China; Tehran, Iran; and Calcutta, India. Respiratory disease is now the leading cause of death for children worldwide. More than 80 percent of these deaths occur in young children (under the age of 5) who live in cities in developing countries.

CONCEPT CHECK STOP

What is the U.S. Clean Air Act and how has it reduced outdoor air pollution?

Where is air pollution worse: in highly developed nations or in developing countries? Why?

Indoor Air Pollution

LEARNING OBJECTIVE

Describe sick building syndrome.

The air in enclosed places, like automobiles, homes, schools, and offices, may have significantly higher levels of air pollutants than the air outdoors. Indoor air pollution is of particular concern to urban residents because they may spend as much as 90 percent to 95 percent of their time in enclosed places.

Because illnesses from indoor air pollution usually resemble common ailments such as colds, influenza, or upset stomachs, they are often not recognized. The most common contaminants of indoor air are radon, cigarette smoke, carbon monoxide, nitrogen dioxide (from gas stoves), formaldehyde (from carpeting, fabrics, and furniture), household pesticides, cleaning solvents, ozone (from photocopiers), and asbestos. In addition, viruses, bacteria, fungi (yeasts, molds, and mildews), dust mites, pollen, and other organisms are often found in heating, air conditioning, and ventilation ducts.

Health officials are paying increasing attention to the **sick building syndrome**. The Labor Department estimates that more than 20 million employees are exposed to health risks from indoor air pollution. The EPA estimates that annual medical costs for treating the health effects of indoor air pollution in the United States exceed $1 billion. When lost work time and diminished productivity are added to health care costs, the total annual cost to the economy may be as much as $50 billion. Fortunately, most building problems are relatively inexpensive to alleviate.

sick building syndrome Eye irritations, nausea, headaches, respiratory infections, depression, and fatigue caused by indoor air pollution.

RADON

The most serious indoor air pollutant is probably radon, a colorless, tasteless radioactive gas produced naturally during the radioactive decay of uranium in Earth's crust. Radon seeps through the ground and enters buildings, where it sometimes accumulates to dangerous levels (FIGURE 8.16). Although radon is also emitted into the atmosphere, it gets diluted and dispersed and is of little consequence outdoors.

Radon and its decay products emit alpha particles, a form of ionizing radiation that is very damaging to tissue but can't penetrate very far into the body. Consequently, only ingested or inhaled radon harms the body. The National Research Council of the National Academy of Sciences estimates that residential exposure to radon causes 12 percent of all lung cancers—between

Detecting and dealing with radon FIGURE 8.17

Radon seeped into the basement of this family's home. Shown are the instruments and instructions they used to reduce indoor radon to a safe level. Photographed in Clinton, New Jersey.

15,000 and 22,000 lung cancers annually. Cigarette smoking exacerbates the risk from radon exposure; about 90 percent of radon-related cancers occur among current or former smokers.

According to the EPA, about 6 percent of U.S. homes have high enough levels of radon to warrant corrective action—a radon level above 4 picocuries per liter of air. (As a standard of reference, outdoor radon concentrations range from 0.1 to 0.15 picocuries per liter of air worldwide.) The highest radon levels in the United States are found in homes across southeastern Pennsylvania into northern New Jersey and New York (FIGURE 8.17). Iowa appears to have the most pervasive radon problem, where 71 percent of the homes tested in 1989 had radon levels high enough to warrant corrective action.

Ironically, efforts to make our homes more energy-efficient have increased the hazard of indoor air pollutants, including radon. Drafty homes waste energy but allow radon to escape outdoors so it does not build up inside. Every home should be tested for radon

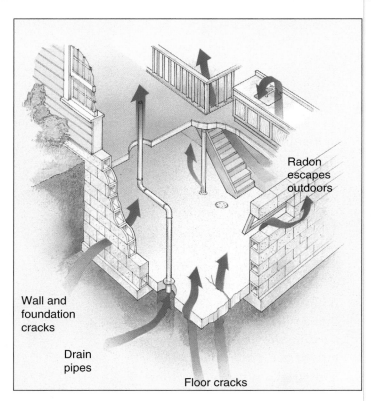

Radon escapes outdoors

Wall and foundation cracks

Drain pipes

Floor cracks

How radon infiltrates a house FIGURE 8.16

Cracks in basement walls or floors, openings around pipes, and pores in concrete blocks provide some of the entries for radon.

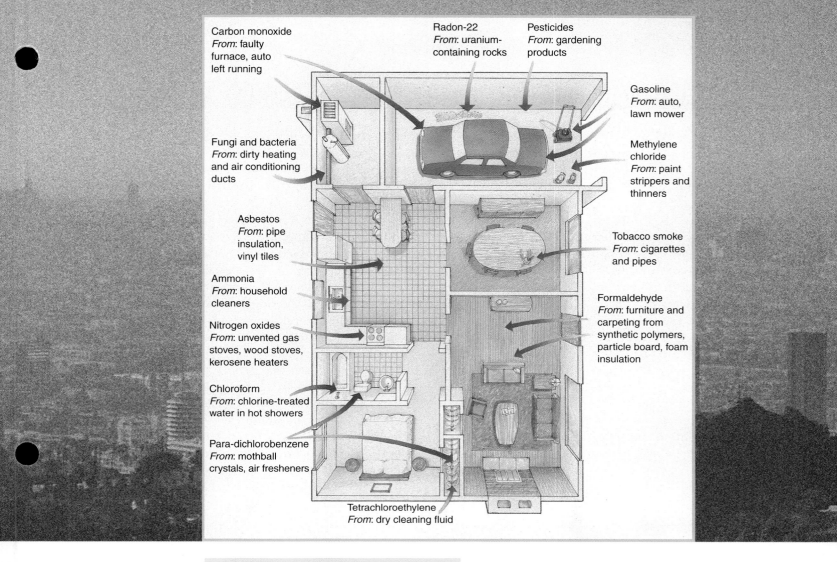

Carbon monoxide *From*: faulty furnace, auto left running

Radon-22 *From*: uranium-containing rocks

Pesticides *From*: gardening products

Gasoline *From*: auto, lawn mower

Fungi and bacteria *From*: dirty heating and air conditioning ducts

Methylene chloride *From*: paint strippers and thinners

Asbestos *From*: pipe insulation, vinyl tiles

Tobacco smoke *From*: cigarettes and pipes

Ammonia *From*: household cleaners

Formaldehyde *From*: furniture and carpeting from synthetic polymers, particle board, foam insulation

Nitrogen oxides *From*: unvented gas stoves, wood stoves, kerosene heaters

Chloroform *From*: chlorine-treated water in hot showers

Para-dichlorobenzene *From*: mothball crystals, air fresheners

Tetrachloroethylene *From*: dry cleaning fluid

Indoor air pollution FIGURE 8.18

Homes may contain higher levels of air pollutants than outside air, even near polluted industrial sites.

because levels vary widely from home to home, even in the same neighborhood. Testing is inexpensive, and corrective actions are reasonably priced.

FIGURE 8.18 summarizes all the possible sources of pollution in our homes.

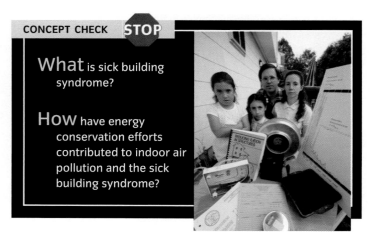

CONCEPT CHECK STOP

What is sick building syndrome?

How have energy conservation efforts contributed to indoor air pollution and the sick building syndrome?

Curbing Air Pollution in Chattanooga

During the 1960s, the federal government gave Chattanooga, Tennessee, the dubious distinction of having the worst air pollution in the United States. The air was so dirty in this manufacturing city that sometimes people driving downtown had to turn on their headlights in the middle of the day. The orange air soiled their white shirts so quickly that many businessmen brought extra ones to work. To compound the problem, the mountains surrounding the city kept the pollutants produced by its inhabitants from dispersing.

Today the air in this scenic midsized city of 200,000 people is clean, and Chattanooga ranks high among U.S. cities in terms of air quality *(see photo)*. City and business leaders are credited with transforming Chattanooga's air. Soon after the passage of the federal Clean Air Act of 1970, the city established an air pollution control board to enforce regulations controlling air pollution. New local regulations allowed open burning by permit only, placed limits on industrial odors and particulate matter, outlawed visible automotive emissions, and set a cap on sulfur content in

fuel, which controlled the production of sulfur oxides. Businesses installed expensive air pollution control devices. The city started an emissions-free electric bus system. Chattanooga also decided to recycle its solid waste rather than build an emissions-producing incinerator.

The measures were so effective that Chattanooga drew national attention. The National Air Pollution Control Association awarded Chattanooga first place in its annual Cleaner Air Week ceremonies. In 1984, the EPA designated Chattanooga in attainment for particulate matter; this designation meant particulate levels were below the federal health limit for one year. The city reached attainment status for ozone in 1989. Since then, the levels for all seven EPA-regulated air pollutants have been lower than federal standards require.

In the early 2000s, Chattanoogans continued to move their city toward *environmental sustainability*. The city plans to convert a run-down business district into a community in which people live near their places of work. Businesses located in this district will form an **industrial ecosystem** in which the wastes of one business are raw materials for another business.

Chattanooga, Tennessee

Chattanooga's air quality has improved dramatically during the past several decades.

CHAPTER SUMMARY

1 The Atmosphere

1. Oxygen (21 percent) and nitrogen (78 percent) are the main gases in the **atmosphere**, the gaseous envelope surrounding Earth. Argon, carbon dioxide, other gases, water vapor, and trace amounts of various air pollutants are also present.

2. The **troposphere**, the layer of atmosphere closest to Earth's surface, extends to a height of approximately 10 km. Temperature decreases with increasing altitude, and weather occurs in the troposphere. In the **stratosphere**, there is a steady wind but no turbulence. The stratosphere contains an ozone layer that absorbs much of the sun's UV radiation. The **mesosphere**, directly above the stratosphere, has the lowest temperatures in the atmosphere. The **thermosphere** has steadily rising temperatures and gases that absorb x-rays and short-wave UV radiation. The thermosphere reflects outgoing radio waves back toward Earth without the aid of satellites. The outermost layer of the atmosphere, the **exosphere** continues to thin until it converges with interplanetary space.

3. The **Coriolis effect** is the tendency of moving air or water to be deflected from its path and swerve to the right in the Northern Hemisphere and to the left in the Southern Hemisphere.

2 Types and Sources of Air Pollution

1. **Air pollution** consists of various chemicals (gases, liquids, or solids) present in the atmosphere in high enough levels to harm humans, other organisms, or materials. **Primary air pollutants** are harmful chemicals that enter the atmosphere directly from either human activities or natural processes; examples include carbon oxides, nitrogen oxides, sulfur dioxide, particulate matter, and hydrocarbons. **Secondary air pollutants** are harmful chemicals that form in the atmosphere when primary air pollutants react chemically with each other or with natural components of the atmosphere; ozone and sulfur trioxide are examples.

2. **Particulate matter**—solid particles and liquid droplets suspended in the atmosphere—corrodes metals, erodes buildings, soils fabrics, and can damage the lungs. **Nitrogen oxides** are gases associated with photochemical smog, acid deposition, global warming, and stratospheric ozone depletion; they also corrode metals and fade textiles. **Sulfur oxides** are gases associated with acid deposition; they corrode metals and damage stone and other materials. **Carbon oxides** include the gases carbon monoxide, which is poisonous, and carbon dioxide, which is linked to global warming. **Hydrocarbons** are solids, liquids, or gases associated with photochemical smog and global warming; some are dangerous to human health. **Ozone** is a secondary air pollutant in the lower atmosphere (troposphere) but an essential part of the stratosphere. Tropospheric ozone reduces visibility, causes health problems, stresses plants, and is associated with global warming. Some air pollutants are called **hazardous air pollutants**, or **air toxics**, because they are potentially harmful and may pose long-term health risks to people who are exposed to them; chlorine, lead, hydrochloric acid, formaldehyde, radioactive substances, and fluorides are examples.

3 Effects of Air Pollution

1. Exposure to low levels of air pollutants irritates the eyes and causes inflammation of the respiratory tract. Many air pollutants suppress the immune system, increasing susceptibility to infection. Exposure to air pollution during respiratory illnesses may result in the development of chronic respiratory diseases, such as emphysema and chronic bronchitis.

2. **Industrial smog** refers to smoke pollution. **Photochemical smog** is a brownish orange haze formed by chemical reactions involving sunlight, nitrogen oxides, and hydrocarbons. A **temperature inversion** is a layer of cold air temporarily trapped near the ground by a warmer, upper layer; during a temperature inversion, polluting gases and particulate matter remain trapped in high concentrations close to the ground. An **urban heat island** is local heat buildup in an area of high population. Urban heat islands affect local air currents and weather conditions and contribute to the buildup of pollutants, especially particulate matter, in the form of a **dust dome**, a dome of heated air that surrounds an urban area and contains a lot of air pollution.

4 Controlling Air Pollutants

1. Improvements in U.S. air quality since 1970 are largely due to the Clean Air Act, which authorizes the EPA to set limits on specific air pollutants. Individual states must meet deadlines to reduce air pollution to acceptable levels and can't mandate weaker limits than those stipulated in the Clean Air Act.

2. Air quality in the United States has slowly improved since passage of the Clean Air Act. The most dramatic improvement is the decline of lead in the air, although levels of sulfur oxides, ozone, carbon monoxide, volatile compounds, and particulate matter have also declined. Air quality is deteriorating in developing nations, the result of rapid industrialization, a growing number of automobiles, and a lack of emissions standards.

5 Indoor Air Pollution

1. The **sick building syndrome** includes eye irritations, nausea, headaches, respiratory infections, depression, and fatigue caused by indoor air pollution.

KEY TERMS

- **atmosphere** p. 186
- **Coriolis effect** p. 189
- **air pollution** p. 190
- **primary air pollutants** p. 190
- **secondary air pollutants** p. 190
- **photochemical smog** p. 196
- **temperature inversion** p. 197
- **urban heat island** p. 198
- **dust dome** p. 198
- **sick building syndrome** p. 203

CRITICAL AND CREATIVE THINKING QUESTIONS

1. What is the global distillation effect? What kinds of air pollutants are involved in the global distillation effect? Where do these pollutants get permanently deposited? Why?

2. The atmosphere of Earth has been compared to the peel covering an apple. Explain the comparison.

3. What basic forces determine the circulation of the atmosphere? Describe the general directions of atmospheric circulation.

4. Distinguish between primary and secondary air pollutants; give examples of each that you are likely to encounter.

5. Distinguish between mobile and stationary sources of air pollution.

6. Why might it be more effective to control photochemical smog in heavily wooded areas like the Atlanta metropolitan area by reducing nitrogen oxides instead of volatile hydrocarbons?

7. One of the most effective ways to reduce the threat of radon-induced lung cancer is to quit smoking. Explain.

8. What air pollutants do the 1990 provisions of the Clean Air Act target?

9. During a formal debate on the hazards of air pollution, one team argues that ozone is helpful to the atmosphere, and the other team argues that it is destructive. Explain why they both are correct.

10. These graphs (in the left column) represent air pollutant measurements taken at two different locations. Which location is indoors, and which is outdoors? Explain your answer.

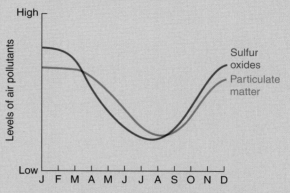

11. The graph above shows air pollutant levels in a city in the Northern Hemisphere, measured throughout a year. Is this city likely to be found in a developing country or in a developed country? Why?

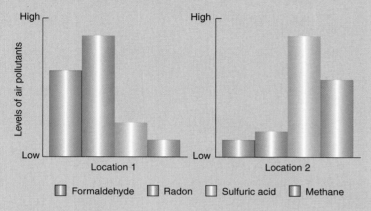

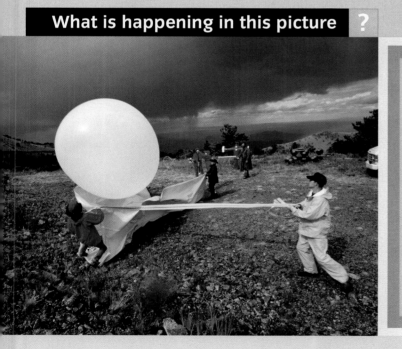

What is happening in this picture ?

- These scientists are about to launch a balloon attached to meteorological instruments *(in far background)*. Judging from the current atmospheric conditions, what processes might they be measuring?

- During a thunderstorm, the energy produced by a lightning strike chemically combines some nitrogen and oxygen molecules to form the reddish brown gas, nitrogen dioxide (NO_2). Nitrogen dioxide is what the scientists are measuring here. Name at least three environmental effects of NO_2.

- What are two human sources of NO_2 and other nitrogen oxides?

Global Atmospheric Changes

MELTING ICE AND RISING SEA LEVELS

In 2002, an iceberg roughly twice the size of Rhode Island broke off from the Antarctic Peninsula. A study of the ice-covered ocean in the Arctic from the 1970s to the 1990s shows that the ice pack there has retreated, and the remaining pack has thinned rapidly, losing 40 percent of its volume in less than three decades. Mountain glaciers around the world are melting at accelerating rates. In Montana's Glacier National Park, 110 of 147 glaciers have melted since 1850; the remaining 27 are retreating so rapidly they may disappear completely by 2030.

During the 20th century, the sea level rose 10 to 20 cm (4 to 8 in), mostly through thermal expansion. Sea-level rise is caused not only by the retreat of glaciers and the thawing of polar ice but also by the expansion of ocean water as it warms. Climate scientists estimate that the sea level will rise an additional 48 cm (19 in) by 2100. Such a rise would flood low-lying coastal areas such as southern Louisiana and South Florida. Coastal areas not inundated will suffer erosion and other damage from more frequent and more intense weather events such as hurricanes.

Small island nations such as the Maldives, a low-elevation chain of 1,200 islands in the Indian Ocean *(see photograph)*, are considered highly vulnerable to a rise in sea level, as storm surges could easily sweep over entire islands. Other vulnerable countries—such as Bangladesh, Egypt, Vietnam, and Mozambique—have dense populations living in low-lying river deltas.

CHAPTER OUTLIN

The Atmosphere and Climate p. 212

Global Warming p. 216

Ozone Depletion in the Stratosphere p. 224

Acid Deposition p. 227

Case Study: International Implications of Global Warming p. 230

The Atmosphere and Climate

LEARNING OBJECTIVES

Distinguish between weather and climate.

Summarize the effects of solar energy on Earth's temperature.

Explain several reasons for regional precipitation differences.

Weather refers to the conditions in the atmosphere at a given place and time; it includes temperature, atmospheric pressure, precipitation, cloudiness, humidity, and wind. Weather changes from one hour to the next and from one day to the next.

The two most important factors that determine an area's overall **climate** are *temperature*—both average temperature and temperature extremes—and both average and seasonal *precipitation*. Latitude, elevation, topography, quantity and quality of vegetation, distance from the ocean, and geographic location all influence temperature, precipitation, and other aspects of climate. Other climate factors include weather conditions like wind, humidity, fog, cloud cover, and, in some areas, lightning. Unlike weather, which changes rapidly, climate changes slowly, over hundreds or thousands of years.

Earth has many different climates, and because each is relatively constant for many years, organisms have adapted to them. The many kinds of organisms on Earth are here in part because of the large number of different climates—from cold, snow-covered polar climates to tropical climates where it is hot and rains almost every day (**FIGURE 9.1**).

> **climate** The average weather conditions that occur in a place over a period of years.

Variations in climate FIGURE 9.1

A A view of the Pacific Ocean from the Na Pali Coast, Kauai, Hawaiian Islands.

B A polar bear and her cub walk across the ice in Nunavut, Canada.

SOLAR RADIATION AND CLIMATE

The sun makes life on Earth possible. It warms the planet, including the atmosphere, to habitable temperatures. Without the sun's energy, all water on Earth would be frozen, including the ocean. Photosynthetic organisms capture the sun's energy and use it to make the food molecules almost all forms of life require. Most of our fuels, such as wood and **fossil fuels** (oil, coal, and natural gas), represent solar energy captured by photosynthetic organisms. The amount of solar energy a region receives (in other words, the amount of sunshine a region gets) is the primary determinant of climate.

Clouds, and to a lesser extent snow, ice, and the ocean, reflect away about 31 percent of the solar radiation that falls on Earth (**FIGURE 9.2**). The remaining 69 percent is absorbed and runs the hydrologic cycle, carbon cycle, and other biogeochemical cycles; drives winds and ocean currents; powers photosynthesis; and warms the planet. Ultimately, all of this energy returns to space as long-wave **infrared radiation** (heat energy).

> **infrared radiation** Electromagnetic radiation with wavelengths longer than visible light but shorter than microwaves; perceived as invisible waves of heat energy.

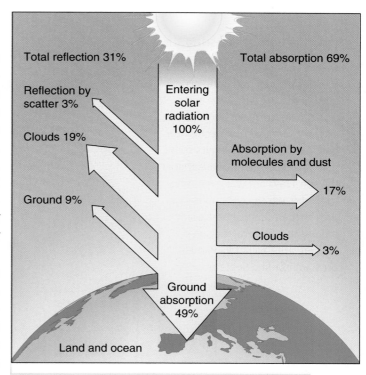

Fate of solar radiation that reaches Earth
FIGURE 9.2

Most of the energy the sun produces never reaches Earth. The solar energy that does reach Earth warms the planet's surface, drives the hydrologic cycle and other biogeochemical cycles, produces our climate, and powers almost all life through the process of photosynthesis.

www.wiley.com/college/berg

Temperature changes with latitude and season

Earth's roughly spherical shape and the tilt of its axis produce a great deal of variation in the exposure of the surface to solar energy (**FIGURE 9.3**). Because the sun's energy does not reach all places uniformly, temperature varies locally.

Solar intensity and latitude FIGURE 9.3

The angle at which the sun's rays strike Earth varies from one geographic location to another owing to Earth's spherical shape and its inclination on its axis. A Sunlight (represented by the flashlight) that shines vertically near the equator is concentrated on Earth's surface. B, C As one moves toward the poles, the light hits the surface more and more obliquely, spreading the same amount of radiation over larger and larger areas.

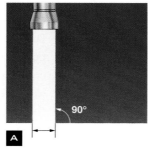

1 unit of surface area

One unit of light is concentrated over one unit of surface area.

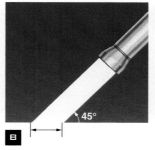

1.4 units of surface area

One unit of light is dispersed over 1.4 units of surface area.

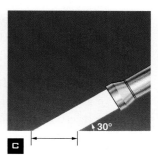

2 units of surface area

One unit of light is dispersed over 2 units of surface area.

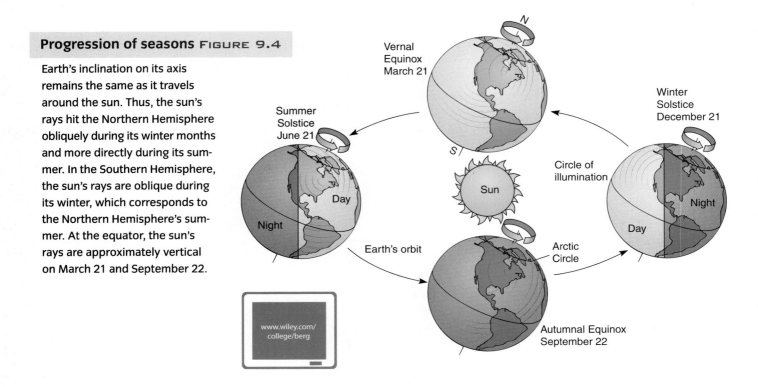

Progression of seasons FIGURE 9.4

Earth's inclination on its axis remains the same as it travels around the sun. Thus, the sun's rays hit the Northern Hemisphere obliquely during its winter months and more directly during its summer. In the Southern Hemisphere, the sun's rays are oblique during its winter, which corresponds to the Northern Hemisphere's summer. At the equator, the sun's rays are approximately vertical on March 21 and September 22.

Vernal Equinox March 21

Summer Solstice June 21

Winter Solstice December 21

Circle of illumination

Sun

Day

Night

Night

Day

Earth's orbit

Arctic Circle

Autumnal Equinox September 22

www.wiley.com/college/berg

Earth's inclination on its axis (23.5 degrees from a line drawn perpendicular to the orbital plane) determines the seasons. During half the year (March 21 to September 22) the Northern Hemisphere tilts toward the sun, and during the other half (September 22 to March 21) it tilts away from the sun (**FIGURE 9.4**). The Southern Hemisphere tilts the opposite way, so that summer in the Northern Hemisphere corresponds to winter in the Southern Hemisphere.

PRECIPITATION

Precipitation refers to any form of water, such as rain, snow, sleet, or hail, that falls from the atmosphere. Differences in precipitation depend on three factors.

1. **The amount of water vapor in the atmosphere.** *Equatorial uplift* of warm, moisture-laden air produces heavy rainfall in some areas of the tropics. High surface water temperatures cause vast quantities of water to evaporate from tropical parts of the ocean. Prevailing winds blow the resulting moist air over landmasses. The moist air continues to rise as it is heated by the sun-warmed land surface. As the air rises, it cools, which decreases its moisture-holding ability. When the air reaches its saturation point—when it can't hold any additional water vapor—clouds form and water is released as precipitation.

2. **Geographic location.** The rising air from the equator eventually descends to Earth near the Tropic of Cancer and Tropic of Capricorn (latitudes 23.5 degrees north and 23.5 degrees south). By then most of its moisture has precipitated, and the dry air returns to the equator. Over land, this dry air produces some of the great tropical deserts, such as the Sahara Desert. Air also dries out as it travels long distances over landmasses. Near the windward coasts of continents (the side from which the wind blows), rainfall may be heavy. However, in the temperate zones (the areas between the tropics and the polar zones), continental interiors are usually dry, because they are far from the ocean that replenishes water in the air passing over it.

3. **Topographic features.** For example, mountains force air to rise and to give up moisture. The air's temperature cools as it gains altitude; clouds form and precipitation occurs, primarily on the mountains' windward slopes. The air mass is warmed as it moves down on the other side of the mountain, reducing the chance of precipitation of the remaining moisture. This situation exists on the West Coast of North America, where precipitation falls on the western slopes of mountains close to the coast (see the What a Scientist Sees figure). The dry land on the side of the mountains away from the prevailing wind—in this case, east of the mountain range—is called a **rain shadow**. The dry conditions of a rain shadow often occur on a regional scale.

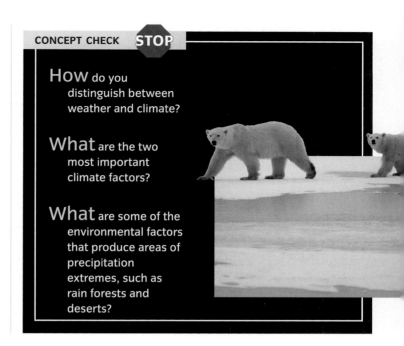

CONCEPT CHECK STOP

How do you distinguish between weather and climate?

What are the two most important climate factors?

What are some of the environmental factors that produce areas of precipitation extremes, such as rain forests and deserts?

Rain Shadow

A Proxy Falls plunges over moss-covered rocks in the Cascade Range, which was named for its many waterfalls. The Cascade Range divides the states of Washington and Oregon into a moist western region and an arid region east of the mountains.

B A rain shadow refers to arid or semi-arid land that occurs on the far side (leeward side) of a mountain. Prevailing winds blow warm, moist air from the windward side. Air temperature cools as it rises, releasing precipitation so dry air descends on the leeward side. Such a rain shadow exists east of the Cascades.

Windward side

Leeward side

Moist air

Dry air

Rain shadow desert

A

B

What a Scientist Sees

Global Warming

LEARNING OBJECTIVES

Describe the enhanced greenhouse effect and list the five main greenhouse gases.

Discuss some of the potential effects of global warming.

Give examples of several ways to mitigate and adapt to global warming.

D ata accumulated from daily measurements indicate that Earth's average surface temperature in 2005 was the highest since the mid-1800s. The last two decades of the 20th century were its warmest (FIGURE 9.5).

Other evidence also suggests an increase in global temperature. Several studies indicate that spring in the Northern Hemisphere now comes about six days earlier than it did in 1959, and autumn comes five days later. Since 1949, the United States has experienced an increased frequency of heat waves, resulting in increased heat-related deaths among elderly and other vulnerable people. In the past few decades, the sea level has risen, glaciers worldwide have retreated, and extreme weather events, such as severe storms, have occurred with increasing frequency in certain regions.

Scientists around the world have been researching global warming for the past 50 years. As the evidence has accumulated, those most qualified to address the issue have reached a strong consensus that the 21st century will experience significant climate change and that human activities are at least partially responsible.

In response to this growing consensus, governments around the world organized the United Nations Intergovernmental Panel on Climate Change (IPCC). With input from hundreds of climate experts, the IPCC provides the definitive scientific statement about global warming. In 2001 the IPCC projected a 1.4° to 5.8°C (2.5° to 10.4°F) increase in global temperature by the year 2100. The IPCC predicts that we will observe higher maximum temperatures and more hot days over nearly all land areas, higher minimum temperatures, fewer frost days, fewer cold days, and an increase in the heat index. We may also experience more intense precipitation events over many areas, an increased risk of drought in the continental interiors in the mid-latitudes, and stronger hurricanes in some coastal areas.

Mean annual global temperature, 1960 to present FIGURE 9.5

Data are presented as surface temperatures (°C) for 1960, 1965, and every year thereafter. The measurements, which naturally fluctuate, show the warming trend of the last several decades. The warmest year on record was 2005, and the second warmest year was 1998. The 10 warmest years occurred from 1990 to 2005. (The dip in global temperatures in the early 1990s was caused by the eruption of Mount Pinatubo in 1991.)

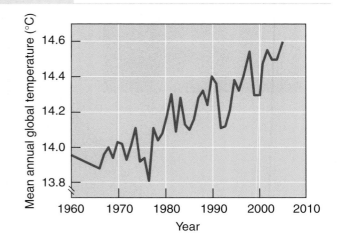

CAUSES OF GLOBAL WARMING

Carbon dioxide (CO_2) and certain other trace gases, including methane (CH_4), nitrous oxide (N_2O), chlorofluorocarbons (CFCs), and tropospheric ozone (O_3), accumulate in the atmosphere as a result of human activities. The concentration of atmospheric carbon dioxide has increased from about 288 parts per million (ppm) approximately 200 years ago (before the Industrial Revolution began) to 380 ppm in 2005 (FIGURE 9.6). Burning carbon-containing fossil fuels accounts for most human-made CO_2. Land conversion, such as when tracts of tropical forests are logged or burned, also releases CO_2. By 2050 the concentration of atmospheric CO_2 may be double what it was in the 1700s.

The combustion of gasoline in your car's engine releases not only CO_2 but also nitrous oxide, which triggers the production of tropospheric ozone. Various industrial processes, land-use conversion, and the use of fertilizers also produce nitrous oxide. CFCs (also discussed later in the chapter as they relate to depletion of the stratospheric ozone layer) are chemicals released into the atmosphere from old, leaking refrigerators and air conditioners. Methane is produced by the decomposition of carbon-containing organic material by anaerobic bacteria in moist places as varied as rice paddies, sanitary landfills, and the intestinal tracts of cattle and other large animals (humans included) (FIGURE 9.7).

Carbon dioxide (CO_2) in the atmosphere, 1958 to present FIGURE 9.6

Note the steady increase in the concentration of atmospheric CO_2 since 1958, when measurements began at the Mauna Loa Observatory in Hawaii. This location was selected because it is far from urban areas where factories, power plants, and motor vehicles emit CO_2. The seasonal fluctuations correspond to winter (a high level of CO_2), when plants are not actively growing and absorbing CO_2, and summer (a low level of CO_2), when plants are growing and absorbing CO_2.

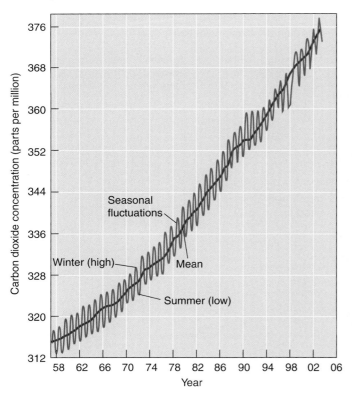

Methane production FIGURE 9.7

The world's population of cattle, which produces huge amounts of methane, is increasing faster than the human population. Bacteria living in the cow's digestive system produce the methane.

greenhouse gases The gases that absorb infrared radiation; include carbon dioxide, methane, nitrous oxide, chlorofluorocarbons, and tropospheric ozone.

enhanced greenhouse effect The additional warming that may be produced by human-increased levels of gases that absorb infrared radiation.

Global warming occurs because these gases absorb infrared radiation—that is, heat energy—given off by Earth's surface. This absorption slows the natural flow of heat into space, warming the lower atmosphere. Some of the heat from the lower atmosphere is transferred to the ocean and raises its temperature as well.

Because CO_2 and other gases trap the sun's infrared radiation somewhat like glass does in a greenhouse, they are called **greenhouse gases**. Greenhouse gases accumulating in the atmosphere as a result of human activities may be causing an **enhanced greenhouse effect** (FIGURE 9.8). TABLE 9.1 summarizes the greenhouse gases.

Other pollutants cool the atmosphere One of the complications that makes the rate and extent of global warming difficult to predict is that other air pol-

lutants, known as atmospheric aerosols, tend to cool the atmosphere. **Aerosols**, which come from both natural and human sources, are particles so small they remain suspended in the atmosphere for days, weeks, or even months. Sulfur haze is an aerosol that reflects sunlight back into space, reducing the amount of solar energy reaching Earth's surface, and thereby cooling the atmosphere. Sulfur haze significantly moderates warming in industrialized parts of the world. Sulfur emissions that produce sulfur haze come from the same smokestacks that emit carbon dioxide. Volcanic eruptions also eject sulfur-containing particles into the atmosphere (FIGURE 9.9).

Despite their cooling effect, human-produced sulfur emissions should not be viewed as a panacea for the enhanced greenhouse effect. Human-produced sulfur emissions remain in the atmosphere for only days, weeks, or months. They do not disperse globally

Changes in selected atmospheric greenhouse gases, preindustrial to present TABLE 9.1

Gas	Estimated pre-1750 concentration	Present concentration
Carbon dioxide	280 ppm*	380 ppm
Methane	730 ppb**	1,847 ppb
Nitrous oxide	270 ppb	319 ppb
Chlorofluorocarbon-12	0 ppt***	545 ppt
Chlorofluorocarbon-11	0 ppt	253 ppt

* ppm = parts per million.
** ppb = parts per billion.
*** ppt = parts per trillion.

Enhanced greenhouse effect FIGURE 9.8

The buildup of carbon dioxide (CO_2) and other greenhouse gases warms the atmosphere by absorbing some of the outgoing infrared (heat) radiation. Some of the heat in the warmed atmosphere is transferred back to Earth's surface, warming the land and ocean.

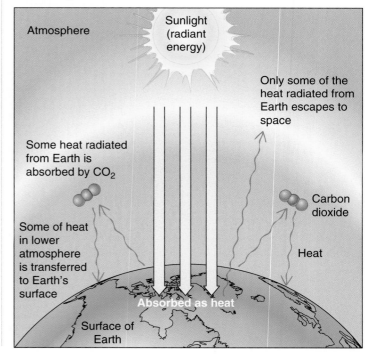

Volcanic eruption FIGURE 9.9

The explosion of Mount Pinatubo in the Philippines in 1991 was the largest volcanic eruption in the 20th century. This eruption injected massive amounts of sulfur into the atmosphere. Because sulfur in the atmosphere reduces the amount of sunlight reaching the surface, this eruption caused Earth to enter a temporary period of global cooling. Compared to temperatures during the rest of the 1990s, 1992 and 1993 global temperatures were relatively cool.

and may cause only regional cooling. Greenhouse gases, on the other hand, remain in the atmosphere for hundreds of years. And carbon dioxide and other greenhouse gases help warm the planet 24 hours a day, whereas sulfur haze cools the planet only during the daytime. In addition, sulfur emissions are a respiratory irritant and cause acid deposition (discussed later in this chapter). Most nations are trying to reduce their sulfur emissions, not maintain or increase them.

EFFECTS OF GLOBAL WARMING

In addition to changes in sea level, already considered in the chapter introduction, some of the observed and potential effects of global warming include changes in precipitation patterns, agriculture, human health, and effects on organisms.

Computer models of weather changes caused by global warming indicate that precipitation patterns will change, causing some areas to have more frequent droughts. At the same time, heavier snow and rainstorms may cause more frequent flooding in other areas. Changes in precipitation patterns could affect the availability and quality of fresh water in many locations, particularly in areas that are currently arid or semi-arid, such as the Sahel region just south of the Sahara Desert.

Global warming will increase problems for agriculture. The rise in sea level will inundate some river deltas, which are fertile agricultural lands. Certain agricultural pests and disease-causing organisms will probably proliferate and reduce crop yields. The likely increase in the frequency and duration of droughts will be a particularly serious problem for countries with limited water resources.

Currently, most evidence linking climate warming to disease outbreaks is circumstantial, so scientists are reluctant to ascribe cause-and-effect relationships. Nonetheless, data linking climate warming and human health problems are accumulating. More frequent and more severe heat waves during summer

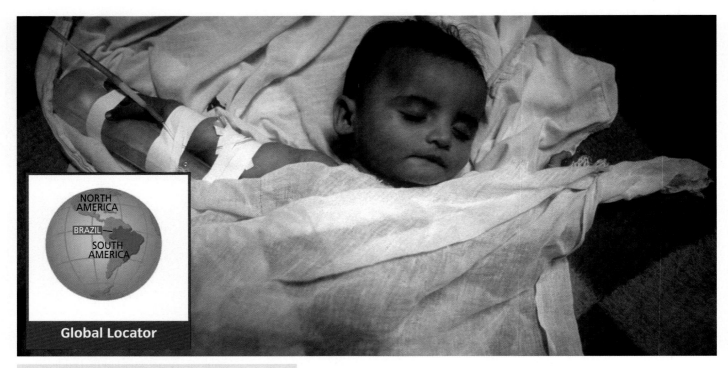

Global Locator

NORTH AMERICA
BRAZIL —
SOUTH AMERICA

Sick with malaria FIGURE 9.10

In a climate-warmed world, the mosquito that spreads malaria could expand into temperate areas. Photographed in Ariquemes, Brazil, where most of the town's inhabitants suffer from malaria.

months may increase the number of heat-related illnesses and deaths. Climate warming may also affect human health indirectly. Mosquitoes and other disease carriers could expand their range into the newly warm areas and spread malaria, dengue fever, schistosomiasis, and yellow fever (**FIGURE 9.10**). According to the World Health Organization, during 1998, the second warmest year on record, the incidence of malaria, Rift Valley fever, and cholera surged in developing countries.

An increasing number of studies report measurable changes in the biology of plant and animal species as a result of climate warming. Climate change also affects populations, communities, and ecosystems. In Visualizing the Effects of Global Warming (on facing page), we report on the results of three of the hundreds of studies conducted thus far.

Rising temperatures in the waters around Antarctica have led to a decline in the population of shrimplike krill, a major food source for Adélie penguins and many other animals. This decline has in turn contributed to a reduction in Adélie penguin populations (**FIGURE 9.11**). Warmer temperatures also cause higher rates of reproductive failure in these penguins by producing wet, snowy conditions that kill the developing chick embryos at egg-laying sites.

The geographic range of one western butterfly (the Edith's Checkerspot butterfly) has shifted considerably northward, as have the ranges of 22 out of 35 European butterfly species examined (**FIGURE 9.12**).

Ecosystems considered at greatest risk of climate-change loss are polar seas, coral reefs, mountain ecosystems, coastal wetlands, and tundra. Water temperature increases of 1° to 2°C cause coral bleaching, which contributes to the destruction of coral reefs. In 1998, when tropical waters were some of the warmest ever recorded, about 10 percent of the world's corals died.

If Earth does warm the projected 1.4° to 5.8°C during the 21st century, the ideal ranges for some temperate tree species may shift northward as much as 483 km (300 mi) (**FIGURE 9.13**).

Adélie penguin and chick FIGURE 9.11

Higher water temperatures threaten this species' food supply and reduce its reproductive success. Photographed in McMurdo Sound, Antarctica.

▼

▲

Climate change and shifting butterfly ranges
FIGURE 9.12

This biologist helped to document that nine species, including the one shown, had died out in southern parts of their ranges but established new colonies farther north. Photographed in the Pyrenees Mountains of Spain. Similar shifts have been documented for butterfly species in other parts of the world.

Climate change and beech trees in North America FIGURE 9.13

A Present geographic range of American beech trees. These large shade trees produce edible beechnuts that are an important food source for squirrels, raccoons, bears, and game birds.

B One projected range of beech trees after global warming occurs. The U.S. Geophysical Fluid Dynamics Laboratory developed this projection.

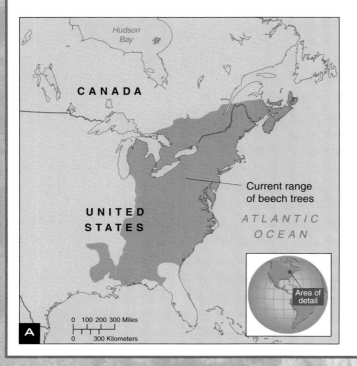

Species such as weeds, insect pests, and disease-carrying organisms already common in a wide range of environments will come out of global warming as winners, with greatly expanded numbers and ranges.

SPECIFIC WAYS TO DEAL WITH GLOBAL WARMING

All greenhouse gases will have to be dealt with as we develop strategies to address global warming, but we focus on CO_2 because it is produced in the greatest quantity and has the largest total effect. Carbon dioxide has an atmospheric lifetime of more than a century, so emissions produced today will still be around in the 22nd century. The extent and severity of global warming will depend on how much additional greenhouse gas emissions we add to the atmosphere. Many studies make the assumption that atmospheric CO_2 will stabilize at 550 ppm, which is roughly twice the concentration of atmospheric CO_2 in the preindustrial world and almost 50 percent higher than the CO_2 currently in the atmosphere.

There are basically two ways to manage global warming: mitigation and adaptation. *Mitigation* is the moderation or postponement of global warming through measures that buy us time to further our understanding of global warming and to pursue more permanent solutions. *Adaptation* is a response to changes caused by global warming. Adopting adaptation strategies for climate warming implies that global warming is unavoidable.

Mitigation of global warming
Climate warming is essentially an energy issue. Developing alternatives to fossil fuels offers a solution to warming caused by CO_2 emissions and addresses the dilemma of dwindling supplies of fossil fuels. Alternatives to fossil fuels include solar energy and nuclear energy.

Increasing the energy efficiency of automobiles and appliances—thus reducing the output of CO_2—would help mitigate global warming. Energy-pricing strategies, such as carbon taxes and the elimination of energy subsidies, are other policies that could mitigate global warming. Most experts think using current technologies and developing such policies would significantly reduce greenhouse gas emissions with little cost to society.

Planting and maintaining forests also mitigates global warming (FIGURE 9.14). Like other green plants, trees remove carbon dioxide from the air and incorporate the carbon into organic matter through photosynthesis. Reasonable estimates suggest that trees could remove 10 percent to 15 percent of the excess CO_2 in the atmosphere, but only through enormous plantings, so such efforts should not be considered a substitute for cutting emissions of greenhouse gases.

Many countries are investigating **carbon management**. Several power plants currently capture the CO_2 in their flue gases, but the technology is very new. Technological innovations that more efficiently trap CO_2 from smokestacks would help mitigate global warming and yet allow us to continue using fossil fuels (while they last) for energy. The carbon could be sequestered in geologic formations or in depleted oil or natural gas wells on land, or it could be injected as a liquid into the ocean depths.

Adaptation to global warming
Because the overwhelming majority of climate experts think human-induced global warming is inevitable, government planners and social scientists are developing strategies to help various regions and sectors of society adapt to climate warming. One of the most pressing issues is rising sea level. People living in coastal areas could be moved inland, away from the dangers of storm surges, although the societal and economic costs would be great.

Another extremely expensive alternative is the construction of dikes and levees to protect coastal land. Rivers and canals that spill into the ocean could be channeled to prevent *saltwater intrusion* into fresh water and agricultural land.

We must also adapt to shifting agricultural zones. Countries with temperate climates are evaluating semitropical crops to determine the best substitutes for traditional crops as the climate warms. Large lumber companies are developing

carbon management
Ways to separate and capture the CO_2 produced during the combustion of fossil fuels and then sequester it from the atmosphere.

Trees help mitigate global warming

FIGURE 9.14

Forests remove carbon dioxide from the air. Photographed on Borneo Island, Malaysia.

drought-resistant strains of trees that will be harvested when global warming may be well advanced. Evaluating such problems and finding and implementing solutions now will ease future stresses of climate warming.

Adaptation to global climate change is under study at 15 locations around the United States. One of the problems identified in the New York City study involves its sewer system. The waterways for storm runoff normally close during high tides. As the sea level rises in response to global climate change, the waterways will have to be shut during many low tides, which will increase the risk of flooding during storms (because excess water will not drain away). City planners will have to rebuild the storm runoff system or find some other way to prevent flooding. Evaluating such problems and implementing solutions now will ease future stresses of climate warming.

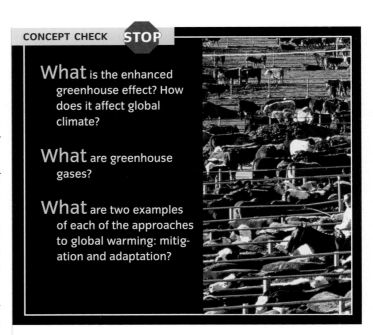

CONCEPT CHECK **STOP**

What is the enhanced greenhouse effect? How does it affect global climate?

What are greenhouse gases?

What are two examples of each of the approaches to global warming: mitigation and adaptation?

Ozone Depletion in the Stratosphere

 lthough ozone (O_3) is a human-made pollutant in the troposphere, it is a naturally produced, essential component in the stratosphere, which encircles our planet some 10 to 45 km (6 to 28 mi) above the surface. The ozone layer shields Earth's surface from much of the **ultraviolet (UV) radiation** coming from the sun (**FIGURE 9.15** in Visualizing the Ozone Layer). Should ozone disappear from the stratosphere, Earth would become uninhabitable for most forms of life.

A slight **ozone thinning** occurs naturally over Antarctica for a few months each year. In 1985, however, the thinning was first observed to be greater than it should be if natural causes were the only factor inducing it. This increased thinning, which occurs each September, is commonly referred to as the "ozone hole" (**FIGURE 9.16**). There, ozone levels decrease as much as 70 percent each year.

During the 1990s the ozone-thinned area continued to grow, and by 2000 it had reached the record size of 29.2 million km^2 (11.3 million mi^2), which is larger than the North American continent. A smaller thinning was also detected in the stratospheric ozone layer over the Arctic. In addition, world levels of stratospheric ozone have been decreasing for several decades. According to the National Center for Atmospheric Research, ozone levels over Europe and North America have dropped almost 10 percent since the 1970s.

CAUSES OF OZONE DEPLETION

The primary chemicals responsible for ozone loss in the stratosphere are a group of compounds called **chlorofluorocarbons (CFCs)**. Chlorofluorocarbons like Freon have been used as propellants for aerosol cans and coolants in air conditioners and refrigerators. Other CFCs have been used as solvents and as foam-blowing agents for insulation and packaging (Styrofoam, for example).

Other compounds that destroy ozone include *halons*, used as fire retardants; *methyl bromide*, a pesticide; *methyl chloroform* and *carbon tetrachloride*, industrial solvents; and *nitrous oxide*, released from the burning of fossil fuels (particularly coal) and from the breakdown of nitrogen fertilizers in the soil.

EFFECTS OF OZONE DEPLETION

With depletion of the ozone layer, higher levels of UV radiation reach the surface of Earth.

Increased levels of UV radiation may disrupt ecosystems. Biologists have documented direct UV damage to natural populations of Antarctic fish. A possible link also exists between the widespread decline of amphibian populations and increased UV radiation. Because organisms are interdependent, the negative effect on one species has ramifications throughout the ecosystem.

■ **ultraviolet (UV) radiation** That part of the electromagnetic spectrum with wavelengths just shorter than visible light; a high-energy form of radiation that can be lethal to organisms at high levels of exposure.

■ **ozone thinning** The removal of ozone from the stratosphere by human-produced chemicals or natural processes.

■ **chlorofluorocarbons** (CFCs) Human-made organic compounds containing chlorine and fluorine; have several industrial and commercial applications but are now banned because they attack the stratospheric ozone layer.

Stratospheric ozone layer FIGURE 9.15

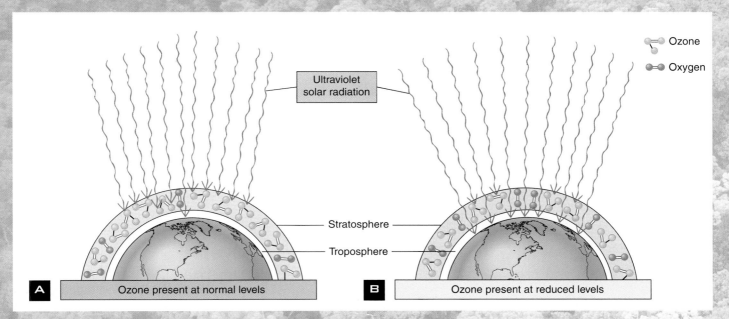

Ozone

Oxygen

Ultraviolet solar radiation

Stratosphere

Troposphere

A Ozone present at normal levels

B Ozone present at reduced levels

A Stratospheric ozone absorbs about 99 percent of incoming solar ultraviolet (UV) radiation, effectively shielding the surface.

B When stratospheric ozone is present at reduced levels, more high-energy UV radiation penetrates the atmosphere to the surface, where its presence harms organisms.

www.wiley.com/college/berg

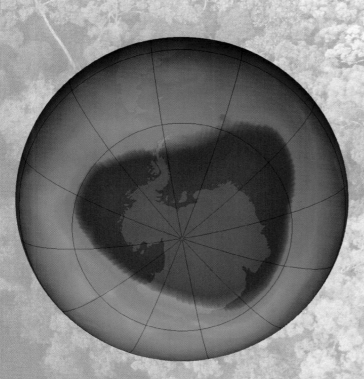

Ozone depletion FIGURE 9.16

A computer-generated image of part of the Southern Hemisphere, taken on October 1, 2005, reveals ozone thinning (the purple area over Antarctica). The ozone-thin area is not stationary but moves about as a result of air currents. It is important to remember that this ozone depletion takes place in the stratosphere, where ozone is produced naturally.

Exposure to UV radiation and skin cancer
FIGURE 9.17

This Australian worked outdoors throughout his career, so it was perhaps inevitable that he contracted skin cancer. Here he attaches an artificial nose after doctors removed a large melanoma.

Excessive exposure to UV radiation is linked to several health problems in humans, including eye cataracts, skin cancer (**FIGURE 9.17**), and weakened immunity. Malignant melanoma, the most dangerous type of skin cancer, is increasing faster than any other type of cancer.

HELPING THE OZONE LAYER RECOVER

In 1978 the United States, the world's largest user of CFCs, banned the use of CFC propellants in products such as antiperspirants and hair sprays. Although this ban was a step in the right direction, it did not solve the problem. Most nations did not follow suit, and besides, propellants represented only a small portion of all CFC use.

In 1987, representatives from many countries met in Montreal to sign the **Montreal Protocol**, an agreement that originally stipulated a 50 percent reduction of CFC production by 1998. Despite this effort, stratospheric ozone continued to thin over the heavily populated middle latitudes of the Northern Hemisphere, and the Montreal Protocol was modified to include even stricter limits on CFC production.

Industrial companies that manufacture CFCs quickly developed substitutes, such as hydrofluorocarbons (HFCs) and hydrochlorofluorocarbons (HCFCs). HFCs do not attack ozone, although they are potent greenhouse gases. HCFCs attack ozone but are less destructive than the chemicals they are replacing.

CFC, carbon tetrachloride, and methyl chloroform production was almost completely phased out in the United States and other highly developed countries in 1996, except for a relatively small amount exported to developing countries. Developing countries were on a different timetable and phased out CFC use by 2005. Methyl bromide was phased out in highly developed countries, which were responsible for 80 percent of its global use, by 2005. HCFCs will be phased out in 2030.

Unfortunately, CFCs are extremely stable and will probably continue to deplete stratospheric ozone for several decades. Human-exacerbated ozone thinning will reappear over Antarctica each year, although the area and degree of thinning will gradually decline over time, until full recovery takes place sometime after 2050.

CONCEPT CHECK **STOP**

What is the stratospheric ozone layer?

What are two harmful effects of stratospheric ozone depletion?

How does the ozone layer protect life on Earth? What is stratospheric ozone thinning? How does it occur?

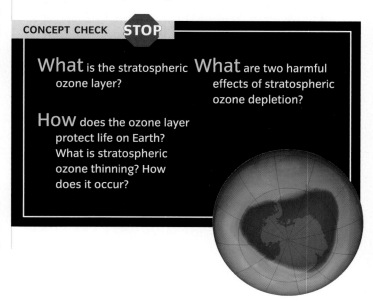

Most environmental studies examine a single issue, such as acid deposition, global climate change, or ozone depletion. In the past few years, however, some researchers have been exploring the interactions of all three problems simultaneously. One recent study of such interactions found that North American lakes may be more susceptible to damage from UV radiation than the thinning of the ozone hole would indicate. The reason: the organic matter in the lakes, which absorbs some UV radiation and protects the lakes' plant and fish life, is affected by acid deposition and global warming. Acid deposition causes organic matter in lakes to settle to the lake floor, so they do not absorb as much of the UV radiation as they once did. And a warmer climate increases evaporation, which reduces the amount of organic matter washed into lakes by streams.

Several studies report a link between human-caused climate warming and polar ozone depletion. Greenhouse gases that warm the troposphere also contribute to stratospheric cooling, presumably because heat trapped in the troposphere is not available to warm the stratosphere. The stratospheric temperature has been dropping for the past several years, and these lower temperatures provide better conditions for ozone-depleting chemicals to attack stratospheric ozone. Record ozone holes over Antarctica are attributed to cooler stratospheric temperatures. Some scientists speculate if the cooling trend in the stratosphere continues, recovery of the ozone layer may be delayed. This means climate warming could prolong ozone depletion in the stratosphere despite the success of the Montreal Protocol.

Scientists now know environmental problems can't be studied as separate issues because they often interact in surprisingly subtle ways. As global warming, ozone depletion, and acid deposition are studied further, it is likely other interactions will be discovered.

Acid Deposition

LEARNING OBJECTIVES

Define acid deposition and explain how acid deposition develops.

Relate some of the effects of acid deposition.

hat do fishless lakes in the Adirondack Mountains, damaged Mayan ruins in southern Mexico, and dead trees in the Czech Republic have in common? All these problems are the result of acid precipitation or, more properly, **acid deposition.** Acid deposition has been around since the Industrial Revolution began. Robert Angus Smith, a British chemist, coined the term *acid rain* in 1872 after he noticed that buildings in areas with heavy industrial activity were being worn away by rain.

> ■ **acid deposition**
> A type of air pollution that includes sulfuric and nitric acids in precipitation as well as dry acid particles that settle out of the air.

Acid precipitation, including acid rain, sleet, snow, and fog, poses a serious threat to the environment. Industrialized countries in the Northern Hemisphere have been hurt the most, especially the Scandinavian countries, central Europe, Russia, and North America. In the United States alone, the annual damage from acid deposition is estimated at $10 billion. Acid deposition is now recognized as a global problem because it also occurs in developing countries as they become industrialized. For example, in 1998 Chinese scientists reported that acid deposition affects 40 percent of their country.

HOW ACID DEPOSITION DEVELOPS

Acid deposition occurs when sulfur dioxide and nitrogen oxides are released into the atmosphere (FIGURE 9.18 in Visualizing the Effects of Acid Deposition). Motor vehicles are a major source of nitrogen oxides. Coal-burning power plants, large smelters, and industrial boilers are the main sources of sulfur dioxide emissions and produce substantial amounts of nitrogen oxides as well. Wind carries sulfur dioxide and nitrogen oxides, released into the air from tall smokestacks, for long distances. Tall smokestacks allow England to "export" its acid deposition problem to the Scandinavian countries, and the midwestern United States to "export" its acid emissions to New England and Canada.

In the atmosphere, sulfur dioxide and nitrogen oxides react with water to produce dilute solutions of sulfuric acid (H_2SO_4), nitric acid (HNO_3), and nitrous acid (HNO_2). Acid deposition returns these acids to Earth's surface in the form of precipitation or particulates.

EFFECTS OF ACID DEPOSITION

Acid deposition corrodes metals and building materials (FIGURE 9.19). It eats away at important monuments, such as the Washington Monument in Washington, D.C., and ancient Mayan ruins in southern Mexico.

The link between acid deposition and declining aquatic animal populations, particularly fish, is well established, but other animals are also adversely affected. Birds living in areas with pronounced acid deposition are more likely to lay eggs with thin, fragile shells that break or dry out before the chicks hatch. The inability to produce strong eggshells is attributed to reduced calcium in the birds' diets. Calcium is less available to the food chain because in acidic soils it becomes soluble and is washed away, with little left for plant roots to absorb.

forest decline
A gradual deterioration and eventual death of many trees in a forest.

Acid deposition also has a serious effect on forest ecosystems. In the Black Forest of Germany, for example, up to 50 percent of trees surveyed are dead or severely damaged. This **forest decline** appears to result from a combination of stressors, including tropospheric ozone, UV radiation (which is more intense at higher altitudes), insect attack, drought, and acid deposition. When one or more stressors weaken a tree, then an additional stressor, such as air pollution, may be decisive in causing its death (FIGURE 9.20).

THE POLITICS OF ACID DEPOSITION

Acid deposition is hard to combat because it does not occur only in the locations where acidic gases are emitted. It is entirely possible for sulfur and nitrogen oxides released in one spot to return to Earth's surface hundreds of kilometers from their source.

The United States has wrestled with this issue. Several states in the Midwest and East—Illinois, Indiana, Missouri, Ohio, Pennsylvania, Tennessee, and West Virginia—produce between 50 percent and 75 percent of the acid deposition that contaminates New England and southeastern Canada. Legislation formulated to deal with acid deposition caused arguments about who should pay for the installation of expensive devices to reduce emissions of sulfur and nitrogen oxides. Should the states emitting the gases be required to pay all the expenses to clean up the air, or should the areas that stand to benefit most from the cleaner air absorb some of the cost?

In international disputes, these issues are magnified even more. For example, gases from coal-burning power plants in England move eastward with prevailing winds and return to the surface as acid deposition in Sweden and Norway. Similarly, emissions from mainland China produce acid deposition in Japan, Taiwan, North Korea, and South Korea. Generally, cases of international air pollution cannot be resolved by local legislation.

FACILITATING RECOVERY FROM ACID DEPOSITION

Although the science and the politics surrounding acid deposition are complex, the basic concept of control is straightforward: Reducing emissions of sulfur and nitrogen oxides curbs acid deposition. Simply stated, if

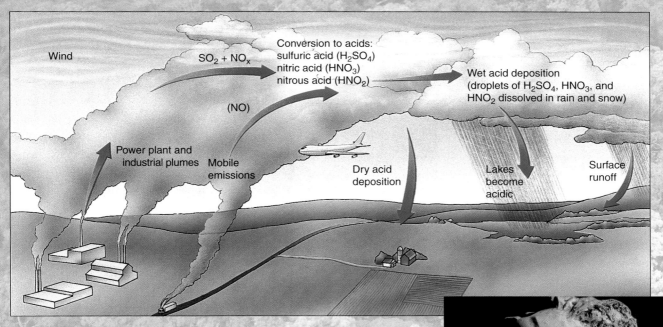

Wind

SO$_2$ + NO$_x$

Conversion to acids:
sulfuric acid (H$_2$SO$_4$)
nitric acid (HNO$_3$)
nitrous acid (HNO$_2$)

(NO)

Wet acid deposition
(droplets of H$_2$SO$_4$, HNO$_3$, and
HNO$_2$ dissolved in rain and snow)

Power plant and
industrial plumes   Mobile
emissions

Dry acid
deposition

Lakes
become
acidic

Surface
runoff

▲ **Acid deposition** FIGURE 9.18

Sulfur dioxide and nitrogen oxide emissions react with water vapor in the atmosphere to form acids that return to the surface as either dry or wet deposition.

▲ **Acid rain damage** FIGURE 9.19

Acid rain reacted with minerals in the bronze to cause the white streaks on this statue. Photographed at the Library of Congress in Washington, D.C.

A

B

◀ **Forest decline** FIGURE 9.20

Acid deposition is one of several stressors that may interact, contributing to the decline and death of trees.
A Healthy Sitka spruce branch.
B Sitka spruce branch exhibiting the effects of forest decline. Photographed in Black Forest, Germany.

sulfur and nitrogen oxides are not released into the atmosphere, they cannot come down as acid deposition. Installing scrubbers in the smokestacks of coal-fired power plants and using clean-coal technologies to burn coal without excessive emissions effectively diminish acid deposition. In turn, a decrease in acid deposition prevents surface waters and soil from becoming more acidic than they already are.

Rainfall in parts of the Midwest, Northeast, and Mid-Atlantic regions is less acidic than it was a few years ago, as a result of cleaner-burning power plants and the use of reformulated gasoline. Many power plants in the Ohio Valley switched from high-sulfur to low-sulfur coal. However, solving one environmental problem often creates others. While the move to low-sulfur coal reduced sulfur emissions, it contributed to the problem of global warming. Because low-sulfur coal has a lower heat value than eastern coal, more of it must be burned—and more CO_2 emitted—to generate a given amount of electricity. Low-sulfur coal also contains higher levels of mercury and other trace metals, so burning it adds more of these hazardous pollutants to the air.

In spite of reduced sulfur emissions, acid precipitation remains a serious problem. Acidified forests and bodies of water have not recovered as quickly as hoped. A likely reason for the slow recovery is that the past 30 or more years of acid rain have profoundly altered soil chemistry in many areas. Because soils take hundreds or even thousands of years to develop, it may be decades or centuries before they recover from the effects of acid rain.

CASE STUDY

International Implications of Global Warming

Various social, economic, and political factors complicate international efforts to deal with global warming. Although highly developed countries are the primary producers of greenhouse gases, many developing countries are rapidly increasing production as they industrialize. But because developing countries have less technical expertise and fewer economic resources, they are less able to respond to the challenges of global warming.

Tensions exist among nations, especially between the highly developed and developing countries, over their differing self-interests. Most developing countries view fossil fuels as their route to industrial development and resist pressure from highly developed nations to decrease fossil fuel consumption. Developing countries also question the need to curb their CO_2 emissions when rich industrialized nations like the United States are the main cause of the problem. Currently, highly developed countries produce about five or six times more CO_2 emissions per person than developing countries (see figure). However, according to the U.S. Department of Energy, CO_2 emissions from developing countries will surpass those from highly developed countries by 2020, assuming current trends in fossil fuel consumption continue.

Despite all the posturing, the international community recognizes it must stabilize CO_2 emissions. At least 174 nations, including the United States, signed the U.N. Framework Convention on Climate Change developed at the 1992 Earth Summit. At the 1996 U.N. Climate Change Convention held in Geneva, Switzerland, highly developed countries agreed to establish

Another factor that may be slowing recovery is that although sulfur emissions have declined, nitrogen oxide emissions have not. In fact, nitrogen oxide emissions increased slightly during the 1990s and early 2000s. Because a substantial portion of nitrogen oxides is produced by motor vehicles, these oxides' emissions are harder to control than those of sulfur oxides. Engine improvements may help reduce nitrogen oxide emissions, but as the human population continues to grow, the increasing number of motor vehicles will probably offset any engineering gains. Thus, dramatic cuts in nitrogen oxide emissions will probably require expensive solutions such as increased mass transit, which would reduce the number of miles people drive, or switching from gasoline-powered cars to hybrids or clean-fuel cars.

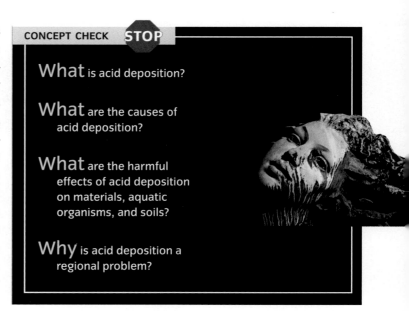

CONCEPT CHECK STOP

What is acid deposition?

What are the causes of acid deposition?

What are the harmful effects of acid deposition on materials, aquatic organisms, and soils?

Why is acid deposition a regional problem?

legally binding timetables to cut emissions of greenhouse gases. In 1997, representatives from 160 countries determined these timetables in Kyoto, Japan.

By 2005, enough countries had ratified the Kyoto Protocol for it to come into force. This international treaty provides operational rules on reducing greenhouse gas emissions. It is noteworthy that the United States has not yet ratified the Kyoto Protocol. Most analysts think the Kyoto Protocol will accomplish little without the full participation of the United States. The United States signed the Kyoto Protocol in 1998, but the administration of President George W. Bush withdrew from the commitment in 2001 because of the perceived economic burden. However, many climate experts think that the costs of climate warming—such as damage to agriculture and human health—make enactment of the Kyoto Protocol economically beneficial.

Per-capita carbon dioxide (CO_2) emission estimates for selected countries

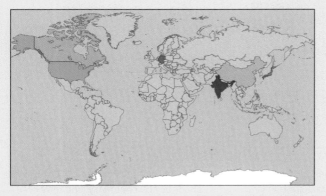

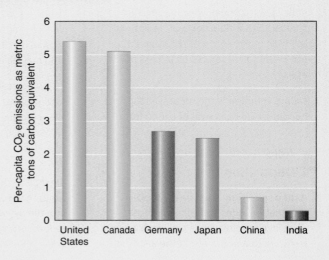

Note that industrialized nations currently produce a disproportionate share of CO_2 emissions. As developing nations such as China and India industrialize, however, their per-capita CO_2 emissions increase. (Map is color-coded with the graph.)

CHAPTER SUMMARY

1 The Atmosphere and Climate

1. **Weather** is the conditions in the atmosphere at a given place and time; it includes temperature, atmospheric pressure, precipitation, cloudiness, humidity, and wind. **Climate** is the average weather conditions that occur in a place over a period of years. The two most important factors that determine an area's climate are temperature and precipitation.

2. Sunlight is the primary (almost sole) source of energy available in the biosphere. The sun's energy runs the hydrologic cycle, drives winds and ocean currents, powers photosynthesis, and warms the planet. Of all the solar energy that reaches Earth, 31 percent is immediately reflected away, and the remaining 69 percent is absorbed. Ultimately, all absorbed solar energy is radiated into space as **infrared radiation**, electromagnetic radiation with wavelengths longer than visible light but shorter than microwaves; it is perceived as invisible waves of heat energy.

3. **Precipitation** is greatest where warm air passes over the ocean, absorbing moisture, and is then cooled, such as when mountains force humid air upward. Deserts develop in the **rain shadows** of mountain ranges or in continental interiors.

2 Global Warming

1. **Greenhouse gases** are the gases that absorb infrared radiation; they include carbon dioxide, methane, nitrous oxide, chlorofluorocarbons, and tropospheric ozone. The **enhanced greenhouse effect** is the additional warming that may be produced by human-increased levels of gases that absorb infrared radiation.

2. Global warming will probably cause a rise in sea level, changes in precipitation patterns, extinction of many species, and problems for agriculture. It could result in the displacement of millions of people, thereby increasing international tensions.

3. Mitigation (slowing down the rate of warming) and adaptation (making adjustments to live with global warming) are two ways to address global warming. Mitigation includes the development of alternatives to fossil fuels; controlling the human population; increasing energy efficiency of automobiles and appliances; planting and maintaining forests; and **carbon management**, ways to separate and capture the CO_2 produced during the combustion of fossil fuels and then sequester it from the atmosphere. Adaptation includes developing strategies to help various regions and sectors of society adapt to climate warming.

3 Ozone Depletion in the Stratosphere

1. Ozone (O_3) is a human-made pollutant in the troposphere but a naturally produced, essential component in the stratosphere. The stratosphere contains a layer of ozone that shields the surface from much of the sun's **ultraviolet (UV) radiation**, that part of the electromagnetic spectrum with wavelengths just shorter than visible light; UV radiation is a high-energy form of radiation that can be lethal to organisms at high levels of exposure.

2. **Ozone thinning** is the natural and human-caused removal of ozone from the stratosphere. The primary chemicals responsible for ozone thinning in the stratosphere are **chlorofluorocarbons (CFCs)**, human-made organic compounds containing chlorine and fluorine. CFCs are now banned because they attack the stratospheric ozone layer. Ozone thinning causes excessive exposure to UV radiation, which may increase cataracts, weaken immunity, and cause skin cancer in humans. Increased levels of UV radiation may also disrupt ecosystems.

3. The Montreal Protocol resulted in an international agreement to phase out CFC production.

4 Acid Deposition

1. **Acid deposition** is a type of air pollution that includes sulfuric and nitric acids in precipitation as well as dry acid particles that settle out of the air. Acid deposition develops when air pollutants (sulfur and nitrogen oxides) are released into the air, where they react with water to form acids and then return to surface waters and soil.

2. Acid deposition kills aquatic organisms, changes soil chemistry, and may contribute to **forest decline**, a gradual deterioration and eventual death of many trees in a forest.

KEY TERMS

- climate p. 212
- infrared radiation p. 213
- greenhouse gases p. 218
- enhanced greenhouse effect p. 218

- carbon management p. 222
- ultraviolet (UV) radiation p. 224
- ozone thinning p. 224
- chlorofluorocarbons p. 224

- acid deposition p. 227
- forest decline p. 228

CRITICAL AND CREATIVE THINKING QUESTIONS

1. How does the sun affect temperature at different latitudes? Why?

2. On the basis of what you know about the nature of science, can we say with absolute certainty that the increased production of greenhouse gases is causing global warming? Why or why not?

3. Biologists who study plants growing high in the Alps found that plants adapted to cold-mountain conditions migrated up the peaks as fast as 3.7 m per decade during the 20th century, apparently in response to climate warming. Assuming that warming continues during the 21st century, what will happen to the plants if they reach the tops of the mountains?

4. Some environmentalists contend the wisest "use" of fossil fuels is to leave them in the ground. How would this affect air pollution? Global warming? Energy supplies?

5. Distinguish between the benefits of the ozone layer in the stratosphere and the harmful effects of ozone at ground level.

6. What is the Montreal Protocol? The Kyoto Protocol?

7. Discuss some of the possible causes of forest decline. How might these factors interact to speed the rate of decline?

This map shows one model of how warmer global temperatures might alter precipitation in the United States in the next 100 years. Use it to answer questions 8–10.

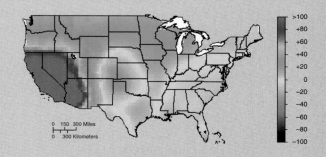

8. Name three states that may experience a significantly wetter climate. Name three states that may experience a significantly drier climate. Explain your answer.

9. Locate the state you live in. Will it be wetter, drier, or about the same?

10. Why are the projected changes in precipitation in the next 100 years a threat to U.S. agriculture? To the U.S. economy?

What is happening in this picture ?

These scientists have drilled into a glacier in Greenland to remove an ice core. Do you think the ice deep within the glacier is old or relatively young? Explain your answer.

Some of the deeper samples were laid down thousands of years ago, when the climate was much cooler. The ice contains bubbles of air. Based on what you have learned in this chapter, do you think the level of carbon dioxide in the air bubbles in the oldest ice is higher or lower than the level in today's atmosphere? Explain your answer.

If the scientists compared CFCs in the air bubbles in the oldest ice with today's levels of CFCs, what do you think they would find?

Freshwater Resources and Water Pollution

10

THE MISSOURI RIVER:
A BATTLE OVER WATER RIGHTS

The Missouri River flows from Montana to St. Louis, Missouri, where it joins the Mississippi River. The longest river in the United States, the Missouri drains about one sixth of the country. The six dams on the Missouri offer both benefits and problems for people living along the river.

Since 1987 the Army Corps of Engineers has increased water flow over the northern dams to protect downstream navigation, including 2 million tons of cargo shipped by boat each year. In addition, people who live downstream count on the river water for irrigation, electrical power, and individual water consumption (see inset, of Kansas City, Missouri). On the other hand, the area along the northern Missouri River depends on the river for its multimillion-dollar fishing and tourism industry *(see larger photo)*. The battle over water rights has intensified as both upstream and downstream parties fight to protect their interests.

Complicating the issue is a growing confrontation between environmentalists and farmers. Farmers want additional dikes and levees to protect their crops on the flood plains from flood damage, whereas environmentalists want the river restored to its natural state, in part to protect the many native fish populations now in decline. Native Americans with claims to water rights want to use the water in a variety of ways, from generating hydroelectric power to irrigating cropland.

The Missouri River Basin Association, a coalition of eight river-basin states and two dozen Native American tribes, works with the Corps to meet the demands of the competing interest groups as they decide the river's future.

■ The Importance of Water p. 236

Water Resource Problems p. 238

Water Management p. 244

Water Pollution p. 248

■ Improving Water Quality p. 254

■ Case Study: Water Pollution in the
Great Lakes p. 258

The Importance of Water

LEARNING OBJECTIVES

Describe the structure of a water molecule and explain how hydrogen bonds form between adjacent water molecules.

List the unique properties of water.

Explain how processes of the hydrologic cycle allow water to circulate through the abiotic environment.

L ife on planet Earth would be impossible without water. All life forms, from unicellular bacteria to multicellular plants and animals, contain water. Humans are composed of approximately 60 percent water by body weight. We depend on water for our survival as well as for our convenience: We drink it, cook with it, wash with it (FIGURE 10.1), travel on it, and use an enormous amount of it for agriculture, manufacturing, mining, energy production, and waste disposal.

Although Earth has plenty of water, about 97% of it is salty and not consumable by most terrestrial organisms. Fresh water is distributed unevenly, resulting in serious regional water supply problems. Water experts predict that by 2025, more than one-third of the human population will live in areas where there isn't enough fresh water for drinking and irrigation.

THE HYDROLOGIC CYCLE AND OUR SUPPLY OF FRESH WATER

In the **hydrologic cycle**, water continuously circulates through the environment, from the ocean to the atmosphere to the land and back to the ocean (see FIGURE 10.2; also see Figure 5.9 on page 106). The result is a balance of the water resources in the ocean, on the land, and in the atmosphere. The hydrologic cycle provides a continual renewal of the supply of fresh water on land, which is essential to terrestrial organisms.

Two important components of the hydrologic cycle FIGURE 10.2 ▼

A Liquid and solid precipitation continuously falls from the atmosphere to the land and ocean.

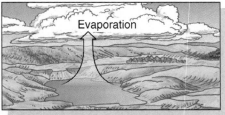

B Evaporation continuously moves water vapor from the land and ocean into the atmosphere.

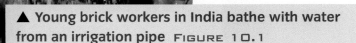

▲ **Young brick workers in India bathe with water from an irrigation pipe** FIGURE 10.1

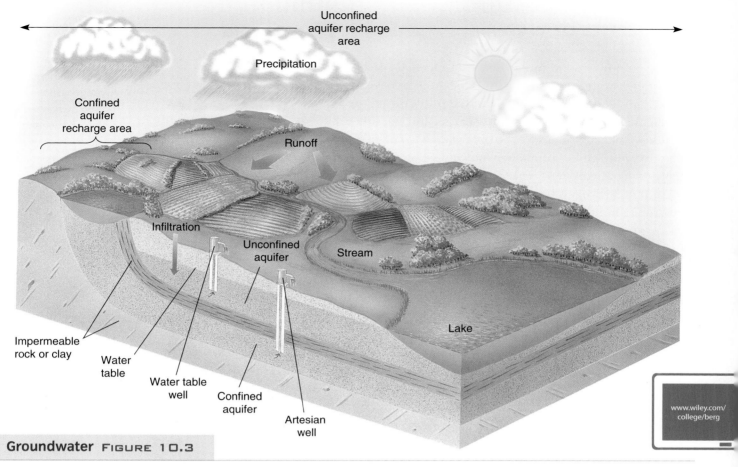

Groundwater FIGURE 10.3

Excess surface water seeps downward through soil and porous rock layers until it reaches impermeable rock or clay. An unconfined aquifer has groundwater recharged by surface water directly above it. In a confined aquifer, groundwater is stored between two impermeable layers and is often under pressure.

Surface water is water found in streams and rivers, lakes, ponds, reservoirs, and **wetlands** (areas of land covered with water for at least part of the year). The **runoff** of precipitation from the land replenishes surface waters and is considered a renewable, although finite, resource. A **drainage basin**, or **watershed**, is the area of land drained by a single river or stream.

Earth contains underground formations that collect and store water. This water originates as rain or melting snow that slowly seeps into the soil. It works its way down through cracks and spaces in sand, gravel, or rock until an impenetrable layer stops it; there it accumulates as **groundwater**. Groundwater is eventually discharged into rivers, wetlands, springs, or the ocean. Thus, surface water and groundwater are interrelated parts of the hydrologic cycle. **Aquifers** are underground reservoirs in which groundwater is stored (FIGURE 10.3).

Most groundwater is considered a nonrenewable resource, because it has taken hundreds or even thousands of years to accumulate and usually only a small portion of it is replaced each year by seepage of surface water.

surface water Precipitation that remains on the surface of the land and does not seep down through the soil.

runoff The movement of fresh water from precipitation and snowmelt to rivers, lakes, wetlands, and, ultimately, the ocean.

groundwater The supply of fresh water under Earth's surface that is stored in underground aquifers.

PROPERTIES OF WATER

Water is composed of molecules of H₂O, each consisting of two atoms of hydrogen and one atom of oxygen. Water molecules are **polar**—that is, one end of the molecule has a positive electrical charge, and the other end has a negative charge (**FIGURE 10.4**). The negative (oxygen) end of one water molecule is attracted to the positive (hydrogen) end of another water molecule, forming a **hydrogen bond** between the two molecules. Hydrogen bonds are the basis for many of water's physical properties, including its high melting/freezing point (0°C, 32°F) and high boiling point (100°C, 212°F). Because most of Earth has a temperature between 0°C and 100°C, most water exists in the liquid form organisms need.

Water absorbs a great deal of solar heat without substantial increases in temperature. This high heat capacity allows the ocean to have a moderating influence on climate, particularly along coastal areas.

Water is a *solvent*, meaning that it can dissolve many materials. In nature, water is never completely pure, because it contains dissolved gases from the atmosphere and dissolved mineral salts from the land. Water's ablities as a solvent have a major drawback: Many of the substances that dissolve in water cause water pollution.

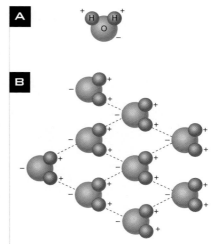

A Water molecules are polar, with positively and negatively charged areas. **B** The polarity causes hydrogen bonds (represented by dashed lines) to form between the positive areas of one water molecule and the negative areas of others. Each water molecule forms up to four hydrogen bonds with other water molecules.

Chemical properties of water FIGURE 10.4

CONCEPT CHECK **STOP**

How do hydrogen bonds form between adjacent water molecules?

What are the unique properties of water?

What is surface water? Groundwater?

Water Resource Problems

LEARNING OBJECTIVES

Relate some of the problems caused by overdrawing surface water, aquifer depletion, and salinization of irrigated soil.

Relate the background behind the water problems of the Ogallala Aquifer and the Colorado River basin.

Briefly describe the role of international cooperation in managing shared water resources.

Water resource problems fall into three categories: too much water, too little water, and poor-quality water. Flooding occurs when a river's discharge cannot be contained within its normal channel. Today's floods are more disastrous in terms of property loss than those of the past because humans often remove water-absorbing plant cover from the soil and construct buildings on flood plains. (A **flood plain** is the area bordering a river channel that has the potential to flood.) These activities increase the likelihood of both floods and flood damage.

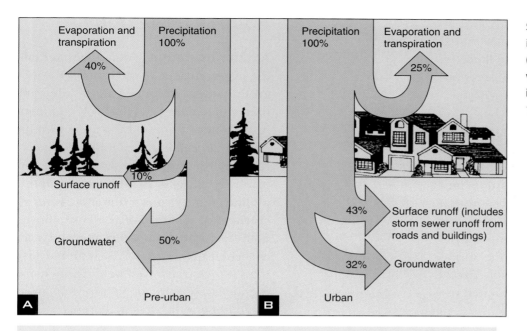

Evaporation and transpiration 40%

Precipitation 100%

Surface runoff 10%

Groundwater 50%

A Pre-urban

Precipitation 100%

Evaporation and transpiration 25%

43% Surface runoff (includes storm sewer runoff from roads and buildings)

32% Groundwater

B Urban

Shown is the fate of precipitation in Ontario, Canada (A) before and (B) after urbanization. After Ontario was developed, surface runoff increased substantially, from 10 percent to 43 percent.

How development changes the natural flow of water FIGURE 10.5

When a natural area—that is, an area undisturbed by humans—is inundated with heavy precipitation, the plant-protected soil absorbs much of the excess water. What the soil cannot absorb runs off into the river, which may then spill over its banks onto the flood plain. Because rivers meander, the flow is slowed, and the swollen waters rarely cause significant damage to the surrounding area. (See Figure 6.12 on page 140 for a diagram of a typical river, including its flood plain.)

When an area is developed for human use, much of the water-absorbing plant cover is removed. Buildings and paved roads don't absorb water, so runoff, usually in the form of storm sewer runoff, is significantly greater in developed areas (FIGURE 10.5). People who build homes or businesses on the flood plain of a river will most likely experience flooding at some point (FIGURE 10.6).

Flooding in California FIGURE 10.6

A house sits in the Santa Clara River after collapsing due to floodwaters. Photographed in Santa Clarita, California, on January 10, 2005.

Arid lands, or deserts, are fragile ecosystems in which plant growth is limited by lack of precipitation. Semiarid lands receive more precipitation than deserts but are subject to frequent and prolonged droughts.

Farmers increase the agricultural productivity of arid and semiarid lands with irrigation. Irrigation of arid and semiarid lands has become increasingly important worldwide in efforts to produce enough food for burgeoning populations (FIGURE 10.7). In fact, irrigation accounts for the highest percentage, 71 percent of the world's total water consumption, industry for 20 percent, and domestic and municipal use for only 9 percent.

aquifer depletion The removal of groundwater faster than it can be recharged by precipitation or melting snow.

saltwater intrusion The movement of seawater into a freshwater aquifer near the coast.

AQUIFER DEPLETION

Aquifer depletion lowers the **water table**, the upper surface of the saturated zone of groundwater. Prolonged aquifer depletion drains an aquifer dry, effectively eliminating it as a water resource. Even areas with high rainfall can experience aquifer depletion if humans remove more groundwater than can be recharged. In addition, aquifer depletion from porous sediments causes **subsidence**, or sinking, of the land above it. **Saltwater intrusion** occurs along coastal areas when groundwater is depleted faster than it recharges. Well water in such areas eventually becomes too salty for human consumption or other freshwater uses.

Agricultural use of water FIGURE 10.7

Center-pivot irrigation produces massive green circles. Each circle is the result of a long irrigation pipe that extends along the radius from the circle's center to its edge and slowly rotates, spraying the crop. Photographed in Nebraska.

The Ogallala Aquifer The High Plains cover 6 percent of U.S. land but produce more than 15 percent of the nation's wheat, corn, sorghum, and cotton and almost 40 percent of its livestock. This productivity requires approximately 30 percent of the irrigation water used in the United States. Farmers on the High Plains rely on water from the **Ogallala Aquifer**, the largest groundwater deposit in the world (FIGURE 10.8).

In some areas farmers are drawing water from the Ogallala Aquifer as much as 40 times faster than nature replaces it. This rapid depletion has lowered the water table more than 30 m (100 ft) in some places. Most hydrologists (scientists who study water supplies) predict groundwater will eventually drop in all areas of the Ogallala to a level uneconomical to pump. Their goal is to postpone that day through water conservation, including the use of water-saving irrigation systems.

OVERDRAWING SURFACE WATERS

Removing too much freshwater from a river or lake can have disastrous consequences in local ecosystems. Humans can remove perhaps 30 percent of a river's flow without greatly affecting the natural ecosystem. In some places considerably more is withdrawn for human use. In the arid American Southwest, is is not unusual to remove 70 percent or more of surface water.

When surface waters are overdrawn, wetlands dry up. **Estuaries**, where rivers empty into seawater, become saltier when surface waters are overdrawn, which reduces their productivity. Wetlands and estuaries, which serve as breeding grounds for many species of birds and other animals, also play a vital role in the hydrologic cycle. When these resources are depleted, the ensuing water shortages and reduced productivity have economic, as well as ecological, ramifications.

The increased use of U.S. surface water for agriculture, industry, and personal consumption has caused many water supply and quality problems. Some regions that have grown in population during this period—for example, California, Nevada, Arizona, and Florida—have had correspondingly greater burdens on their water supplies. If water consumption in these and other areas continues to increase, the availability of surface waters could become a serious regional problem, even in places that have never experienced water shortages.

Nowhere in the country are water problems as severe as they are in the West and Southwest. Much of this large region is arid or semiarid. With the rapid expansion of the population during the past 25 years, municipal, commercial, and industrial uses now compete heavily with irrigation for available water.

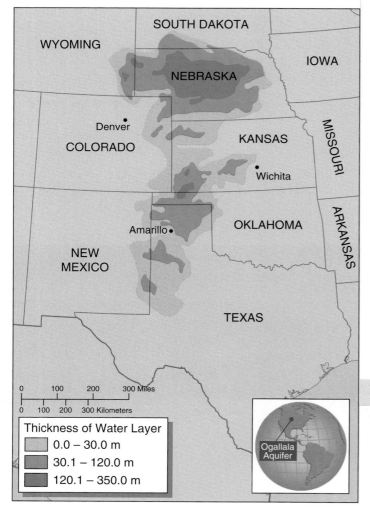

Ogallala Aquifer FIGURE 10.8

This massive deposit of groundwater lies under eight Midwestern states, with extensive portions in Texas, Kansas, and Nebraska. Water in the Ogallala Aquifer takes hundreds or even thousands of years to renew after it is withdrawn to grow crops and raise cattle.

The Colorado River basin One of the most serious water-supply problems in the United States is in the Colorado River basin. The Colorado River provides water for 25 million people, including the cities of Denver, Las Vegas, Salt Lake City, Albuquerque, Phoenix, Los Angeles, and San Diego. It supplies irrigation water for 3.5 million acres of fruit, vegetable, and field crops worth $1.5 billion per year. The Colorado River has 49 dams, 11 of which produce electricity by hydropower.

An international agreement with Mexico, along with federal and state laws, severely restricts the use of the Colorado's waters. The most important of all the state treaties is the 1922 **Colorado River Compact**. It stipulates an annual allotment of 7.5 million acre-feet of water each to the upper Colorado (Colorado, Utah, and Wyoming) and the lower Colorado (California, Nevada, Arizona, and New Mexico). Each acre-foot equals 326,000 gal, enough for about eight people for one year. However, the Colorado River Compact overestimated the average annual flow of the Colorado River, and it locked that estimate into the multistate agreement.

Population growth in the upper Colorado region threatens the lower Colorado region's water supply. Further, people in the states through which the lower Colorado flows take so much water that the remainder is insufficient to meet Mexico's needs as stipulated by international treaty (**FIGURE 10.9**). To com-

> **salinization** The gradual accumulation of salt in a soil, often as a result of improper irrigation methods.

pound the problem, as more and more water is used, the lower Colorado becomes increasingly salty—in some places saltier than the ocean—as it flows toward Mexico.

Some positive steps have been taken. In 2003, California agreed to limit its water withdrawals from the Colorado River to quantities specified in the Colorado River Compact. Also, some California farmers agreed to sell some water they would normally use for irrigation, then use the money earned to update their irrigation systems so they make more efficient use of the water they use.

SALINIZATION OF IRRIGATED SOIL

Although irrigation improves the agricultural productivity of arid and semiarid lands, it often causes salt to accumulate in the soil, a phenomenon called **salinization**. Irrigation water contains small amounts of dissolved salts. Normally, as a result of precipitation runoff, rivers carry salt away. Irrigation water, however, normally soaks into the soil and does not run off the land into rivers. The continued application of such water, season after season, year after year, leads to the gradual accumulation of salt in the soil. Given enough time, the salt concentration can rise to such a high level that plants are poisoned or their roots dehydrated. Thus, salt hurts soil productivity and, in extreme cases, renders soil unfit for crop production.

Colorado River bed in San Luis Rio Colorado FIGURE 10.9

As a result of diversion for irrigation and other uses in the United States, the Colorado River often dries up before reaching the Gulf of California in Mexico.

GLOBAL WATER ISSUES

As the world's population continues to increase, global water problems will become more serious. In India, where 17 percent of the world's population has access to 4 percent of the world's fresh water, approximately 8,000 villages have no local water supply. Water supplies are precarious in much of China, owing to population pressures: water table levels are dropping, wells have gone dry, and much of the water in the Yellow River is diverted for irrigation, depriving downstream areas of water. Mexico is facing the most serious water shortages of any country in the Western Hemisphere. The main aquifer supplying Mexico City is dropping rapidly, and the water table is falling fast in Guanajuato, an agricultural state.

As the needs of the growing human population deplete freshwater supplies, less water will be available for crops. Local famines from water shortages are a possibility.

Sharing water resources among countries

In the 1950s, the then Soviet Union began diverting water that feeds into the Aral Sea to irrigate desert areas surrounding the lake. Since 1960 the Aral Sea has declined more than 50 percent in area (**FIGURE 10.10**). Its total volume is down 80 percent, and much of its biological diversity has disappeared. Millions of people living in the Aral Sea's watershed have developed serious health problems, probably due to toxic salt storms caused by winds that lift the salt on the receding shoreline into the air.

Following the breakup of the Soviet Union in 1991, responsibility for plans to save the Aral Sea shifted from Moscow to the five central Asian countries that share the Aral basin—Uzbekistan, Kazakstan, Kyrgyzstan, Turkmenistan, and Tajikistan. Despite recent cooperative restoration efforts made by these nations and backed by the World Bank and the U.N. Environment Program, it is improbable that the Aral Sea will ever return to its former size and economic importance.

Aral Sea FIGURE 10.10

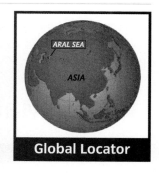

Global Locator

The satellite images show the Aral Sea in (**A**) 1976 and (**B**) 1997. As water was diverted for irrigation, the sea level subsided.

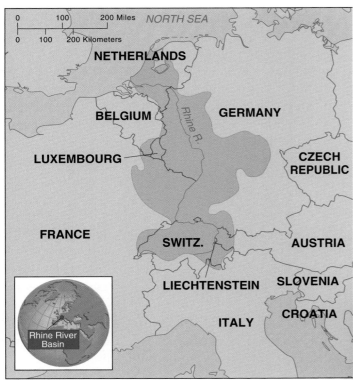

Rhine River basin FIGURE 10.11

The Rhine River drains five European countries. (The green area represents the drainage basin.) Water management of such a river requires international cooperation.

Three-fourths of the world's 200 or so major watersheds are shared between at least two nations. International cooperation is required to manage rivers that cross international borders. The heavily populated drainage basin for the Rhine River in Europe is in five countries—Switzerland, Germany, France, Luxembourg, and the Netherlands (FIGURE 10.11). All five nations recognize that international cooperation is essential to conserve and protect the supply and quality of the Rhine River. Their efforts have paid off: the main sources of pollution have been eliminated and water in the Rhine River today is almost as pure as drinking water; long-absent fishes have returned; and projects are under way to restore riverbanks, control flooding, and clean up remaining pollutants.

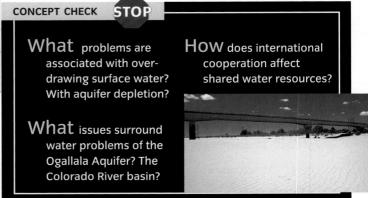

CONCEPT CHECK STOP

What problems are associated with over-drawing surface water? With aquifer depletion?

What issues surround water problems of the Ogallala Aquifer? The Colorado River basin?

How does international cooperation affect shared water resources?

Water Management

NATIONAL GEOGRAPHIC

LEARNING OBJECTIVES

Define sustainable water use.

Contrast the benefits and drawbacks of dams and reservoirs, using the Columbia River to provide specific examples.

Give examples of water conservation in agriculture, industry, and individual homes and buildings.

The main goal of water management is to provide a sustainable supply of high-quality water. **Sustainable water use** means humans use water resources carefully, so water is available for future generations.

Water supplies are obtained by building dams, diverting water, or removing salt from seawater or salty groundwater. Conservation, which includes reusing water, recycling water, and improving water-use efficiency, augments water supply and is an important aspect of sustainable water use. Economic policies are also important in managing water sustainably: When water is inexpensive, it tends to be wasted.

> **sustainable water use** The wise use of water resources, without harming the essential functioning of the hydrologic cycle or the ecosystems on which humans depend.

DAMS AND RESERVOIRS: MANAGING THE COLUMBIA RIVER

The Columbia River, the fourth largest river in North America, illustrates the impact of dams on natural fish communities. There are more than 100 dams in the Columbia River system, 19 of which are major generators of inexpensive hydroelectric power (FIGURE 10.12). The Columbia River system supplies municipal and industrial water to several major urban areas in the northwestern United States and irrigation water for more than 1.2 million hectares (3 million acres) of agricultural land.

As is often the case in natural resource management, one particular use of the Columbia River system may have a negative impact on other uses. The dam impoundments along the Columbia River generate electricity and control floods but were expensive to build and have adversely affected fish populations. The salmon population in the Columbia River system is only a fraction of what it was before the watershed was developed. The many dams that impede salmon migrations are widely considered the most significant factor in salmon decline.

Various efforts to assist migrating salmon have not proved particularly successful (FIGURE 10.13). Conservationists and biologists support using a natural approach of releasing water to flush young salmon downstream; they also support adopting a controversial proposal to tear down several dams on the lower Snake River, a tributary of the Columbia River. Farmers and the hydroelectric companies strongly oppose these plans, which they fear would threaten their water supplies.

Grand Coulee Dam on the Columbia River
FIGURE 10.12

Shown are the dam and part of its reservoir, Lake Roosevelt. The many beneficial uses of dams include electricity generation and flood control, and their reservoirs are often popular locations for water sports and other outdoor activities. But dams are expensive to build, and they disrupt or destroy natural river habitats.

Fish ladder FIGURE 10.13

This ladder is located at the Bonneville Dam along the Oregon side of the Columbia River. Fish ladders help migratory fishes to bypass dams in their migration upstream.

WATER CONSERVATION

Today there is more competition than ever among water users with different priorities, and water conservation measures are necessary to guarantee sufficient water supplies.

Reducing agricultural water waste

Irrigation generally makes inefficient use of water. Traditional irrigation methods involve flooding the land or diverting water to fields through open channels. Plants absorb about 40 percent of the water that flood irrigation applies to the soil; the rest of the water usually evaporates into the atmosphere or seeps into the ground.

One of the most important innovations in agricultural water conservation is **microirrigation**, also called **drip** or **trickle irrigation**, in which pipes with tiny holes bored in them convey water directly to individual plants (**FIGURE 10.14** in Visualizing Water Conservation). Microirrigation substantially reduces the water needed to irrigate crops—usually by 40 percent to 60 percent compared to traditional irrigation—and also reduces the amount of salt that irrigation water leaves in the soil.

> **microirrigation**
>
> A type of irrigation that conserves water by piping it to crops through sealed systems.

Other measures that could save irrigation water include using lasers to level fields, which allows a more even water distribution, and making greater use of recycled wastewater. A drawback of such techniques is their cost, which makes them unaffordable for most farmers in highly developed countries, let alone subsistence farmers in developing nations.

Reducing water waste in industry

Electric power generators and many industries require water. In the United States, five major industries—chemical products, paper and pulp, petroleum and coal, primary metals, and food processing—consume almost 90 percent of industrial water.

Stricter pollution control laws provide some incentive for industries to conserve water. Industries usu-ally recapture, purify, and reuse water to reduce their water use and their water treatment costs. The U.S. Steel Corporation plant in Granite City, Illinois, for example, recycles approximately two-thirds of the water it uses daily. The potential for industries to conserve water by recycling is enormous.

Reducing municipal water waste

Like industries, regions and cities recycle or reuse water to reduce consumption. For example, homes and other buildings can be modified to collect and store gray water. *Gray water* is water that was already used in sinks, showers, washing machines, and dishwashers. Gray water is recycled to flush toilets, wash the car, or sprinkle the lawn (**FIGURE 10.15**). In contrast to water *recycling*, wastewater *reuse* occurs when water is collected and treated before being redistributed. The reclaimed water is generally used for irrigation.

Cities also decrease water consumption by providing consumer education, requiring water-saving household fixtures, developing economic incentives to save water, and repairing leaky water supply systems. Also, increasing the price of water to reflect its true cost promotes water conservation.

The average person in the United States uses 295 L (78 gal) of water per day at home. As a water user, you have a responsibility to use water carefully and wisely. The cumulative effect of many people practicing personal water conservation measures has a significant impact on overall water consumption. You can practice these yourself (**TABLE 10.1**).

CONCEPT CHECK **STOP**

What are the benefits of dams on the Columbia River? The drawbacks?

How can individuals conserve and manage water resources?

How do migratory salmon in the Columbia River cause conflicts over use of water resources?

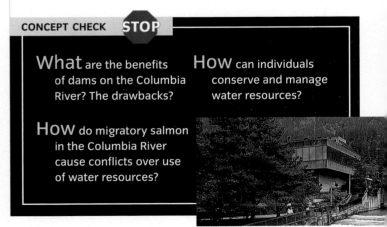

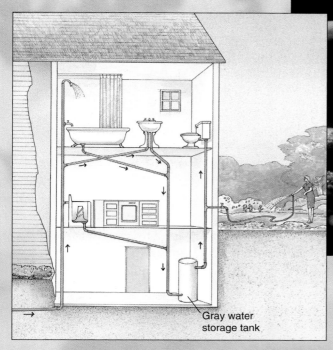

Recycling water FIGURE 10.15

Gray water storage tank

Microirrigation FIGURE 10.14

Close-up of a drip irrigation pipe system about to release a drop of water directly over a seedling, eliminating much of the waste associated with traditional methods of irrigation. Photographed in the Negev Desert, Israel.

◀ Individual homes and buildings can be modified to collect and store "gray water," water already used in sinks, showers, washing machines, and dishwashers. This "gray water" is used when clean water is not required—for example, in flushing toilets, washing the car, and sprinkling the lawn.

Conserving water at home TABLE 10.1

Location	What you can do
Bathroom	Install water-saving showerheads and faucets and low-flush toilets. Or, use water displacement device in tank of conventional toilet (but not a brick—water erodes brick, and resulting sediment can damage the porcelain and the pipes). Fix leaky fixtures. Modify personal habits: Avoid leaving faucet running while shaving or brushing teeth. Take shorter showers.
Kitchen	Use a dishwasher. It requires less water than washing dishes by hand with tap running—but only if you run the dishwasher with a full load.
Laundry room	If you are in the market for a washing machine, choose high-efficiency model. These models not only use less water, but they spin more water out of the clothes, which saves on energy costs when using a clothes dryer.

www.wiley.com/college/berg

Water Pollution

LEARNING OBJECTIVES

Define water pollution.

Discuss how sewage is related to eutrophication, biochemical oxygen demand (BOD), and dissolved oxygen.

Distinguish between the two types of pollution sources and give examples of each.

Water pollution is a global problem that varies in magnitude and type of pollutant from one region to another. In many locations, particularly in developing countries, the main water pollution issue is providing individuals with disease-free drinking water.

TYPES OF WATER POLLUTION

Water pollutants are divided into eight categories: sewage, disease-causing agents, sediment pollution, inorganic plant and algal nutrients, organic compounds, inorganic chemicals, radioactive substances, and thermal pollution. Causes and examples of each of these types of water pollution are summarized in **TABLE 10.2** (on facing page). Here we explore the pollution threats associated with sewage.

■ **water pollution**
Any physical or chemical change in water that adversely affects the health of humans and other organisms.

■ **sewage** Wastewater from drains or sewers (from toilets, washing machines, and showers); includes human wastes, soaps, and detergents.

Sewage The release of **sewage** into water causes several pollution problems. First, because sewage may carry disease-causing agents, water polluted with sewage poses a threat to public health (see Chapter 4). Sewage also generates two serious environmental problems, enrichment and oxygen demand. Enrichment of a body of water is due to the presence of high levels of plant and algal nutrients such as nitrogen and phosphorus, both of which are sewage products. When an aquatic ecosystem contains high levels of sewage or other organic material, decomposing microorganisms use up most of the dissolved oxygen, leaving little available for fishes or other aquatic animals.

Sewage and other organic wastes are measured in terms of their **biochemical oxygen demand (BOD)**, or **biological oxygen demand**. A large amount of sewage in water generates a high BOD, which robs the water of dissolved oxygen (**FIGURE 10.16**). When dissolved oxygen levels are low, anaerobic (without oxygen) microorganisms produce compounds with unpleasant odors, further deteriorating water quality.

■ **biochemical oxygen demand (BOD)**
The amount of oxygen that microorganisms need to decompose biological wastes into carbon dioxide, water, and minerals.

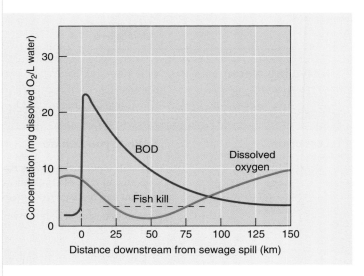

Effect of sewage on dissolved oxygen and biochemical oxygen demand (BOD) FIGURE 10.16

Note the initial oxygen depletion (blue line) and increasing BOD (red line) close to the sewage spill (at distance 0). The stream gradually recovers as the sewage is diluted and degraded. As indicated by the dashed line, fishes can't live in water that contains less than 4 mg of dissolved oxygen per liter of water.

Type of water pollution TABLE 10.2

Type of pollution	Source	Examples	Effects
Sewage	Wastewater from drains or sewers	Human wastes, soaps, detergents	Threatens public health; causes enrichment and high biochemical oxygen demand (BOD)
Disease-causing agents	Wastes of infected individuals	Bacteria, viruses, protozoa, parasitic worms	Spread infectious diseases (cholera, dysentery, typhoid, infectious hepatitis, poliomyelitis, etc.)
Sediment pollution	Erosion of agricultural lands, forest soils exposed by logging, degraded stream banks, overgrazed rangelands, strip mines, construction	Clay, silt, sand, and gravel, suspended in water and eventually settling out	Reduces light penetration, limiting photosynthesis and disrupting food chain; clogs gills and feeding structures of aquatic animals; carries and deposits disease-causing agents and toxic chemicals
Inorganic plant and algal nutrients	Human and animal wastes, plant residues, atmospheric deposition, fertilizer runoff from agricultural and residential land	Nitrogen and phosphorus	Stimulate growth of excess plants and algae, which disrupts natural balance between producers and consumers and causes enrichment, bad odors, and a high BOD; suspected of causing red tides, explosive blooms of toxic pigmented algae that threaten the health of humans and aquatic animals in coastal areas
Organic compounds	Landfills, agricultural runoff, industrial wastes	Synthetic chemicals: pesticides, cleaning solvents, industrial chemicals, plastics	Contaminate groundwater and surface water; threaten drinking water supply; some are suspected endocrine disrupters
Inorganic chemicals	Industries, mines, irrigation runoff, oil drilling, urban runoff from storm sewers	Acids, salts, heavy metals such as lead, mercury, and arsenic	Contaminate groundwater and surface water; threaten drinking water supply; don't easily degrade or break down
Radioactive substances	Nuclear power plants, nuclear weapons industry, medical and scientific research facilities	Unstable isotopes of radioactive minerals such as uranium and thorium	Contaminate groundwater and surface water; threaten drinking water supply
Thermal pollution	Industrial runoff	Heated water produced during industrial processes, then released into waterways	Depletes water of oxygen and reduces amount of oxygen that water can hold; reduced oxygen threatens fish

Oligotrophic and Eutrophic Lakes

A and B The average person looking at these two photographs would notice the dramatic differences between them but wouldn't understand the environmental conditions responsible for the differences. A shows Crater Lake, an oligotrophic lake in Oregon; B shows a small eutrophic lake in western New York.

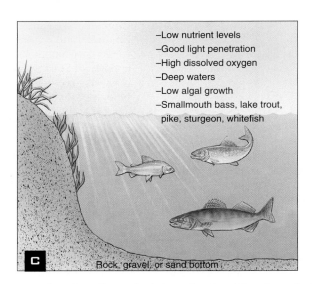

–Low nutrient levels
–Good light penetration
–High dissolved oxygen
–Deep waters
–Low algal growth
–Smallmouth bass, lake trout, pike, sturgeon, whitefish

Rock, gravel, or sand bottom

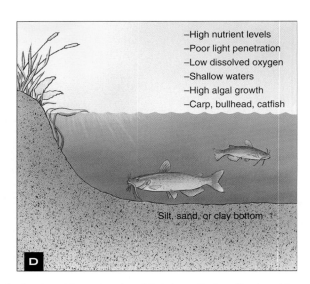

–High nutrient levels
–Poor light penetration
–Low dissolved oxygen
–Shallow waters
–High algal growth
–Carp, bullhead, catfish

Silt, sand, or clay bottom

C and D Aquatic ecologists understand the characteristics of oligotrophic and eutrophic lakes. C An oligotrophic lake has a low level of inorganic plant and algal nutrients. D A eutrophic lake has a high level of these nutrients.

Eutrophication: an enrichment problem

Lakes, estuaries, and slow-flowing streams that have minimal levels of nutrients are considered unenriched, or oligotrophic. An **oligotrophic** lake has clear water and supports small populations of aquatic organisms (see What a Scientist Sees, parts A and C). **Eutrophication** is the enrichment of a lake, estuary, or slow-flowing stream by inorganic plant and algal nutrients such as phosphorus; an enriched body of water is said to be **eutrophic**. The enrichment of water results in an increased photosynthetic productivity. The water in a eutrophic lake is cloudy and usually resembles pea soup because of the presence of vast numbers of algae and cyanobacteria (see parts B and D).

Over vast periods, oligotrophic lakes, estuaries, and slow-moving streams become eutrophic naturally. However, some human activities greatly accelerate eutrophication. This fast, human-induced process is usually called **artificial eutrophication** to distinguish it from natural eutrophication. Artificial eutrophication results from enrichment of aquatic ecosystems by nutrients found predominantly in fertilizer runoff and sewage.

SOURCES OF WATER POLLUTION

Water pollutants come from both natural sources and human activities. Natural sources of pollution like mercury and arsenic tend to be local concerns, but human-generated pollution is generally more widespread.

The sources of water pollution are classified into two types, point source pollution and nonpoint source pollution. **Point source pollution** is discharged into the environment through pipes, sewers, or ditches from specific sites such as factories or sewage treatment plants (**FIGURE 10.17**).

Pollutants that enter bodies of water over large areas rather than at a single point cause **nonpoint source pollution**, also called *polluted runoff*. Nonpoint source pollution occurs when precipitation moves over and through the soil, picking up and carrying away pollutants that are eventually deposited in lakes, rivers, wetlands, groundwater, estuaries, and the ocean. Nonpoint source pollution includes agricultural runoff (such as fertilizers, pesticides, livestock wastes, and salt from irrigation), mining wastes (such as acid mine drainage), municipal wastes (such as inorganic plant and algal nutrients), construction sediments, and soil erosion (from fields, logging operations, and eroding stream banks). Although nonpoint sources are diffuse, their cumulative effect can be huge.

Point source pollution FIGURE 10.17

Industrial runoff pours into a waste pit in the Amazon River basin. Natural gas burns from an adjacent pipe.

According to the EPA, agriculture is the leading source of water quality impairment of surface waters nationwide and is responsible for 72 percent of the water pollution in U.S. rivers. Agricultural practices produce several types of pollutants that contribute to nonpoint source pollution. Fertilizer runoff causes water enrichment. Animal wastes and plant residues in waterways produce high BODs and high levels of suspended solids as well as water enrichment. Highly toxic chemical pesticides may leach into the soil and from there into water or may find their way into waterways by adhering to sediment particles. Soil erosion from fields and rangelands causes sediment pollution in waterways.

Although sewage is the main pollutant produced by cities and towns, municipal water pollution also has a nonpoint source: urban runoff from storm sewers (**FIGURE 10.18**). The water quality of urban runoff from city streets is often worse than that of sewage. Urban runoff carries salt from roadways, untreated garbage, construction sediments, and traffic emissions (via rain that washes pollutants out of the air). It often may contain such contaminants as asbestos, chlorides, copper, cyanides, grease, hydrocarbons, lead, motor oil, organic wastes, phosphates, sulfuric acid, and zinc.

Different industries generate different types of pollutants. Food processing industries produce organic wastes that decompose quickly but have a high BOD. Pulp and paper mills also release wastes with a high BOD and produce toxic compounds and sludge. The paper industry, however, has begun to adopt new manufacturing methods, such as eliminating chlorine as a bleaching agent, that produce significantly less toxic effluents.

Many industries in the United States treat their wastewater with advanced treatment methods. The electronics industry, for example, produces wastewater containing high levels of heavy metals such as copper, lead, and manganese but uses special techniques such as ion exchange and electrolytic recovery to reclaim those heavy metals.

Urban runoff FIGURE 10.18

These pollutants may be carried from storm drains on streets to streams and rivers.

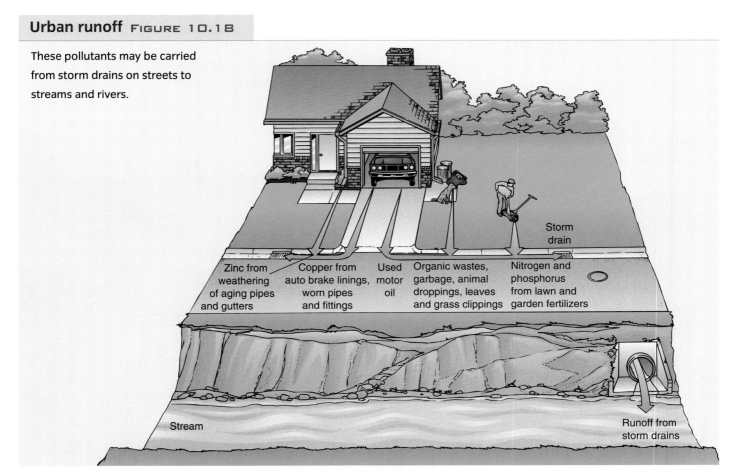

Zinc from weathering of aging pipes and gutters

Copper from auto brake linings, worn pipes and fittings

Used motor oil

Organic wastes, garbage, animal droppings, leaves and grass clippings

Nitrogen and phosphorus from lawn and garden fertilizers

Storm drain

Stream

Runoff from storm drains

GROUNDWATER POLLUTION

Roughly half the people in the United States obtain their drinking water from groundwater, which is also withdrawn for irrigation and industry. In recent years, the quality of the nation's groundwater has become a concern. The most common pollutants, such as pesticides, fertilizers, and organic compounds, seep into groundwater from municipal sanitary landfills, underground storage tanks, backyards, golf courses, and intensively cultivated agricultural lands (FIGURE 10.19).

Currently, most of the groundwater supplies in the United States are of good quality and don't violate standards established to protect human health. However, those areas that do experience local groundwater contamination face quite a challenge. Cleanup of polluted groundwater is costly, takes years, and in some cases is not technically feasible.

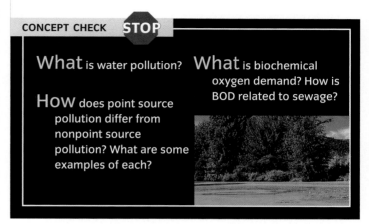

CONCEPT CHECK STOP

What is water pollution?

How does point source pollution differ from nonpoint source pollution? What are some examples of each?

What is biochemical oxygen demand? How is BOD related to sewage?

Sources of groundwater contamination FIGURE 10.19

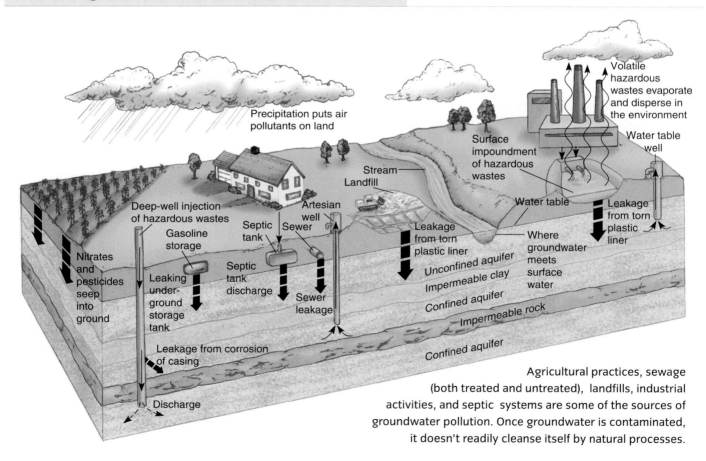

Precipitation puts air pollutants on land

Volatile hazardous wastes evaporate and disperse in the environment

Water table well

Surface impoundment of hazardous wastes

Stream
Landfill

Water table

Leakage from torn plastic liner

Deep-well injection of hazardous wastes

Artesian well

Septic tank   Sewer

Gasoline storage

Leakage from torn plastic liner

Where groundwater meets surface water

Unconfined aquifer

Impermeable clay

Confined aquifer

Nitrates and pesticides seep into ground

Leaking underground storage tank

Septic tank discharge

Sewer leakage

Impermeable rock

Confined aquifer

Leakage from corrosion of casing

Discharge

Agricultural practices, sewage (both treated and untreated), landfills, industrial activities, and septic systems are some of the sources of groundwater pollution. Once groundwater is contaminated, it doesn't readily cleanse itself by natural processes.

Water Pollution   253

Improving Water Quality

LEARNING OBJECTIVES

Describe how most drinking water is purified in the United States.

Distinguish among primary, secondary, and tertiary treatments for wastewater.

Compare the goals of the Safe Drinking Water Act and the Clean Water Act.

ater quality is improved by removing contaminants from the water supply before and after it is used. Technology assists in both processes.

PURIFICATION OF DRINKING WATER

Most U.S. municipal water supplies are treated before being used so the water is safe to drink (**FIGURE 10.20**). Turbid water is treated with a chemical coagulant that causes the suspended particles to clump to-gether and settle out. The water is then filtered through sand to remove remaining suspended materials as well as many microorganisms.

In the final purification step before distribution in the water system, the water is disinfected to kill any remaining disease-causing agents. The most common way to disinfect water is to add chlorine. A small amount of chlorine is left in the water to provide protection during its distribution through many kilometers of pipes. Other disinfection systems use ozone or ultraviolet (UV) radiation in place of chlorine.

Process Diagram

Water treatment for municipal use FIGURE 10.20

www.wiley.co
college/be

A The water supply for a town may be stored in a reservoir, as shown, or obtained from groundwater.

B The water is treated before use so it is safe to drink.

D The quality of the wastewater is fully or partially restored by the sewage treatment before the treated effluent is dispersed into a nearby body of water.

C After use, municipal sewer lines collect the wastewater.

MUNICIPAL SEWAGE TREATMENT

Wastewater, including sewage, usually undergoes several treatments at a sewage treatment plant to prevent environmental and public health problems. The treated wastewater is then discharged into rivers, lakes, or the ocean.

Primary treatment removes suspended and floating particles, such as sand and silt, by mechanical processes such as screening and gravitational settling (FIGURE 10.21, left side). The solid material that settles out at this stage is called **primary sludge**. **Secondary treatment** uses microorganisms (aerobic bacteria) to decompose the suspended organic material in wastewater (FIGURE 10.21, right side). After several hours of processing, the particles and microorganisms are allowed to settle out, forming **secondary sludge**, a slimy mixture of bacteria-laden solids. Water that has undergone primary and secondary treatment is clear and free of organic wastes like sewage.

Even after primary and secondary treatments, wastewater still contains pollutants, such as dissolved minerals, heavy metals, viruses, and organic compounds.

Primary and secondary sewage treatment FIGURE 10.21

www.wiley.com/college/berg

A Raw sewage enters from the municipal sewage system.
B Large debris are removed, and sand settles to the bottom.
C In the primary sedimentation tank, suspended solids sink to the bottom.
D Aeration tanks mix the wastewater with air (oxygen) to support bacteria that consume the suspended organic wastes.
E The cleanest water is taken from the surface and placed in a secondary sedimentation tank, where remaining suspended particles settle.

F The cleanest water is taken from the surface and disinfected by chlorination or ultraviolet light.
G The treated water is discharged to a river or other natural water source.
H Sludge from the primary and secondary sedimentation tanks is pumped to a digester, where bacteria consume the organic wastes.
I The digested sewage sludge is disposed of in a sanitary landfill, incinerated, or converted into fertilizer.

Process Diagram

Advanced wastewater treatment methods, or **tertiary treatment**, include a variety of biological, chemical, and physical processes. Tertiary treatment reduces phosphorus and nitrogen, the nutrients most commonly associated with enrichment, and purifies wastewater for reuse in communities where water is scarce.

Disposal of primary and secondary sludge presents a major problem associated with wastewater treatment. Sludge is generally handled by application to soil as fertilizer, incineration, disposal in a sanitary landfill, or anaerobic digestion. (In anaerobic digestion, bacteria break down the organic material in sludge in the absence of oxygen.)

Some communities have adopted an environmentally innovative and economical approach to wastewater treatment (**FIGURE 10.22**). Beginning in 1978, the small town of Arcata, California, restored and constructed a series of freshwater wetlands in a former industrial area and then routed the town's wastewater through these wetlands. The marshes absorb and assimilate contaminants normally removed through more expensive treatment methods. The highly productive marsh ecosystem—the Arcata Marsh and Wildlife Sanctuary—also provides wildlife habitat for many organisms and opportunities for human recreation.

CONTROLLING WATER POLLUTION

Many governments have passed legislation to control water pollution. Point source pollutants lend themselves to effective control more readily than nonpoint source pollutants.

The two U.S. laws that have the most impact on water quality today are the Safe Drinking Water Act and the Clean Water Act.

The **Safe Drinking Water Act**, passed in 1974, set uniform federal standards for drinking water, to guarantee safe public water supplies throughout the United States. This law required the EPA to determine the **maximum contaminant level**, which is the maximum permissible amount of any water pollutant that might adversely affect human health. The EPA oversees the states to ensure they adhere to the maximum contaminant levels for specific water pollutants. A 1996 amendment to the Safe Drinking Water Act requires municipal water suppliers to tell consumers what contaminants are present in their city's water and whether these contaminants pose a health risk.

The **Clean Water Act** affects the quality of rivers, lakes, aquifers, estuaries, and coastal waters in the United States. Originally passed as the Water Pollution

Checking plant life in the wastewater treatment system in Arcata, California FIGURE 10.22

This constructed wetland is a successful way to treat sewage in a small community like Arcata.

Control Act of 1972, it was amended and renamed the Clean Water Act of 1977; additional amendments were made in 1981 and 1987. The Clean Water Act has two basic goals: to eliminate the discharge of pollutants in U.S. waterways and to attain water quality levels that make these waterways safe to fish and swim in. Under the provisions of this act, the EPA is required to set up and monitor **national emission limitations**, the maximum permissible amounts of water pollutants that can be discharged from a sewage treatment plant, factory, or other point source.

Overall, the Clean Water Act has effectively improved the quality of water from point sources. According to the EPA, nonpoint source pollution is a major cause of water pollution yet is much more difficult and expensive to control than point source pollution. The 1987 amendments to the Clean Water Act expanded regulations on nonpoint sources.

The United States has improved its water quality in the past several decades, thereby demonstrating that the environment recovers once pollutants are eliminated. Much remains to be done, however. The EPA's 2002 *National Water Quality Inventory* indicated that water pollution has increased in U.S. rivers, lakes, estuaries, and coastal areas in recent years. According to the report, 39 percent of the nation's rivers, 45 percent of its lakes, and 51 percent of its estuaries are too polluted for swimming, fishing, or drinking.

Preventing water pollution at home

Although individuals produce little water pollution, the collective effect of municipal water pollution, even in a small neighborhood, can be quite large. There are many things you can do to protect surface waters and groundwater from water pollution (see TABLE 10.3).

Preventing water pollution at home   TABLE 10.3

Location	What you can do
Bathroom	Never throw unwanted medicines down the toilet.
Kitchen	Use smallest effective amount of toxic household chemicals like oven cleaners, mothballs, drain cleaners, and paint thinners. Substitute less hazardous chemicals wherever possible. Dispose of unwanted hazardous household chemicals at hazardous waste collection centers.
Driveway/Car	Never pour used motor oil or antifreeze down storm drains or on ground. Recycle these chemicals at service station or local hazardous waste collection center. Clean up spilled oil, brake fluid, and antifreeze, and sweep sidewalks and driveways instead of hosing them off. Dispose of dirt properly; don't sweep it into gutters or storm drains. Drive less: air pollution emissions from automobiles eventually get into groundwater and surface water. Toxic metals and oil byproducts deposited on road by automobiles are washed into surface waters by precipitation.
Lawn and garden	Pick up pet waste and dispose of it in garbage or toilet. If left on ground, it eventually washes into waterways where it can contaminate shellfish and enrich water. Replace some of grass lawn with trees, shrubs, and ground covers, which absorb up to 14 times more precipitation and require little or no fertilizer. To reduce erosion, use mulch to cover bare ground. Use fertilizer sparingly; excess fertilizer leaches into groundwater or waterways. Never apply fertilizer near surface water. Make sure that gutters and downspouts drain onto water-absorbing grass or graveled areas instead of paved surfaces.

Water Pollution in the Great Lakes

North America's Great Lakes—Superior, Michigan, Huron, Erie, and Ontario—collectively hold about one-fifth of the world's fresh surface water (FIGURE A). More than 37 million people live in the Great Lakes watershed, an area home to agriculture, trade, industry, and tourism, and at least 26 million people obtain their drinking water from the Great Lakes.

Industrial wastes, sewage, fertilizers, and other pollutants have contaminated the Great Lakes since the mid-1800s. By the 1960s, pollution in the Great Lakes became a highly visible problem. Thousands of toxic chemicals polluted the lakes, eutrophication was pronounced, high bacterial counts caused health hazards, and fish kills were common. Birth defects were observed in almost 50 percent of the animal species studied.

Canada and the United States have cooperated to improve the condition of the Great Lakes. Since a joint pollution control program was enacted in 1972, $20 billion has been spent on cleanup. As of the early 2000s, the Great Lakes are in better shape than they have been at any other time in recent memory. Levels of dichlorodiphenyltrichloroethane (DDT) in women's breast milk have declined, as have levels of polychlorinated biphenyls (PCBs) in trout, Coho salmon, and herring gulls. Some affected animal populations, such as double-crested cormorants and bald eagles, have rebounded.

However, many problems remain. Zebra mussels, sea lampreys, and at least 160 other invasive aquatic species have proliferated and threaten native species. Shoreline development continues to encroach on natural habitats and contributes to flooding and shoreline erosion during storms. Pesticides and fertilizers from suburban lawns pollute the water.

Persistent toxic compounds, such as mercury and PCBs, remain in the lakes, many of them coming from air pollution or from contaminated lake sediments. The presence of certain toxic chemicals is causing hormonal changes linked to reproductive failures, abnormal development, and abnormal behaviors in some fishes, birds, and mammals (recall the discussion of endocrine disrupters in Chapter 4). Because persistent toxic chemicals have accumulated in food webs, health advisories warn people of the potential risks associated with eating Great Lakes fishes (FIGURE B).

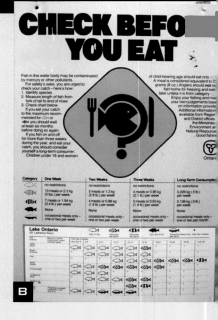

B

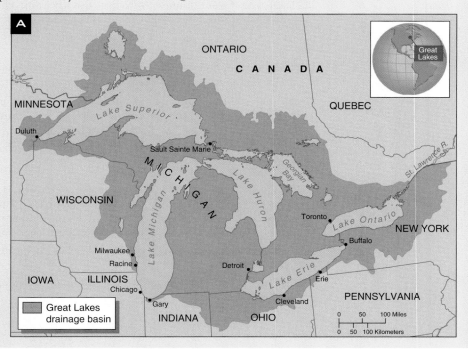

A

Water pollution in developing countries

According to the World Health Organization, an estimated 1.4 billion people don't have access to safe drinking water, and about 2.9 billion people don't have access to adequate sanitation systems; most of these people live in rural areas of developing countries. Worldwide, at least 250 million cases of water-related illnesses occur each year, with 5 million or more of these resulting in death.

Municipal water pollution from sewage is a greater problem in developing countries, many of which lack water treatment facilities, than in highly developed nations. Sewage from many densely populated cities in Asia, Latin America, and Africa is dumped directly into rivers or coastal harbors (**FIGURE 10.23**). Other serious sources of water pollution in developing countries include industrial wastes, agricultural chemicals, and even human remains.

Ganges River FIGURE 10.23

Bathing and washing clothes in the Ganges River are common practices in India. The river is contaminated by raw sewage discharged directly into the river at many different locations.

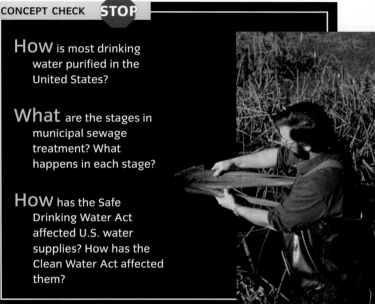

CONCEPT CHECK STOP

HOW is most drinking water purified in the United States?

What are the stages in municipal sewage treatment? What happens in each stage?

HOW has the Safe Drinking Water Act affected U.S. water supplies? How has the Clean Water Act affected them?

CHAPTER SUMMARY

1 The Importance of Water

1. Water molecules are **polar**: the negatively charged (oxygen) end of one molecule is attracted to the positively charged (hydrogen) end of another molecule, forming a **hydrogen bond**.

2. Hydrogen bonds are the basis for many of water's properties, including its high melting point, high boiling point, high heat capacity, and dissolving ability.

3. In the **hydrologic cycle**, water continuously circulates through the abiotic environment. **Surface water** is precipitation that remains on the surface. **Runoff** is the movement of fresh water from precipitation and snowmelt to rivers, lakes, wetlands, and the ocean. **Groundwater** is the supply of freshwater that is stored in **aquifers**, underground reservoirs.

CHAPTER SUMMARY

2 Water Resource Problems

1. Overdrawing surface water causes **wetlands** to dry up and **estuaries** to become saltier. **Aquifer depletion** is the removal of groundwater faster than it can be recharged. **Saltwater intrusion** is the movement of seawater into a freshwater aquifer near the coast. **Salinization** is the gradual accumulation of salt in soil, often due to improper irrigation.

3 Water Management

1. **Sustainable water use** is the wise use of water resources, without harming the hydrologic cycle or the ecosystems on which humans depend.

2. Dams and reservoirs allow rivers to be tapped for hydroelectric power and

5 Improving Water Quality

1. Most U.S. municipal water supplies are treated so the water is safe to drink. A chemical coagulant traps suspended particles, filtration removes suspended materials and microorganisms, and disinfection kills disease-causing agents.

2. Wastewater usually undergoes several treatments at a sewage treatment plant. **Primary treatment** removes suspended

2. Farmers on the U.S. High Plains are depleting water from the **Ogallala Aquifer** much faster than nature replaces it. In the Colorado River basin, rapid population growth upstream threatens the water supply of users downstream.

3. Most of the world's major watersheds are shared between at least two nations. International cooperation is often required to manage shared water use.

used to supply municipal and industrial water, but they are expensive to build and alter the natural environment.

3. **Microirrigation** is an innovative type of irrigation that conserves water by piping it to crops through sealed systems. Industries and cities can employ measures to recapture, purify, and reuse water in homes and buildings.

and floating particles from wastewater by mechanical processes. **Secondary treatment**, which reduces water's biochemical oxygen demand, treats wastewater biologically to decompose suspended organic material. **Tertiary treatment** reduces phosphorus and nitrogen.

3. The **Safe Drinking Water Act** protects the safety of the nation's drinking water. The **Clean Water Act** affects the quality of U.S. rivers, lakes, aquifers, estuaries, and coastal waters.

4 Water Pollution

1. **Water pollution** is any physical or chemical change in water that adversely affects the health of humans and other organisms.

2. **Sewage** is the release of wastewater from drains or sewers. It carries disease-causing agents and causes **enrichment**, the fertilization of a body of water due to high levels of nutrients such as nitrogen and phosphorus. **Eutrophication** is the enrichment of a lake, estuary, or slow-flowing stream. **Artificial eutrophication** is overnourishment of an aquatic ecosystem by nutrients due to human activities such as agriculture and sewage discharge. Sewage in water also raises the **biochemical oxygen demand (BOD)**, the amount of oxygen needed by microorganisms to decompose biological wastes. A high BOD decreases water quality by reducing levels of dissolved oxygen.

3. Sources of water pollution are classified into two types. **Point source pollution** is water pollution that can be traced to a specific spot, such as wastewater released from a factory or sewage treatment plant. **Nonpoint source pollution** consists of pollutants that enter bodies of water over large areas rather than at a single point of entry. Examples include agricultural runoff, mining wastes, municipal wastes, construction sediments, and soil erosion.

KEY TERMS

- surface water p. 237
- runoff p. 237
- groundwater p. 237
- aquifer depletion p. 240
- saltwater intrusion p. 240
- salinization p. 242
- sustainable water use p. 244
- microirrigation p. 246
- water pollution p. 248
- sewage p. 248
- biochemical oxygen demand (BOD) p. 248
- artificial eutrophication p. 251
- point source pollution p. 251
- nonpoint source pollution p. 251
- primary treatment p. 255
- secondary treatment p. 255
- tertiary treatment p. 256

1. What issues complicate water rights in the Missouri River?

2. Are our water supply problems largely the result of too many people? Give reasons for your answer.

3. Briefly describe the complexity of international water use, using the Rhine River or the Aral Sea as an example.

4. Outline a brief water conservation plan for your own daily use. How could you use water more sustainably?

5. What is water pollution? Why is wastewater treatment an important part of sustainable water use?

6. Explain why untreated sewage may kill fishes when it is added directly to a body of water.

7. Tell whether each of the following represents point or nonpoint source pollution: fertilizer runoff from farms, thermal pollution from a power plant, urban runoff, sewage from a ship, and erosion sediments from deforestation. Why is nonpoint pollution more difficult to control than point source pollution?

8. Is the Clean Water Act related in any way to the quality of U.S. public drinking water? Explain your answer.

9–10. The graph reflects the monitoring of dissolved oxygen concentrations at six stations along a river. The stations are located 20 m apart, with A the farthest upstream and F the farthest downstream.

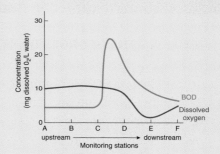

9. Where along the river did a sewage spill occur?

10. At which station would you most likely discover dead fish?

- Rain soaks the streets of New York City. Where does the rainwater go?

- Many materials dissolve in water. What types of pollution might be dissolved in or carried by the rainwater?

- Is street runoff an example of point source pollution or nonpoint source pollution?

The Ocean and Fisheries

11

CLOSING THE GEORGES BANK FISHERY

In 1994, the U.S. Commerce Department closed two large portions of Georges Bank, a vast area off the coast of New England in the North Atlantic Ocean, which was once one of the world's richest fishing grounds *(see inset)*. These closures did not surprise any observer of the fishing industry. Catches of cod, haddock, and yellowtail flounder had been steadily declining there for several decades, primarily due to *overfishing*, the harvesting of fishes faster than they can reproduce *(see larger photo)*. The closure of the Georges Bank fishery resulted in the loss of jobs for thousands of people. Since the closures, fish populations have increased. Fishery experts, however, do not know how long it will take for Georges Bank to recover enough to be reopened for large commercial harvests.

The overfishing of Georges Bank is not unique. According to the National Research Council, 80 percent of nearly 200 commercially important fish stocks in the United States are fully exploited or overfished. Worldwide, about 30 percent of fish species have been overfished. Overfishing occurs because the world demand for fish has grown. To meet the demand, there are more large fishing boats than ever before, often doing their work in large international fleets. In addition, more and more fishermen are using high-tech methods to find and harvest greater quantities of fishes.

The closing of the Georges Bank fishery teaches us that as our technologies (like more effective fishing techniques) advance, so do the impacts we have on our environment. As a result, we must regulate these impacts (like the overharvesting of fishes) more closely than ever.

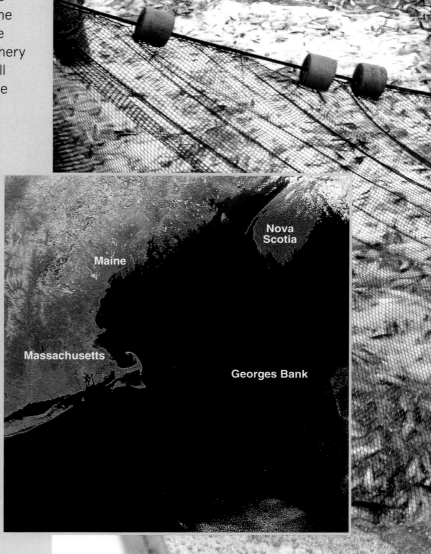

262

CHAPTER OUT

The Global Ocean p. 264

Major Ocean Life Zones p. 268

Human Impacts on the Ocean p.

Addressing Ocean Problems p.

Case Study: Humans and the Antarctic Food Web p. 283

The Global Ocean

The ocean is a vast wilderness, much of it unknown. It teems with life—from warm-blooded mammals like whales to soft-bodied invertebrates like jellyfish. The ocean is essential to the hydrologic cycle that provides us with water. It affects cycles of matter on land, influences our climate and weather, and provides foods that enable millions of people to survive. The ocean dominates Earth, and its condition determines the future of life on our planet. If the ocean dies, then we do as well.

The global ocean is a huge body of salt water that surrounds the continents and covers almost three-fourths of Earth's surface. It is a single, continuous body of water, but geographers divide it into four sections separated by the continents: the Pacific, Atlantic, Indian, and Arctic Oceans. The Pacific is the largest: It covers one-third of Earth's surface and contains more than half of Earth's water.

PATTERNS OF CIRCULATION IN THE OCEAN

The persistent prevailing winds blowing over the ocean produce currents, mass movements of surface–ocean water (**FIGURE 11.1** in Visualizing Ocean Currents). The prevailing winds generate **gyres**, circular ocean currents. In the North Atlantic Ocean, the tropical trade winds tend to blow toward the west, whereas the westerlies in the midlatitudes blow toward the east. This helps establish a clockwise gyre in the North Atlantic.

The **Coriolis effect** (also see Chapter 8, page 189) influences the paths of surface, or shallow, ocean currents just as it does the winds. Earth's rotation from west to east causes surface ocean currents to swerve to the right in the Northern Hemisphere, helping establish the circular, clockwise pattern of water currents. In the Southern Hemisphere, ocean currents swerve to the left, thereby moving in a circular, counterclockwise pattern.

gyres Large, circular ocean current systems that often encompass an entire ocean basin.

Vertical mixing of ocean water

Variations in the **density** (mass per unit volume) of seawater affect deep ocean currents. Cold, salty water is denser than warmer, less salty water. Colder, salty ocean water sinks and flows under warmer, less salty water, generating currents far below the surface. Deep ocean currents often travel in different directions and at different speeds than do surface currents, in part because the Coriolis effect is more pronounced at greater depths. **FIGURE 11.2** shows the present circulation of shallow and deep currents—the **ocean conveyor belt**—that moves cold, salty deep-sea water from higher to lower latitudes, where it warms up. Note that the Atlantic Ocean gets its cold deep water from the Arctic Ocean, whereas the Pacific and Indian Oceans get theirs from the water surrounding Antarctica.

As the Gulf Stream and North Atlantic Drift push into the North Atlantic (see Figure 11.1), they deliver an immense amount of heat from the tropics to Europe. As this shallow current transfers its heat to the atmosphere, the water becomes denser and sinks. The deep current flowing southward in the North Atlantic is, on average, 8°C cooler than the shallow current flowing northward.

Evidence from seafloor sediments and Greenland ice indicates that the ocean conveyor belt is not unchanging but shifts from one equilibrium state to another in a relatively short period (a few years to a few decades).

Surface ocean currents FIGURE 11.1

Winds largely cause the basic pattern of ocean currents. The main ocean current flow—clockwise in the Northern Hemisphere and counterclockwise in the Southern Hemisphere—results partly from the Coriolis effect.

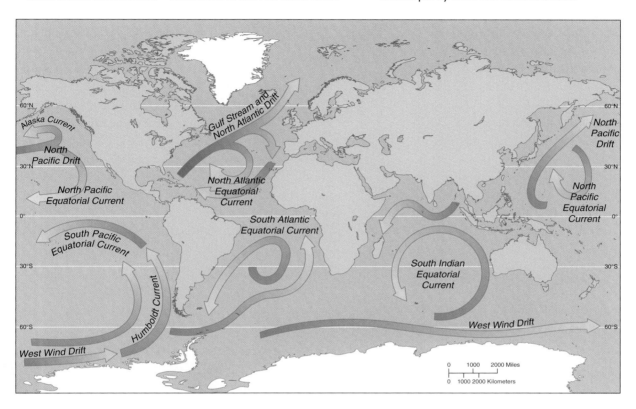

Ocean conveyor belt
FIGURE 11.2

This loop consists of both warm, shallow water and cold, deep water. The ocean conveyor belt affects regional and possibly global climate.

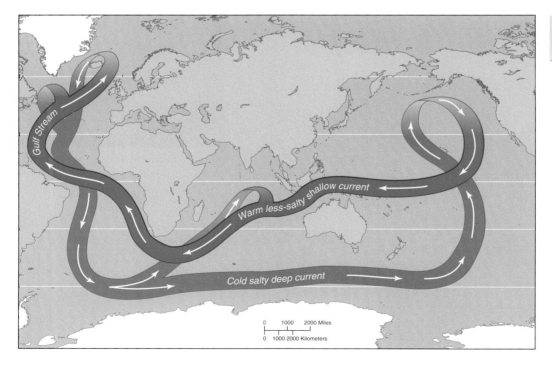

OCEAN–ATMOSPHERE INTERACTION

The ocean and the atmosphere are strongly linked, with wind from the atmosphere affecting the ocean currents and heat from the ocean affecting atmospheric circulation. One of the best examples of the interaction between ocean and atmosphere is the **El Niño–Southern Oscillation (ENSO)** event, which is responsible for much of Earth's interannual (from one year to the next) climate variability (FIGURE 11.3A). Normally, westward-blowing trade winds restrict the warmest waters to the western Pacific near Australia. Every 3 to 7 years, however, the trade winds weaken, and the warm mass of water expands eastward to South America, increasing surface temperatures in the usually cooler east Pacific (FIGURE

El Niño–Southern Oscillation (ENSO) A periodic, large-scale warming of surface waters of the tropical eastern Pacific Ocean that temporarily alters both ocean and atmospheric circulation patterns.

11.3B). Ocean currents, which normally flow westward in this area, slow down, stop altogether, or even reverse and go eastward. The name for this phenomenon, El Niño (in Spanish, "the boy child"), refers to the Christ child: the warming usually reaches the fishing grounds off Peru just before Christmas. Most ENSOs last between 1 and 2 years.

ENSO devastates the fisheries off South America. Normally, the colder, nutrient-rich deep water is about 40 m (130 ft) below the surface and **upwells** (comes to the surface) along the coast, partly in response to strong trade winds (FIGURE 11.4A). During an ENSO event, however, the deep water is about 152 m (500 ft) below the surface, and the warmer surface temperatures and weak trade winds prevent upwelling (FIGURE 11.4B). The lack of

ENSO FIGURE 11.3

A El Niño–Southern Oscillation (ENSO) events drastically alter the climate, even in many areas far from the Pacific Ocean. As a result of ENSO, some areas are drier, some wetter, some cooler, and some warmer than usual. Typically, northern areas of the contiguous United States are warmer during the winter, whereas southern areas are cooler and wetter.

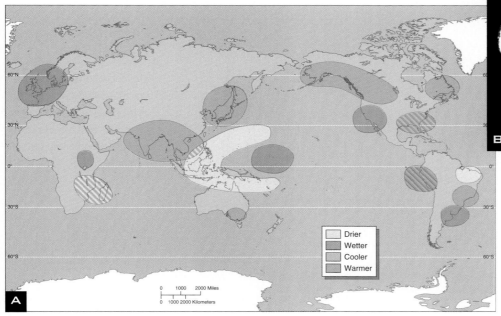

Drier
Wetter
Cooler
Warmer

B ENSO-driven warmer waters off western South America.

nutrients in the water results in a severe decrease in the populations of anchovies and many other fishes. During the 1982–83 El Niño, one of the worst ever recorded, the anchovy population decreased 99%. Other species such as shrimp and scallops thrive during an ENSO event.

ENSO alters global air currents, directing unusual, sometimes dangerous, weather to areas far from the tropical Pacific where it originates. By one estimate, the 1997–98 ENSO, the strongest on record, caused more than 20,000 deaths and $33 billion in property damage worldwide. It resulted in heavy snows in parts of the western United States; ice storms in eastern Canada; torrential rains that flooded Peru, Ecuador, California, Arizona, and Western Europe; and droughts in Texas, Australia, and Indonesia. An ENSO-caused drought—the worst in 50 years—particularly hurt In-

donesia. Fires that had been deliberately set to clear land for agriculture got out of control and burned an area in Indonesia the size of New Jersey.

Climate scientists observe and monitor sea surface temperatures and winds to better understand and predict the timing and severity of ENSO events. The **TAO/TRITON array** consists of 70 moored buoys in the tropical Pacific Ocean. These instruments collect oceanic and weather data during normal conditions and El Niño events. The data are transmitted to scientists onshore by satellite.

Scientists at the National Oceanic and Atmospheric Administration's Climate Prediction Center forecasted the 1997–98 ENSO six months in advance using data from TAO/TRITON. Such forecasts give governments time to prepare for the extreme weather changes associated with ENSO.

Upwelling FIGURE 11.4

A Coastal upwelling, where deeper waters come to the surface, occurs in the Pacific Ocean along the South American coast. Upwelling provides nutrients for microscopic algae, which in turn support a complex food web.

B Coastal upwelling weakens considerably during years with El Niño–Southern Oscillation (ENSO) events, temporarily reducing fish populations.

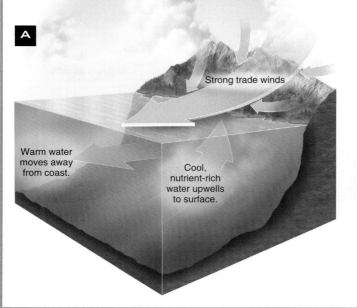

A

Strong trade winds

Warm water moves away from coast.

Cool, nutrient-rich water upwells to surface.

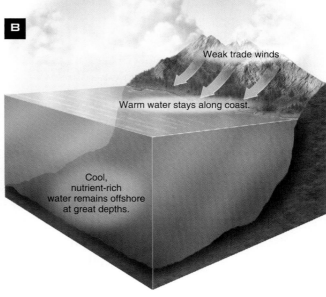

B

Weak trade winds

Warm water stays along coast.

Cool, nutrient-rich water remains offshore at great depths.

La Niña   El Niño isn't the only periodic ocean temperature event to affect the tropical Pacific Ocean. **La Niña** (in Spanish, "the girl child") occurs when the surface water temperature in the eastern Pacific Ocean becomes unusually cool and westbound trade winds become unusually strong. La Niña often occurs after an El Niño event and is considered part of the natural oscillation of ocean temperature.

During the spring of 1998, the surface water of the eastern Pacific cooled 6.7°C (12°F) in just 20 days. Like ENSO, La Niña affects weather patterns around the world, but its effects are more difficult to predict. In the contiguous United States, La Niña typically causes wetter-than-usual winters in the Pacific Northwest, warmer weather in the Southeast, and drought conditions in the Southwest. Atlantic hurricanes are stronger and more numerous during a La Niña event.

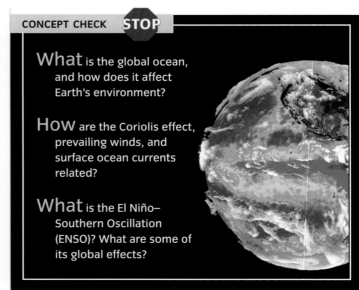

CONCEPT CHECK   **STOP**

What is the global ocean, and how does it affect Earth's environment?

How are the Coriolis effect, prevailing winds, and surface ocean currents related?

What is the El Niño–Southern Oscillation (ENSO)? What are some of its global effects?

Major Ocean Life Zones

LEARNING OBJECTIVE

Describe and distinguish among the four main ocean life zones.

The immense marine environment is subdivided into several zones (**FIGURE 11.5**):

- The intertidal zone.

- The benthic (ocean floor) environment.

- The two provinces—neritic and oceanic—of the pelagic (ocean water) environment.

The *neritic province* is that part of the pelagic environment from the shore to where the water reaches a depth of 200 m (650 ft). The *oceanic province* is that part of the pelagic environment where the water depth is greater than 200 m.

intertidal zone
The area of shoreline between low and high tides.

THE INTERTIDAL ZONE: TRANSITION BETWEEN LAND AND OCEAN

Although high levels of light, nutrients, and oxygen make the **intertidal zone** a biologically productive habitat, it is a stressful one. On sandy intertidal beaches, inhabitants must contend with a constantly shifting environment that threatens to engulf them and gives them little protection against wave action.

Rocky shores provide fine anchorage for seaweeds and marine animals, but they are exposed to wave action when submerged during high tides and exposed to temperature changes and drying out when in contact with the air during low tides (**FIGURE 11.6**).

A rocky-shore inhabitant generally has some way of sealing in moisture, perhaps by closing its shell (if it has one), and a means of anchoring itself to the rocks. For example, mussels have tough, threadlike anchors secreted by a gland in the foot, and barnacles secrete a tightly bonding glue that hardens under water. Some organisms hide in burrows or under rocks or crevices at low tide. Some small crabs run about the splash line, following it up and down the beach.

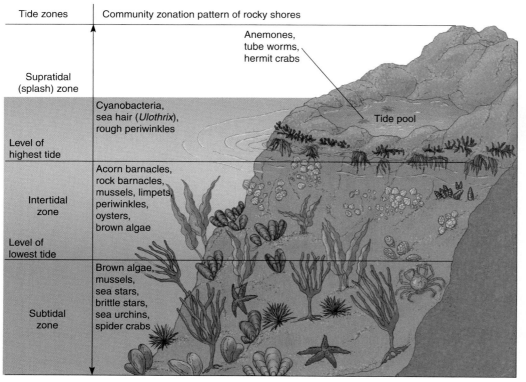

Zonation in the ocean FIGURE 11.5

The intertidal zone, the benthic environment, and the pelagic environment make up the ocean. The pelagic environment consists of the neritic and oceanic provinces. (The slopes of the ocean floor aren't as steep as shown; they are exaggerated to save space.)

Intertidal zone:

Rockweed (brown algae)

Shallow benthic zone:

American lobster

Neritic province:

Bottlenose dolphins

Oceanic province:

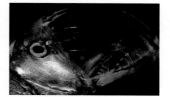

Sabertooth viperfish

www.wiley.com/college/berg

Zonation along a rocky shore FIGURE 11.6

Three zones are shown: the supratidal, or "splash" zone, which is never fully submerged; the intertidal zone, which is fully submerged at high tide; and the subtidal zone (part of the benthic environment), which is always submerged. Representative organisms are given for each of these zones.

Coral reefs FIGURE 11.7

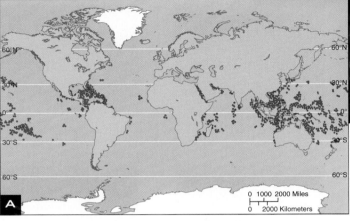

A This map shows the distribution of coral reefs around the world. There are more than 6,000 of them worldwide.

B A coral reef in Fiji has a variety of soft corals as well as basslets and yellow butterfly fish.

THE BENTHIC ENVIRONMENT

Most of the **benthic environment** consists of sediments (mainly sand and mud) where many bottom-dwelling animals, such as worms and clams, burrow. Bacteria are common in marine sediments and have even been reported in ocean sediments more than 500 m (1625 ft) below the ocean floor. The communities in the relatively shallow benthic zone that are particularly productive include coral reefs, seagrass beds, and kelp forests.

Corals are small, soft-bodied animals similar to jellyfish and sea anemones. Corals live in hard cups, or shells, of limestone (calcium carbonate) that they produce using the minerals dissolved in ocean water. When the coral animals die, the tiny cups remain, and a new generation of coral animals grows on top of these. Over thousands of generations, a **coral reef** forms from the accumulated layers of limestone.

Coral reefs are found in warm (usually greater than 21°C), shallow seawater (**FIGURE 11.7A**). The living portions of coral reefs grow in shallow waters where light penetrates. The tiny coral animals require light for **zooxanthellae** (symbiotic algae) that live and photosynthesize in their tissues. In addition to obtaining food from the zooxanthellae that live inside them, coral animals capture food at night with stinging tentacles that paralyze **plankton** (small or microscopic organisms carried by currents and waves) and small animals that drift nearby. The waters where coral reefs grow are often poor in nutrients, but other factors are favorable for high productivity, including the presence of zooxanthellae, favorable temperatures, and year-round sunlight.

Coral reef ecosystems are the most diverse of all marine environments (**FIGURE 11.7B**). They contain

> **benthic environment**
> The ocean floor, which extends from the intertidal zone to the deep ocean trenches.

Sea grass bed FIGURE 11.8

Turtle grasses *(Thalassia testudinum)* form underwater meadows that are ecologically important for shelter and food for many organisms. Photographed in the Caribbean Sea off the coast of Belize.

hundreds of species of fishes and invertebrates, such as giant clams, snails, sea urchins, sea stars, sponges, flatworms, brittle stars, sea fans, shrimp, and spiny lobsters. The Great Barrier Reef occupies only 0.1 percent of the ocean's surface, but 8 percent of the world's fish species live there. The multitude of relationships and interactions that occur at coral reefs is comparable only to those of the tropical rain forest. As in the rain forest, competition is intense, particularly for light and space to grow.

Coral reefs are ecologically important because they both provide habitat for many kinds of marine organisms and protect coastlines from shoreline erosion. They provide humans with seafood, pharmaceuticals, and recreation and tourism dollars.

Sea grasses are flowering plants adapted to complete submersion in salty ocean water. They only occur in shallow water (to depths of 10 m, or 33 ft) where they receive enough light to photosynthesize efficiently. Extensive beds of sea grasses occur in quiet temperate, subtrop-

ical, and tropical waters. Eelgrass is the most widely distributed sea grass along the coast of North America; the largest eelgrass bed in the world is in Izembek Lagoon on the Alaska Peninsula. The most common sea grasses in the Caribbean Sea are manatee grass and turtle grass (FIGURE 11.8). Sea grasses have a high primary productivity and are ecologically important in shallow marine areas. Their roots and rhizomes help stabilize sediments, thereby reducing erosion, and they provide food and habitat for many marine organisms.

In temperate waters, ducks and geese eat sea grasses, whereas in tropical waters, manatees, green turtles, parrot fish, sturgeon fish, and sea urchins eat them. These herbivores consume only about 5% of the sea grasses. The remaining 95% eventually enters the detritus food web and is decomposed when the sea grasses die. The decomposing bacteria are in turn consumed by a variety of animals such as mud shrimp, lugworms, and mullet (a type of fish).

Kelps, which may reach lengths of 60 m (200 ft), are the largest and most complex of all algae commonly called seaweeds (FIGURE 11.9). Kelps, which are brown algae, are common in cooler temperate marine waters of both the Northern Hemisphere and Southern Hemisphere. They are especially abundant in relatively shallow waters (depths of about 25 m, or 82 ft) along rocky coastlines. Kelps are photosynthetic and are the primary food producers for the kelp "forest" ecosystem. Kelp forests provide habitats for many marine animals, such as tubeworms, sponges, sea cucumbers, clams, crabs, fishes, and sea otters.

Some animals eat the fronds, but kelps are mainly consumed in the detritus food web. Bacteria that decompose kelp provide food for sponges, tunicates, worms, clams, and snails. The diversity of life supported by kelp beds almost rivals that found in coral reefs.

Kelp forest FIGURE 11.9

These underwater forests are ecologically important because they support many kinds of aquatic organisms. Photographed off the coast of California.

EnviroDiscovery   Otters in Trouble

Sea otters play an important role in their environment. They feed on sea urchins, thereby preventing the urchins from eating kelp, which allows kelp forests to thrive. Now scientists have uncovered an alarming decline in sea otter populations in western Alaska's Aleutian Islands—a stunning 90 percent crash since 1990—that in turn poses wide-ranging threats to the coastal ecosystem there. The population of sea urchins in these areas is exploding, and kelp forests are being devastated. Strong evidence identifies killer whales, or orcas, as the culprits. Recently, orcas were observed for the first time preying on sea otters. Orcas generally feed on sea lions, seals, and fishes of all sizes. In comparison, sea otters, the smallest marine mammal species, are more like a snack rather than a desirable meal. So why are the orcas now choosing sea otters? Most likely the reason is that seal and sea lion populations have collapsed across the north Pacific.

In a scenario partly documented and partly speculative, the starting point of this disastrous chain of events is a drop in fish stocks, possibly caused by overfishing or climate change. With their food fish in decline, seal and sea lion populations have suffered, and orcas have looked elsewhere for food. Such a change in the orcas' feeding behavior transforms the food chain of kelp forests, putting orcas rather than otters at the top, and linking coastal and oceanic ecosystems off the Alaskan coast.

Otters in Alaskan waters

THE NERITIC PROVINCE: FROM THE SHORE TO 200 METERS

Organisms that live in the pelagic environment's **neritic province** are all floaters or swimmers. The upper level of the pelagic environment is the **euphotic zone**, which extends from the surface to a maximum depth of 150 m (488 ft) in the clearest open ocean water. Sufficient light penetrates the euphotic zone to support photosynthesis.

> **neritic province**
> The part of the pelagic environment that overlies the ocean floor from the shoreline to a depth of 200 m (650 ft).

Large numbers of phytoplankton (microscopic algae) produce food by photosynthesis and are the base of food webs. Zooplankton, including tiny crustaceans, jellyfish, comb jellies, and the larvae of barnacles, sea urchins, worms, and crabs, feed on phytoplankton. Zooplankton are in turn consumed by plankton-eating nekton (any marine organism that swims freely), such as herring, sardines, squid, baleen whales, and manta rays (Figure 11.10). These in turn become prey for carnivorous nekton such as sharks, tuna, porpoises, and toothed whales. Nekton are mostly confined to the shallower neritic waters (less than 60 m, or 195 ft, deep), near their food.

Neritic province FIGURE 11.10

A manta ray swims slowly through the water, swallowing vast quantities of microscopic plankton as it swims. The wingspan of a mature manta ray can reach about 6 m (20 ft) across. Note the remoras that are hitching a ride.

THE OCEANIC PROVINCE: MOST OF THE OCEAN

The **oceanic province** is the largest marine environment, representing about 75 percent of the ocean's water. Most of the oceanic province is loosely described as the "deep sea." (The average depth of the ocean is 4000 m, more than 2 mi.) All but the surface waters of the oceanic province have cold temperatures, high pressure, and an absence of sunlight. These environmental conditions are uniform throughout the year.

> **oceanic province** The part of the pelagic environment that overlies the ocean floor at depths greater than 200 m (650 ft).

Fishes of the deep waters of the oceanic province are strikingly adapted to darkness and scarcity of food (**FIGURE 11.11**). Adapted to drifting or slow swimming, animals of the oceanic province

Oceanic province FIGURE 11.11

The gulper eel is highly specialized for its deep-sea habitat. It uses its "trap-door" jaws to swallow prey as large as itself. Gulper eels grow to 1.8 m (6 ft). Shown is a live specimen, photographed in an aquarium aboard ship after being captured at a depth of 1,500 m (4,921 ft) off Southern California.

often have reduced bone and muscle mass. Many of these animals have light-producing organs to locate one another for mating or food capture.

Most organisms of the deep waters of the oceanic province depend on **marine snow**, organic debris that drifts down into their habitat from the upper, lighted regions of the oceanic province. Organisms of this little-known realm are filter feeders, scavengers, or predators. Many are invertebrates, some of which attain great sizes. The giant squid measures up to 18 m (59 ft) in length, including its tentacles.

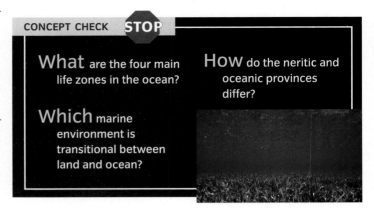

CONCEPT CHECK STOP

What are the four main life zones in the ocean?

Which marine environment is transitional between land and ocean?

How do the neritic and oceanic provinces differ?

Human Impacts on the Ocean

LEARNING OBJECTIVES

Contrast fishing and aquaculture and relate the environmental challenges of each activity.

Identify the human activities that contribute to marine pollution and describe their effects.

Explain how global warming could potentially alter the ocean conveyor belt.

The ocean is so vast it's hard to imagine that human activities could harm it. Such is the case, however. Fisheries and aquaculture, marine shipping, marine pollution, coastal development, offshore mining, and global warming all contribute to the degradation of marine environments.

MARINE POLLUTION AND DETERIORATING HABITAT

One of the great paradoxes of human civilization is that the same ocean that provides food to a hungry world is used as a dumping ground. Coastal and marine ecosystems receive pollution from land, from rivers emptying into the ocean, and from atmospheric contaminants that enter the ocean via precipitation. Offshore mining and oil drilling pollute the neritic province with oil and other contaminants. Pollution increasingly threatens the world's fisheries. Everything from accidental oil spills to the deliberate dumping of litter pollutes the water. The World Resources Institute estimates that about 80 percent of global ocean pollution comes from human activities on land. In 2003 the Pew Oceans Commission, composed of scientists, economists, fishermen, and other experts, verified the seriousness of ocean problems in a series of studies. Some of their findings are shown in **FIGURE 11.12**.

WORLD FISHERIES

The ocean contains a valuable food resource. About 90 percent of the world's total marine catch is fishes, with clams, oysters, squid, octopus, and other mollusks representing an additional 6 percent of the total catch. Crustaceans, including lobsters, shrimp, and crabs, make up about 3 percent, and marine algae constitute the remaining 1 percent.

Fleets of deep-sea fishing vessels obtain most of the world's marine harvest. Numerous fishes are also captured in shallow coastal waters and inland waters. According to the U.N. Food and Agricultural Organiza-

Overfishing
Example: The populations of many commercial fish species are severely depleted.

Bycatch
Example: Fishermen unintentionally kill dolphins, sea turtles, and sea birds.

Aquaculture
Example: Produces wastes that can pollute ocean water and harm marine organisms.

Point Source Pollution
Example: Passenger cruise ships dump sewage, shower and sink water, and oily bilge water.

Habitat Destruction
Example: Trawl nets (fishing equipment pulled along the ocean floor) destroy habitat.

Invasive Species
Example: Release of ships' ballast water, which contains foreign crabs, mussels, worms, and fishes.

Nonpoint Source Pollution (runoff from land)
Example: Agricultural runoff (fertilizers, pesticides, and livestock wastes) pollutes water.

Coastal Development
Example: Developers destroy important coastal habitat, such as salt marshes and mangrove swamps.

Climate Change
Example: Coral reefs and polar seas are particularly vulnerable to increasing temperatures.

An overview of major threats to the ocean FIGURE 11.12

www.wiley.com/college/berg

tion (FAO), the world annual fish harvest increased substantially from 19 million tons in 1950 to 133 million tons in 2003, the latest year for which data are available.

Problems and challenges for the fishing industry

No nation lays legal claim to the open ocean. Consequently, resources in the ocean are more susceptible to overuse and degradation than land resources, which individual nations own and for which they feel responsible (see the Tragedy of the Commons section in Chapter 2, page 24).

The most serious problem for marine fisheries is that many species, particularly large predatory fish, have been harvested to the point that their numbers are severely depleted. This generally causes the fishery to become unusable for commercial or sport fishermen, let alone the other marine species that rely on it as part of the food web. Scientists have found that dramatically depleted fish populations recover only slowly. Some show no real increase in population size up to 15 years after the fishery has collapsed.

According to the FAO, 62 percent of the world's fish stocks are in urgent need of management action. Fisheries have experienced such pressure for two reasons. First, the growing human population requires protein in its diets, leading to a greater demand for fish. Second, technological advances allow us to fish so efficiently that every single fish is often removed from an area (see What a Scientist Sees).

Fishermen tend to concentrate on a few fish species with high commercial value, such as menhaden, salmon, tuna, and flounder, while other species, collectively called **bycatch**, are unintentionally caught and then discarded. The FAO reports that about

bycatch The fishes, marine mammals, sea turtles, seabirds, and other animals caught unintentionally in a commercial fishing catch.

25 percent of all marine organisms caught—some 27 million metric tons (30 million tons)—are dumped back into the ocean. Most of these unwanted animals are dead or soon die because they are crushed by the fishing gear or are out of the water too long. The United States and other countries are trying to significantly reduce the amount of bycatch and develop uses for the bycatch that remains.

Modern Commercial Fishing Methods

A A full fishnet is pulled on board a fishing vessel off the coast of Alaska.

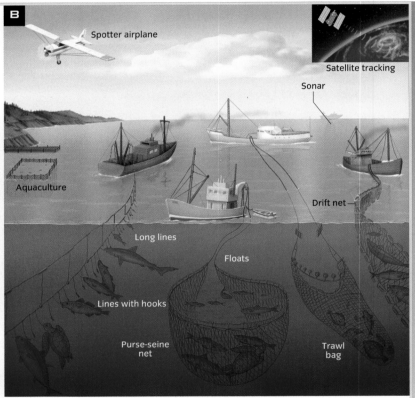

B Scientists know modern methods of harvesting fish are so effective that many fish species have become rare. Sea turtles, dolphins, seals, whales, and other aquatic organisms are accidentally caught and killed in addition to the target fish. The depth of longlines is adjusted to catch open-water fishes like sharks and tuna or bottom fishes like cod and halibut. Purse seines catch anchovies, herring, mackerel, tuna, and other fishes that swim near the water's surface. Trawls catch cod, flounder, red snapper, scallops, shrimp, and other fishes and shellfish that live on or near the ocean floor. Drift nets catch salmon, tuna, and other fishes that swim in ocean waters.

In response to harvesting, many nations extended their limits of jurisdiction to 320 km (200 mi) offshore. This action removed most fisheries from international use, because more than 90 percent of the world's fisheries are harvested in relatively shallow waters close to land. This policy was supposed to prevent overharvesting by allowing nations to regulate the amounts of fishes and other seafood harvested from their waters. However, many countries also have a policy of **open management**, in which all fishing boats of that country are given unrestricted access to fishes in national waters.

Aquaculture: fish farming

Aquaculture is more closely related to agriculture on land than it is to the fishing industry (**FIGURE 11.13**). Aquaculture is carried out both in fresh and marine water; the cultivation of marine organisms is sometimes called **mariculture**. According to the FAO, world aquaculture production has increased substantially, from 600,000 tons in 1950 to 55 million tons in 2003.

Aquaculture differs from fishing in several respects. For one thing, although highly developed nations harvest more fishes from the ocean, developing nations produce much more seafood by aquaculture. Developing nations have an abundant supply of cheap labor, which is a requirement of aquaculture because it is labor-intensive like land-based agriculture. Another difference between fishing and aquaculture is that the limit on the size of a catch in fishing is the size of the natural population, whereas the limit on aquacultural production is primarily the size of the area in which organisms can be grown.

In aquacultural "fish farms," fish populations are concentrated in a relatively small area and produce higher-than-normal concentrations of waste that pollute the adjacent water and harm other organisms. Aquaculture also causes a net loss of wild fish because many of the fishes farmed are carnivorous. Sea bass and salmon, for example, eat up to 5 kg of wild fish to gain 1 kg of weight.

The National Oceanic and Atmospheric Administration (NOAA) is investigating prospects for fish farming in the offshore deep waters of the U.S. Exclusive Economic Zone. Opponents are concerned about the potential for pollution, the spread of disease, and the accidental release of caged species into the deep-water environment.

aquaculture The growing of aquatic organisms (fishes, shellfish, and seaweeds) for human consumption.

An underwater clam farm in Micronesia
FIGURE 11.13

Cultivated giant clams are measured as part of a propagation project that is launching clam farms around various islands of Micronesia to reinstate clams as a food and industry source.

SHIPPING, OCEAN DUMPING, AND PLASTIC DEBRIS

Millions of ships dump oily ballast and other wastes overboard in the neritic and oceanic provinces. The U.N. International Maritime Organization's International Convention for the Prevention of Pollution from Ships (MARPOL) bans marine pollution arising from the shipping industry. MARPOL regulations specifically address six types of marine pollution caused by shipping: oil, noxious liquids, harmful packaged substances, sewage, garbage, and air pollution released by ships. The 2004–2006 revisions to MARPOL regulations included stricter controls on oil tankers and added certain marine sites to the list of special protected areas. Unfortunately, MARPOL is not well enforced in the open ocean.

In the past, U.S. coastal cities such as New York dumped their sewage sludge into the ocean. Disease-causing viruses and bacteria from human sewage contaminated shellfish and other seafood and posed an increasing threat to public health. In 1988 the U.S. Congress passed the **Ocean Dumping Ban Act**, which barred ocean dumping of sewage and industrial waste, beginning in 1991.

Huge quantities of trash containing plastics are released into the ocean from coastal communities or, sometimes accidentally, from cargo ships. Plastics don't biodegrade; they photodegrade, which means that they break down into smaller and smaller pieces yet still exist for an indefinite period. This trash collects in certain areas of the open ocean defined by atmospheric pressure systems. For example, in the north Pacific gyre—halfway between Hawaii and the U.S. mainland—researchers found a continuous array of floating plastics estimated at 1 million pieces per square mile and covering an area the size of Texas.

Not only are marine mammals and birds susceptible to being entangled and/or strangled by the larger pieces of plastic (**FIGURE 11.14**), but the many filter-feeding organisms near the bottom of the ocean food chain constantly ingest the smaller degraded pieces. These plastic pieces may absorb and transport hazardous chemicals such as PCBs. Scientists have yet to determine whether these substances are released into marine food webs when organisms ingest the plastic.

Plastic pollution in the ocean FIGURE 11.14

Note the plastic entangled on the spinner dolphin's fin.

COASTAL DEVELOPMENT

Development of resorts, cities, industries, and agriculture along coasts alters or destroys many coastal ecosystems, including mangrove forests, salt marshes, sea grass beds, and coral reefs. Many coastal areas are overdeveloped, highly polluted, and overfished. Although more than 50 countries have coastal management strategies, their goals are narrow and usually deal only with the economic development of the thin strips of land that directly border the oceans. Coastal management plans generally don't integrate the management of both land and water, nor do they take into account the main cause of coastal degradation—sheer human numbers.

Perhaps as many as 3.8 billion people—about two-thirds of the world's population—live within 150 km (93 mi) of a coastline. Demographers project that three-fourths of all humans—perhaps as many as 6.4 billion—will live in that area by 2025. If the world's natural coastal areas aren't to become urban sprawl or continuous strips of tourist resorts during the 21st century, coastal management strategies must be developed that take into account projections of human population growth and distribution.

Humans and the Antarctic Food Web

Although the icy waters around Antarctica seem inhospitable, they shelter a complex food web. The base of the web is microscopic algae, which live in vast numbers in the well-lit nutrient-rich water. A huge population of herbivores—tiny shrimp-like **krill**—eat these marine algae *(see photo)*. Krill in turn support a variety of larger animals.

Baleen whales—including blue, humpback, and right whales—are major consumers of krill. Until a 1986 global ban, whaling steadily reduced the numbers of baleen whales in Antarctic waters. As a result, more krill became available for other krill-eating animals, whose populations then increased. Seals, penguins, and smaller baleen whales replaced the large baleen whales as the main eaters of krill. Now that commercial whaling is regulated, it is hoped the number of large baleen whales will increase, although it is impossible to say whether they will return to their former dominance in the food web in terms of krill consumption.

Scientists are concerned that ozone thinning over Antarctica may damage the algae that form the base of the food web. Increased ultraviolet radiation is penetrating the surface waters around Antarctica, and algal productivity has declined. (The problem of stratospheric ozone depletion is discussed in detail in Chapter 9.)

Another human-induced change that may be responsible for declines in certain Antarctic populations is global warming. As the water around Antarctica warms, less pack ice forms during winter months. Large numbers of marine algae are found in and around the pack ice, providing a critical supply of food for the krill. Below-average pack ice cover means less algae, which mean less krill. (Global warming, including its effect on Adélie penguins in Antarctica, is discussed in Chapter 9.)

To complicate matters, commercial fishermen have started harvesting krill to make fish-meal for aquaculture operations. Scientists caution that the human harvest of krill may endanger the marine animals that depend on it for food.

CHAPTER SUMMARY

1 The Global Ocean

1. The global ocean is a huge body of salt water that surrounds the continents. It affects the hydrologic cycle and other cycles of matter, influences climate and weather, and provides food to millions of people.

2. Prevailing winds over the ocean generate **gyres**, large, circular ocean current systems that often encompass an entire ocean basin. The **Coriolis effect** is a force resulting from Earth's rotation that influences the paths of surface ocean currents, which move in a circular, clockwise pattern in the Northern Hemisphere and in a circular, counter-clockwise pattern in the Southern Hemisphere.

3. The ocean and the atmosphere are strongly linked. The **El Niño–Southern Oscillation (ENSO)** event, which is responsible for much of Earth's interannual climate variability, is a periodic, large-scale warming of surface waters of the tropical eastern Pacific Ocean that temporarily alters both ocean and atmospheric circulation patterns.

CHAPTER SUMMARY

2 Major Ocean Life Zones

1. The vast ocean is subdivided into major life zones. The biologically productive **intertidal zone** is the area of shoreline between low and high tides. The **benthic environment** is the ocean floor, which extends from the intertidal zone to the deep ocean trenches. Most of the benthic environment consists of sediments where many animals burrow. Common benthic habitats include seagrass beds, kelp forests, and coral reefs. The **pelagic environment** is divided into two provinces. The **neritic province** is the part of the pelagic environment from the shore to where the water reaches a depth of 200 m. Organisms that live in the neritic province are all floaters or swimmers. The **oceanic province**, "the deep sea," is the part of the pelagic environment where the water depth is greater than 200 m. The oceanic province is the largest marine environment, comprising about 75 percent of the ocean's water.

3 Human Impacts on the Ocean

1. The most serious problem for marine fisheries is the overharvesting of many species to the point that their numbers are severely depleted. Fishermen usually concentrate on a few fish species with high commercial value. In doing so, they also catch **bycatch**, those fishes, marine mammals, sea turtles, seabirds, and other animals caught unintentionally in a commercial fishing catch and then discarded. **Aquaculture** is the growing of aquatic organisms (fishes, shellfish, and seaweeds) for human consumption. Aquaculture is common in developing nations with abundant cheap labor, and it is limited by the size of the space dedicated to cultivation. Aquaculture produces wastes that pollute the adjacent water and also causes a net loss of wild fish because many of the fishes farmed are carnivorous.

2. **Marine pollution** is generated by many human activities, including the release of trash or contaminants through commercial shipping, ocean dumping of sludge and industrial wastes, and discarding of plastics that are potentially harmful to marine organisms. Marine environments are also deteriorated by coastal development and the extraction of offshore minerals.

3. The **ocean conveyor belt** moves cold, salty, deep-sea water from higher to lower latitudes, affecting regional and possibly global climate. Global warming associated with human activities may alter the link between the ocean conveyor belt and global climate.

4 Addressing Ocean Problems

1. International initiatives aimed at protecting the global ocean include the **U.N. Convention on the Law of the Sea (UNCLOS)**, a "constitution for the ocean" that protects ocean resources, and the **U.N. Fish Stocks Agreement**, the first international treaty to regulate marine fishing.

2. Long-term goals for halting and reversing destruction of the ocean focus on adopting an ecosystem-based approach to management of ocean environments. Consolidating ocean programs, funding research on marine ecosystems, and enhancing ocean education to instill citizens with a stewardship ethic could improve U.S. ocean policy.

KEY TERMS

- gyres p. 264
- El Niño–Southern Oscillation (ENSO) p. 266
- intertidal zone p. 268
- benthic environment p. 270
- neritic province p. 273
- oceanic province p. 273
- bycatch p. 276
- aquaculture p. 277

1. How do ocean currents affect climate on land? In particular, describe the role of the ocean conveyor belt.

2. Compare the different global effects of the El Niño–Southern Oscillation (ENSO) event with those of La Niña. How are the two events similar? How are they different?

3. Identify which of the ocean life zones would be home to each of the following organisms: giant squid, kelp, tuna, and mussels. Explain your answers.

4. Explain how human activities impact coral reefs.

5. How might the production of plastic shopping bags contribute to ocean pollution?

6. A 2003 study reports that in recent years tropical ocean waters have become saltier, whereas polar ocean waters have become less salty. What do you think is a possible explanation for these changes?

7. Compare and contrast the environmental stresses faced by fish populations on Georges Bank and by krill in Antarctica. Explain the role of human activities in each case.

8. Imagine you live in a small community along the Atlantic coast where a company wants to set up an aquaculture facility in a salt marsh. What are the benefits? What are the environmental drawbacks? Would you support or oppose this proposal? Explain your answer.

9–10. Examine the graph, which shows the global fish catch, including both wild fish and farmed fish (aquaculture).

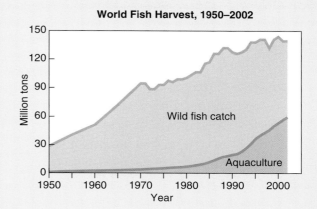

World Fish Harvest, 1950–2002

(Y-axis: Million tons, 0 to 150; X-axis: Year, 1950 to 2000)

Wild fish catch

Aquaculture

9. From 1950 to 1980, did most of the growth in the world fish harvest occur in the wild fish catch or in aquaculture?

10. From 1980 to 2002 (most recent data available), did most of the growth in the world fish harvest occur in the wild fish catch or in aquaculture?

11. In 2003 the wild fish catch was 78 million tons, and the harvest from aquaculture was 55 million tons. What percent of the global fish catch was from aquaculture?

What is happening in this picture ?

These divers in Florida are scientists conducting underwater experiments to learn more about damage to coral reefs. What climate change are they likely studying?

What controlled differences might there be in the enclosed habitats containing the miniature coral reefs?

Why do you suppose the scientists are conducting their experiments in this habitat rather than in a laboratory?

Environmental Geology: Mineral and Soil Resources

12

COPPER BASIN, TENNESSEE

Copper Basin, Tennessee, provides an example of environmental degradation caused by smelting, a stage of mineral processing. During the 19th century, mining companies in southeastern Tennessee extracted copper ore—rock containing copper—from the ground and dug vast pits to serve as open-air smelters. They cut down the surrounding trees to fire the smelters, producing the high temperatures needed for the separation of copper from other substances in the ore. One of these substances, sulfur, reacted with oxygen in the air to form sulfur dioxide. This sulfur dioxide entered the atmosphere, reacted with water vapor there, and became sulfuric acid that returned to Copper Basin as acid precipitation.

Ecological ruin took only a few short years *(see larger photograph)*. Acid precipitation killed plants. Without plants to hold the soil in place, erosion cut gullies in the rolling hills. The forest animals disappeared along with the plants, their food and shelter destroyed.

State and federal reclamation efforts were only marginally successful until the 1970s, when specialists began using new replanting techniques. The new plants had a greater survival rate, and as they became established, their roots held the soil in place *(see inset)*. Birds and field mice slowly began to return.

Today, reclamation of Copper Basin continues; the goal is to have the entire area under plant cover early in the 21st century. The return of the original forest ecosystem will take at least a century or two. Plant scientists and land reclamation specialists have learned a lot from Copper Basin, and they will put this knowledge to use in future reclamation projects around the world.

CHAPTER OUTLIN

■ Plate Tectonics and Shifting Continents p. 288

■ Economic Geology: Useful Minerals p. 292

■ Environmental Implications of Mineral Use p. 296

■ Soil Properties and Processes p. 299

■ Soil Problems and Conservation p. 302

■ Case Study: Industrial Ecosystems p. 307

Plate Tectonics and Shifting Continents

LEARNING OBJECTIVE

Define plate tectonics and explain its relationship to earthquakes and volcanic eruptions.

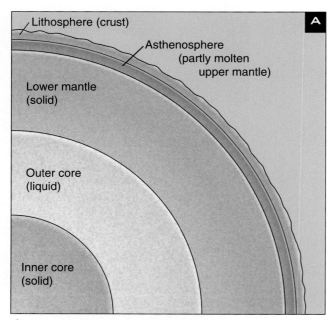

A The main layers of planet Earth.

eology is an essential part of environmental science. To better understand the environmental effects of humans on mineral and soil resources, you must first know something about the geologic properties of the lithosphere.

Earth's outermost rigid rock layer (the *lithosphere*) is composed of seven large plates, plus a few smaller ones, that float on the *asthenosphere*, the region of the mantle where rocks become hot and soft (**FIGURE 12.1**). Continents and landmasses are situated on some of these plates. As the plates move across Earth's surface, the continents change their relative positions. **Plate tectonics** is the study of the movement of these plates.

Any area where two plates meet—a **plate boundary**—is a site of intense geologic activity (**FIG-URE 12.2**). Earthquakes and volcanoes are common in

> **plate tectonics**
> The study of the processes by which the lithospheric plates move over the asthenosphere.

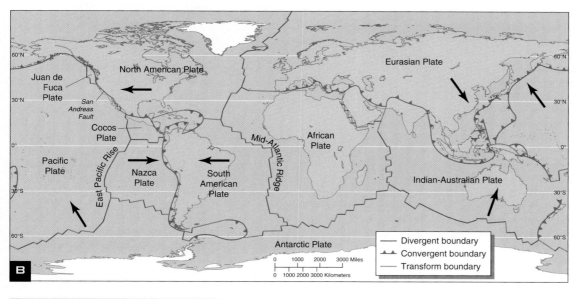

B Plates and plate boundary locations. There are seven major independent plates that move horizontally across Earth's surface. Arrows show the directions of plate movements. The three types of plate boundaries are explained in Figure 12.2.

Earth's layers FIGURE 12.1

these regions. San Francisco, California (noted for its earthquakes), and the volcano Mount Saint Helens in Washington State are both situated on plate boundaries. Where landmasses meet on the boundary between two plates, mountains may form: when the plate carrying India rammed into the plate carrying Asia, the resulting pressure pushed up the Himalayas. When two plates grind together, one of them is sometimes forced under the other, in the process of **subduction**. When two plates move apart, a ridge of molten rock from the mantle wells up between them, continually expanding as the plates move farther apart. The Atlantic Ocean is growing as a result of the buildup of lava along the Mid-Atlantic ridge, where two plates are separating.

VOLCANOES

The movement of tectonic plates on the hot, soft rock of the asthenosphere causes most volcanic activity. In places where the asthenosphere is close to the surface, heat from this part of Earth's mantle melts the surrounding rock, forming pockets of **magma**. When one plate slides under or away from another, this magma may rise to the surface, often forming volcanoes. Magma that reaches the surface is called **lava**.

Volcanoes occur at three kinds of locations: subduction zones, spreading centers, and above hot spots. Subduction zones around the Pacific Basin have given rise to hundreds of volcanoes around Asia and the Americas known as the "ring of fire." Iceland is a volcanic island that formed along the Mid-Atlantic ridge as the adjoining plates there spread apart. The volcanic Hawaiian Islands formed as the Pacific plate moved over a **hot spot**, a rising plume of magma that flowed from an undersea opening in Earth's crust.

The largest volcanic eruption in the 20th century occurred in 1991 when Mount Pinatubo in the Philippines exploded (see Figure 9.9 on page 219). Despite the evacuation of more than 200,000 people, several hundred deaths occurred, mostly from the collapse of buildings under the thick layer of wet ash that blanketed the area. The lava and ash ejected into the atmosphere by the eruption blocked much of the sun's warmth and caused a slight cooling of global temperatures for a year or so.

Plate boundaries FIGURE 12.2

All three types of plate boundaries occur both in the ocean and on land.

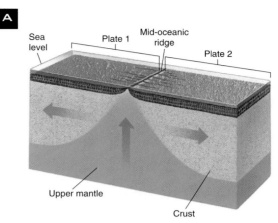

A Two plates move apart at a divergent plate boundary.

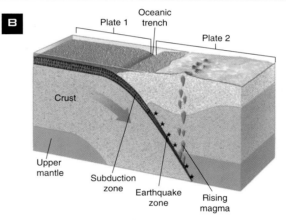

B When two plates collide at a convergent plate boundary in the seafloor, subduction may occur. Convergent collision can also form a mountain range (not shown).

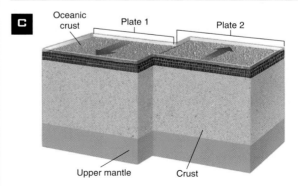

C At a transform plate boundary, plates move horizontally in opposite but parallel directions. On land, such a boundary is often evident as a long, thin valley due to erosion along the fault line.

EARTHQUAKES

Forces inside Earth sometimes push and stretch rocks in the lithosphere. The rocks absorb this energy for a time, but eventually, as the energy accumulates, the stress is too great and the rocks suddenly shift or break. The energy—released as **seismic waves**, vibrations that rapidly spread through rock in all directions—causes one of the most powerful events in nature, an earthquake. Most earthquakes occur along **faults**, fractures in the crust where rock moves forward and backward, up and down, or from side to side. Fault zones are often found at plate boundaries.

The site where an earthquake begins, often far below the surface, is the **focus**. Directly above the focus, at Earth's surface, is the earthquake's **epicenter**. When seismic waves reach the surface, they cause the ground to shake. Buildings and bridges may collapse, and roads may break. One of the instruments used to measure seismic waves is a seismograph, which helps seismologists (scientists who study earthquakes) determine where an earthquake started, how strong it was, and how long it lasted.

Seismologists record more than 1 million earthquakes each year. Some of these are major, but most are too small to feel, equivalent to readings of about 2 on the *Richter scale,* a measure of the magnitude of energy released by an earthquake. In populated areas, a magnitude 5 earthquake usually causes property damage, and quakes of 8 or higher cause massive property destruction and kill large numbers of people. In 2005, a 7.6-magnitude earthquake devastated a mountainous region in northern Pakistan, killing as many as 87,000 people and leaving 2 million homeless. The earthquake occurred along the convergent boundary of the Indian and Eurasian plates (Figure 12.1B), a region known for its high tectonic activity. The Indian plate is thrusting northward and being subducted under the Eurasian plate; this process is the same that formed the Himalayan mountain range.

Not all earthquakes occur at plate boundaries. Some, like the major earthquakes that damaged Northridge, California, in 1994 and Kobe, Japan, in 1995, occur on smaller faults that crisscross the large plates. Such earthquakes pose a seismic hazard because they are difficult to predict. Major quakes may occur on larger fault lines every century or so, but along any given small fault line, like those in Northridge and Kobe, only every 1,000 to 5,000 years.

Side effects of earthquakes include landslides and tsunamis. A *landslide* is an avalanche of rock, soil, and other debris that slides swiftly down a mountainside. A 1970 earthquake in Peru resulted in a landslide that buried the town of Yungay and killed 17,000 people.

A *tsunami*, a giant sea wave caused by an underwater earthquake or volcanic eruption, may sweep across the ocean at more than 750 km (450 mi) per hour. Although a tsunami may be only about 1 m (3 ft) high in deep ocean water, it can build to a wall of water 30.5 m (100 ft)—as high as a 10-story building—when it comes ashore, often far from where the original earthquake triggered it. Tsunamis have caused thousands of deaths, particularly along the Pacific coast. Although the Pacific Tsunami Warning System monitors submarine earthquakes and warns people of approaching tsunamis, deaths still occur because there is so little time to respond and not all nations are part of the network.

One of the deadliest natural disasters in modern history was the tsunami generated by an Indian Ocean earthquake in 2004 that killed more than 225,000 people along the coasts of South India, Thailand, Sri Lanka, Indonesia, and eastern Africa (FIGURE 12.3A). The earthquake was recorded as a magnitude 9 or higher, and the waves it generated reached heights of 30.5 m (100 ft) (FIGURE 12.3B). Unfortunately, no gauges or buoys were in place in the Indian Ocean to detect the tsunami as it developed there.

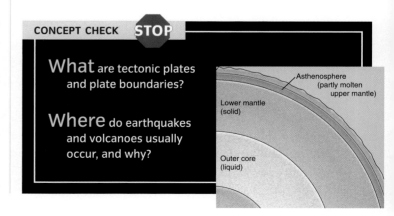

CONCEPT CHECK STOP

What are tectonic plates and plate boundaries?

Where do earthquakes and volcanoes usually occur, and why?

Asthenosphere (partly molten upper mantle)

Lower mantle (solid)

Outer core (liquid)

A The earthquake that triggered the Indian Ocean tsunami originated off the western coast of northern Sumatra, Indonesia.

B Waves crash onto the shores of a Thailand town; across the region, the tsunami generated high waves.

Economic Geology: Useful Minerals

LEARNING OBJECTIVES

Define minerals, and explain the differences between high-grade ores and low-grade ores.

Distinguish between surface mining and subsurface mining, using the terms overburden and spoil bank in your answer.

Briefly describe the process of smelting.

E arth's outermost layer, the crust, contains many kinds of rocks and minerals that are of economic importance. We now focus on the economic and environmental impacts of extracting and using mineral resources. We then consider soil, that part of the crust where biological and physical processes meet.

Minerals are such an integral part of our lives that we often take them for granted (**FIGURE 12.4**). Steel, an essential building material, is a blend of iron and other metals. Beverage cans, aircraft, automobiles, and buildings all contain aluminum. Copper, which readily conducts electricity, is used for electrical and communications wiring. The concrete used in buildings and roads is made from sand and gravel, as well as cement, which contains crushed limestone. Sulfur, a component of sulfuric acid, is an indispensable industrial mineral. It is used to make plastics and fertilizers and to refine oil. Other important minerals include platinum, mercury, manganese, and titanium.

> **minerals**
> Elements or compounds of elements that occur naturally in Earth's crust.

Earth's minerals are elements or (usually) compounds of elements and have precise chemical compositions. **Sulfides** are mineral compounds in which certain elements are combined chemically with sulfur, and **oxides** are mineral compounds in which elements are combined chemically with oxygen. Minerals are metallic or nonmetallic (**FIGURE 12.5**). **Metals** are minerals such as iron, aluminum, and copper, which are malleable, lustrous, and good conductors of

Examples of minerals FIGURE 12.4

A Concrete highways are made of sand, gravel, and crushed limestone.

B Table salt is a nonmetallic mineral.

C Copper, a metallic mineral, is often shaped into wire for electrical equipment or sheets for roofing, gutters, and downspouts.

heat and electricity. **Nonmetallic minerals**, such as sand, stone, salt, and phosphates, lack these characteristics.

Rocks are naturally formed mixtures of minerals that have varied chemical compositions. **Ore** is rock that contains a large enough concentration of a particular mineral to be profitably mined and extracted.

Aluminum (Al)	Borax ($Na_2B_4O_7$)	Chromium (Cr)	Cobalt (Co)
Planes, automobiles, packaging, fireworks	Manufacturing, soaps, antiseptics	Chrome plate, pigments, steel alloys	Alloys, medicines, pigments, varnishes
Copper (Cu)	Gold (Au)	Gypsum ($CaSO_4–2H_2O$)	Iron (Fe)
Gold jewelry, pipes, electric wiring	Jewelry, money, dentistry	Plaster of Paris, soil treatments	Steel (buildings and machinery)
Lead (Pb)	Magnesium (Mg)	Manganese (Mn)	Mercury (Hg)
Pipes, solder, battery electrodes	Alloys, firecrackers, bombs, flashbulbs	Steel, alloys, batteries, chemicals	Electric switches, medicines
Molybdenum (Mo)	Nickel (Ni)	Phosphorus (P)	Platinum (Pt)
Lamp filaments, boiler plates, rifle barrels	Money, alloys, metal plating	Medicine, fertilizers, detergents	Jewelry, delicate instruments
Potassium (K)*	Common salt (NaCl)	Sand (largely SiO_2)	Silicon (Si)
Fertilizers, glass, photography, medicine	Food additive, raw material for synthetics	Glass, concrete (buildings, roads)	Electronics, solar batteries, ceramics
Silver (Ag)	Sulfur (S)	Titanium (Ti)	Zinc (Zn)
Jewelry, silverware, photography	Insecticides, rubber tires, matches, paint	Paints, satellites, chemical equipment	Brass, electrodes in batteries, medicine

Metal ☐, Nonmetal ☐

* Potassium is very reactive chemically and is never found free in nature.

Some important minerals and their uses FIGURE 12.5

High-grade ores contain relatively large amounts of particular minerals, whereas **low-grade ores** contain lesser amounts. Although some minerals are abundant, all minerals are nonrenewable resources that are not replenished by natural processes on a human timescale.

HOW MINERALS ARE EXTRACTED AND PROCESSED

The process of making mineral deposits available for human consumption occurs in several steps. First, a particular mineral deposit is located. Geologic knowledge of Earth's crust and how minerals are formed is used to estimate locations of possible mineral deposits. Once these sites are identified, geologists drill or tunnel for mineral samples and analyze their composition. Second, mining extracts the mineral from the ground. Third, the mineral is processed, or refined, by concentrating it and removing impurities. Last, the purified mineral is used to make a product.

Extracting minerals   The depth of a particular mineral deposit determines whether surface or subsurface mining will be used. In **surface mining**, minerals are extracted near the surface. Surface mining is more common because it is less expensive than subsurface mining. Because even surface mineral deposits occur in rock layers beneath Earth's surface, the overlying soil and rock layers, called **overburden**, must first be removed, along with the vegetation growing in the soil. Then giant power shovels scoop the minerals out.

There are two kinds of surface mining, **open-pit surface mining** and strip mining. Iron, copper, stone, and gravel are usually extracted by open-pit surface mining, in which a giant hole, called a quarry, is dug in the ground to extract the minerals (**FIGURE 12.6A**). In **strip mining**, a trench is dug to extract

surface mining
The extraction of mineral and energy resources near Earth's surface by first removing the soil, subsoil, and overlying rock strata.

overburden Soil and rock overlying a useful mineral deposit.

Types of mining operations FIGURE 12.6

A Copper is extracted from this open-pit surface mine near Tucson, Arizona.

B Strip mining removes overburden along narrow strips to reach the ore beneath.

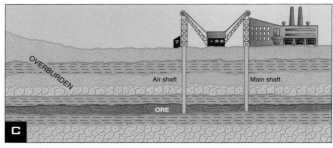

C In a shaft mine, a hole is dug straight through the overburden to the ore, which is removed up through the shaft in buckets.

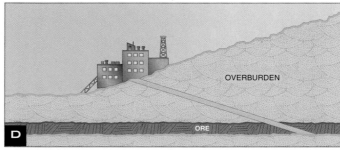

D In a slope mine, an entry to the ore is dug at an angle so that the ore can be hauled out in carts.

the minerals (**Figure 12.6B**). Then a new trench is dug parallel to the old one, and the overburden from the new trench is put into the old one, creating a hill of loose rock called a **spoil bank**.

spoil bank A hill of loose rock created when the overburden from a new trench is put into the already excavated trench during strip mining.

subsurface mining The extraction of mineral and energy resources from deep underground deposits.

Subsurface mining extracts minerals too deep in the ground to be removed by surface mining. It disturbs the land less than surface mining, but it is more expensive and more hazardous for miners. There is always a risk of death or injury from explosions or collapsing walls, and prolonged breathing of dust in subsurface mines can result in lung disease.

Subsurface mining may be done with underground shaft mines or slope mines. A **shaft mine**, often used for mining coal, is a direct vertical shaft to the vein of ore (**Figure 12.6C**). The ore is broken up underground and then hoisted through the shaft to the surface in buckets. A **slope mine** has a slanting passage that makes it possible to haul the broken ore out of the mine in cars rather than hoisting it up in buckets (**Figure 12.6D**). Sump pumps keep the subsurface mine dry, and a second shaft is usually installed for ventilation.

Processing minerals

Processing minerals often involves **smelting**. Purified copper, tin, lead, iron, manganese, cobalt, or nickel smelting is done in a blast furnace. **Figure 12.7** shows a blast furnace used to smelt iron. The iron ore reacts with coke (modified coal) to form molten iron and carbon dioxide. The limestone reacts with impurities in the ore to form a molten mixture called **slag**. Note the vent near the top of the iron smelter for exhaust gases. If air pollution control devices are not installed, many dangerous gases are emitted during smelting.

smelting The process in which ore is melted at high temperatures to separate impurities from the molten metal.

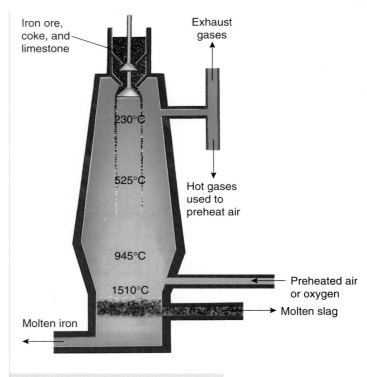

Blast furnace Figure 12.7

Such towerlike furnaces separate metal from impurities in the ore. The energy for smelting comes from a blast of heated air.

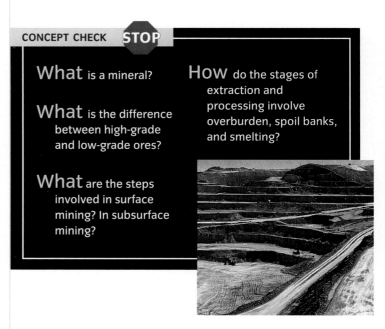

CONCEPT CHECK **STOP**

What is a mineral?

What is the difference between high-grade and low-grade ores?

What are the steps involved in surface mining? In subsurface mining?

How do the stages of extraction and processing involve overburden, spoil banks, and smelting?

Economic Geology: Useful Minerals   295

Environmental Implications of Mineral Use

LEARNING OBJECTIVES

Relate the environmental impacts of mining and refining minerals. Include a brief description of acid mine drainage.

Explain how mining lands can be restored.

The economies of industrialized nations require the extraction and processing of large amounts of minerals to make products. Most of these countries rely on the mineral deposits found in developing countries, having long since exhausted their own supplies. As developing countries become more industrialized, their own mineral requirements increase, placing further pressure on a nonrenewable resource. In fact, humans have consumed more minerals since World War II than in the previous 5,000 years.

Clearly, the extraction, processing, and disposal of minerals harm the environment. Mining disturbs and damages the land, and the processing and disposal of minerals pollute the air, soil, and water. Although pollution can be controlled and damaged lands can be restored, these remedies are costly.

MINING AND THE ENVIRONMENT

Mining, particularly surface mining, disturbs large areas of land. In the United States, functioning and abandoned metal and coal mines occupy an estimated 9 million hectares (22 million acres). Because mining destroys existing vegetation, this land is particularly prone to erosion, with wind erosion causing air pollution, and water erosion polluting nearby waterways and damaging aquatic habitats.

Open-pit mining of gold and other minerals uses huge quantities of water. As miners dig deeper, they eventually hit the water table and pump out the water to keep the pit dry. Farmers and ranchers in open-pit mining areas are concerned about depletions in the groundwater they need for irrigation. Environmentalists and others would like the mining op-

erations to reinject the water into the ground after pumping it out.

Mining has contaminated thousands of kilometers of streams and rivers in the United States. Rocks rich in minerals often contain high concentrations of heavy metals like arsenic and lead. Rainwater seeping through the sulfide minerals in mine waste produces sulfuric acid, which dissolves the heavy metals and other toxic substances in the spoil banks. These acids, called **acid mine drainage**, are highly toxic and are washed into soil and water—including groundwater—by precipitation runoff (**FIGURE 12.8**). When such acids and toxic compounds make their way into nearby lakes and streams, particularly through "toxic pulses" of thunderstorms or spring snowmelt, they adversely affect aquatic life.

> ■ **acid mine drainage** Pollution caused when sulfuric acid and dangerous dissolved materials such as lead, arsenic, and cadmium wash from mines into nearby lakes and streams.

ENVIRONMENTAL IMPACTS OF REFINING MINERALS

Approximately 80 percent of mined ore consists of impurities that become wastes after processing. These wastes, called **tailings**, are usually left in giant piles on the ground or in ponds near the processing plants (**FIGURE 12.9**). The tailings contain toxic materials such as cyanide, mercury, and sulfuric acid. Left exposed, they contaminate the air, soil, and water.

Smelting plants may emit large quantities of air pollutants during mineral processing, particularly sulfur. Unless expensive pollution control devices are added to smelters, the sulfur escapes into the atmosphere, where it forms sulfuric acid. (The environmental implications of the resulting acid precipitation are discussed in Chapter 9.)

Pollution control devices for smelters are the same as those used for the burning of sulfur-containing coal—scrubbers and electrostatic precipitators.

Annual global gold production in 2003 was about 2,600 metric tons, up from 1,300 metric tons in 1980. Worldwide demand for gold is increasing, and the environment is suffering from the increased mining. The waste from mining and processing ore is enormous: Six tons of wastes are produced to yield enough gold to make two wedding rings. One technology, *cyanide heap leaching*, allows profitable mining when minuscule amounts of gold are present, but this process produces up to 3 million pounds of waste for every pound of gold produced. The world's largest gold mine, located in Indonesia but owned by a U.S. company,

dumps more than 100,000 tons of cyanide-contaminated waste into the local river each day. The highly toxic cyanide threatens waterfowl and fishes, as well as underground drinking water supplies.

Small-scale miners use other extraction techniques with destructive side effects: soil erosion, production of silt that clogs streams and threatens aquatic organisms, and contamination from mercury used to extract the gold. The environmental hazards of gold mining do not end when the gold is carried away: If not disposed of properly, mining wastes cause long-term problems like acid mine drainage and heavy-metal contamination.

Acid mine drainage FIGURE 12.8

For more than 100 years, the Leadville mining district in the central Colorado Rockies has been mined for silver, gold, lead, and zinc, which occur as sulfide deposits. Shown is the characteristic orange-red acid runoff that contains sulfuric acid contaminated with lead, arsenic, cadmium, silver, and zinc. The runoff here drains into the Arkansas River from snowmelt and precipitation runoff.

Copper ore tailings from a strip-mining operation in southern Illinois FIGURE 12.9

Toxic materials from mine tailings left in mountainous heaps pollute the air, soil, and water.

Other contaminants in ores include the heavy metals lead, cadmium, arsenic, and zinc. These toxic elements may pollute the atmosphere during the smelting process and cause harm to humans. In addition to airborne pollutants, smelters emit hazardous liquid and solid wastes that can pollute the soil and water.

One of the most significant environmental impacts of mineral production is the large amount of energy required to mine and refine minerals, particularly if they are being refined from low-grade ore. Most of this energy is obtained by burning fossil fuels, which depletes nonrenewable energy reserves and produces carbon dioxide and other air pollutants.

RESTORATION OF MINING LANDS

When a mine is no longer profitable to operate, the land can be reclaimed, or restored to a seminatural condition, as has been done to most of the Copper Basin in Tennessee (see the chapter introduction). Reclamation prevents further degradation and erosion of the land, eliminates or neutralizes local sources of toxic pollutants, and makes the land productive for purposes other than mining (FIGURE 12.10). Restoration also makes such areas visually attractive.

Restoring lands degraded by mining—called **derelict lands**—involves filling in and grading the area to the shape of its natural contours, then planting vegetation to hold the soil in place. Often the topsoil is completely gone or contains toxic levels of metals, so special types of plants that tolerate such a challenging environment must be used.

The **Surface Mining Control and Reclamation Act** of 1977 requires reclamation of areas that were surface mined for coal. However, no federal law is in place to require restoration of derelict lands produced by other kinds of mines. As a result, restoration of mining lands often does not occur.

CONCEPT CHECK STOP

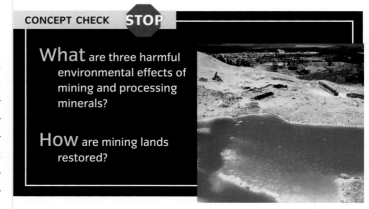

What are three harmful environmental effects of mining and processing minerals?

How are mining lands restored?

Restoration of mining lands
FIGURE 12.10

Part of a phosphate mine near Fort Meade, Florida, was reclaimed and is currently used as a pasture (background). The unrestored area that remains is in the foreground. Restoration of mining lands makes them usable once again, or at least stabilizes them so that further degradation does not occur.

Soil Properties and Processes

Soil is the relatively thin surface layer of Earth's crust. It consists of mineral and organic matter modified by the natural actions of agents such as weather, wind, water, and organisms. It is easy to take soil for granted. We walk on and over it throughout our lives but rarely stop to think about how important it is to our survival. Vast numbers and kinds of organisms, mainly microorganisms, inhabit soil and depend on it for shelter, food, and water. Plants anchor themselves in soil, and from it they receive essential minerals and water. Terrestrial plants could not survive without soil, and because we depend on plants for our food, humans could not exist without soil either (**FIGURE 12.11**).

> **soil** The uppermost layer of Earth's crust, which supports terrestrial plants, animals, and microorganisms.

SOIL FORMATION AND COMPOSITION

Soil is formed from *parent material*, rock that is slowly broken down, or fragmented, into smaller and smaller particles by biological, chemical, and physical **weathering processes**. It takes a long time, sometimes thousands of years, for rock to disintegrate into finer and finer mineral particles. Time is also required for organic material to accumulate in the soil. Soil formation is a continuous process that involves interactions between Earth's solid crust and the biosphere. The weathering of parent material beneath already formed soil continues to add new soil.

Topography, a region's surface features (such as the presence or absence of mountains and valleys), is also involved in soil formation. Steep slopes often have little or no soil on them because soil and rock are continually transported down the slopes by gravity. Runoff from precipitation tends to amplify erosion on steep slopes. Moderate slopes and valleys, on the other hand, may encourage the formation of deep soils.

A farmer plows his soil prior to planting crops
FIGURE 12.11

Soil is an important natural resource that humans and countless soil organisms rely on.

Soil Profile

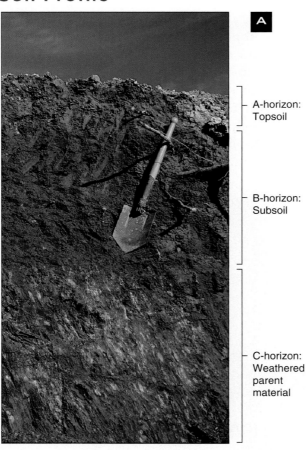

A

A-horizon: Topsoil

B-horizon: Subsoil

C-horizon: Weathered parent material

B

O-horizon: Mostly organic matter and humus; plant litter accumulates and decays.

A-horizon (topsoil): dark; high concentration of organic matter.

B-horizon (subsoil): light-colored; litter and nutrient minerals leached from A-horizon accumulate here.

C-horizon (weathered parent material): below roots, often saturated with groundwater.

Consolidated bedrock (parent material).

A This soil, located on a farm in Virginia, has no O-horizon because it is used for agriculture; the surface litter that would normally comprise the O-horizon was plowed into the A-horizon. The shovel gives an idea of the relative depths of each horizon.

B A "typical" soil profile as it appears to the trained eye of a soil scientist. Each horizon has its own chemical and physical properties.

Soil is composed of four distinct parts: mineral particles, organic matter, water, and air. The mineral portion, which comes from parent material, is the main component of soil. It provides anchorage and essential nutrient minerals for plants, as well as pore space for water and air. Litter (dead leaves and branches on the soil's surface), animal dung, and the remains of plants, animals, and microorganisms constitute the organic portion of soil. Organisms like bacteria and fungi gradually decompose this material. The black or dark brown organic material remaining after extended decomposition is called **humus** (FIGURE 12.12). Humus, which is a mix of many organic compounds, binds to nutrient mineral ions and holds water. Many soils are organized into distinctive horizontal layers called **soil horizons**. A **soil profile** is a vertical section from surface to parent material, showing the soil horizons (see What a Scientist Sees). The topsoil (or A-horizon) is somewhat nutrient-poor due to the leaching of many nutrients into deeper soil layers. **Leaching** is the removal of dissolved materials from the soil by water percolating downward.

soil horizons The horizontal layers into which many soils are organized, from the surface to the underlying parent material.

Soil rich in humus Figure 12.12

Humus is partially decomposed organic material, primarily from plant and animal remains. Soil rich in humus has a loose, somewhat spongy structure with several properties, such as increased water-holding capacity, that are beneficial for plants and other organisms living in it.

SOIL ORGANISMS

Although soil organisms are usually hidden underground, their numbers are huge. Organisms that colonize the soil ecosystem include plant roots, insects like termites and ants, earthworms, moles, snakes, and groundhogs (Figure 12.13). Most numerous in soil are bacteria, which number in the hundreds of millions per gram of soil. Other microorganisms that are abundant in soil ecosystems include fungi, algae, microscopic worms such as nematodes, and protozoa.

In a balanced ecosystem, the relationship between soil and the organisms that live in and on it ensure soil fertility. Soil organisms provide several essential **ecosystem services**, such as maintaining soil fertility, preventing soil erosion, breaking down toxic materials, and cleansing water.

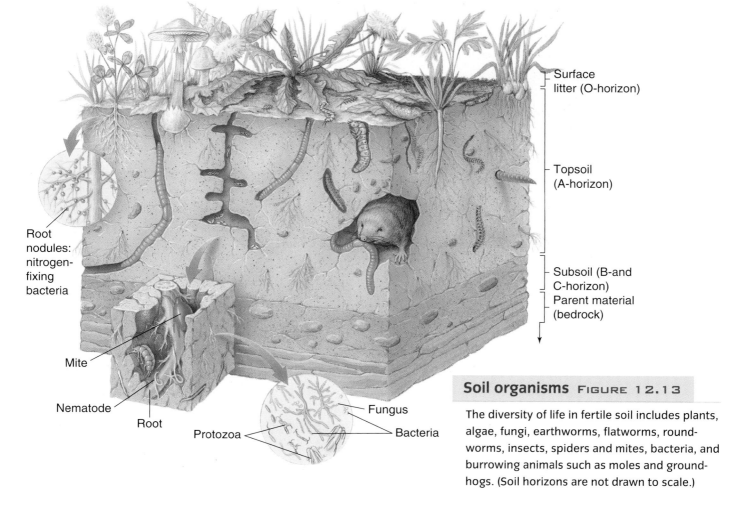

Surface litter (O-horizon)

Topsoil (A-horizon)

Root nodules: nitrogen-fixing bacteria

Subsoil (B-and C-horizon)
Parent material (bedrock)

Mite

Nematode

Root

Protozoa

Fungus

Bacteria

Soil organisms Figure 12.13

The diversity of life in fertile soil includes plants, algae, fungi, earthworms, flatworms, roundworms, insects, spiders and mites, bacteria, and burrowing animals such as moles and groundhogs. (Soil horizons are not drawn to scale.)

nutrient cycling
The pathway of various nutrient minerals or elements from the environment through organisms and back to the environment.

Decomposition, another ecosystem service, is part of **nutrient cycling**. Bacteria and fungi decompose plant and animal detritus and wastes, transforming large organic molecules into small inorganic molecules, including carbon dioxide, water, and nutrient minerals; the nutrient minerals are released into the soil to be reused (**FIGURE 12.14**; also see Chapter 5). Nonliving processes are also involved in nutrient cycling: The weathering of a parent material replaces some nutrient minerals lost by erosion or agricultural practices.

Nutrient cycling Figure 12.14

In a balanced ecosystem, nutrient minerals cycle from the soil to organisms and then back to the soil.

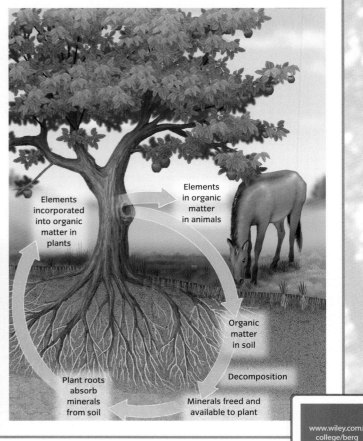

Elements incorporated into organic matter in plants

Elements in organic matter in animals

Organic matter in soil

Decomposition

Plant roots absorb minerals from soil

Minerals freed and available to plant

www.wiley.com/college/berg

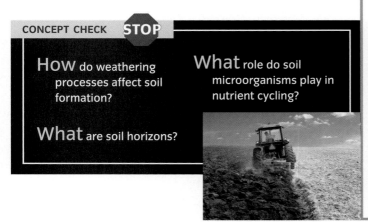

CONCEPT CHECK STOP

How do weathering processes affect soil formation?

What are soil horizons?

What role do soil microorganisms play in nutrient cycling?

Soil Problems and Conservation

LEARNING OBJECTIVES

Define sustainable soil use.

Explain the impacts of soil erosion on plant growth and on other resources such as water.

Identify and summarize the major soil conservation methods.

S oil is as important as air and water for human survival. Yet, humans have had a harmful impact on soil resources worldwide, particularly by intensifying agricultural use. Increasing the production of crops on soil and other human activities often cause or exacerbate soil problems like erosion, mineral depletion, and soil pollution, all of which occur worldwide. Such activities do not promote **sustainable soil use**. Soil used in a sustainable way renews itself by natural processes year after year.

sustainable soil use The wise use of soil resources, without a reduction in the amount or fertility of soil, so it is productive for future generations.

SOIL EROSION

Water, wind, ice, and other agents promote **soil erosion**, a natural process often accelerated by human activities. Water and wind are particularly effective in moving soil from one place to another. Rainfall loosens soil particles, which are transported by moving water (**FIGURE 12.15**). Wind loosens soil and blows it away, particularly if the soil is barren and dry. Erosion reduces the amount of soil in an area and therefore limits the growth of plants.

Humans often accelerate soil erosion with poor soil management. Poor agricultural practices are partly to blame, as is the removal of natural plant communities during road and building construction, and unsound logging practices like clear-cutting. Soil erosion has an impact on other natural resources as well. Sediment that gets into streams, rivers, and lakes affects water quality and fish habitats (see Chapter 10). If the sediments contain pesticide and fertilizer residues, they further pollute the water.

> ■ **soil erosion** The wearing away or removal of soil from the land.

Sufficient plant cover limits soil erosion. Leaves and stems cushion the impact of rainfall, and roots help to hold the soil in place. Although soil erosion is a natural process, the abundant plant cover in many natural ecosystems makes it negligible.

SOIL POLLUTION

Soil pollution is any physical or chemical change in soil that adversely affects the health of plants and other organisms living in or on the soil. Soil pollution is important not only in its own right but because so many soil pollutants tend to also pollute surface water, groundwater, and the atmosphere. For example, selenium, an extremely toxic natural element found in many western soils, leaches off irrigated farmlands and poisons nearby lakes, ponds, and rivers, causing death and deformity in thousands of migratory birds and other organisms every year. Except for salts, petroleum products, and heavy metals, most soil pollutants originate as agricultural chemicals like fertilizers and pesticides.

Irrigation of agricultural fields often results in their becoming increasingly saline, an occurrence known as **salinization** (**FIGURE 12.16**). In time, salt concentrations in soil can rise to such a high level that plants are poisoned or their roots dehydrated.

Soil erosion caused by water FIGURE 12.15

The branching gullies shown here are the most serious form of erosion and will continue to grow unless checked by some type of erosion control. Photographed in Colorado.

FIGURE 12.16 **Salinized soil**

This irrigated soil has become too salty for plants to tolerate.

SOIL CONSERVATION AND REGENERATION

Only 11 percent of the world's soil is suitable for agriculture (FIGURE 12.17 in Visualizing Soil Conservation). We therefore need to protect the soils we do use for agriculture. Although agriculture may cause or accelerate soil degradation, good soil conservation practices promote sustainable soil use. Conservation tillage, crop rotation, contour plowing, strip cropping, terracing, and shelterbelts minimize erosion and mineral depletion of the soil. Badly eroded and depleted land can be restored, but restoration is costly and time-consuming.

Conservation tillage and crop rotation

Conventional methods of tillage, or working the land, include spring plowing, in which the soil is cut and turned in preparation for planting seeds. Although conventional tillage prepares the land for crops, it greatly increases the likelihood of soil erosion. Conventionally tilled fields contain less organic material and generally hold less water than undisturbed soil.

Conservation tillage is one of the fastest-growing trends in U.S. agriculture (FIGURE 12.18). More than one-third of U.S. farmland is currently planted using conservation tillage. In addition to reducing soil erosion, conservation tillage increases the organic material in the soil, which improves the soil's water-holding capacity. Decomposing organic matter releases nutrient minerals more gradually than when conventional tillage methods are employed. However, use of conservation tillage requires new equipment, new techniques, and greater use of herbicides to control weeds. Research in developing alternative methods of weed control for use with conservation tillage is underway. (Chapter 14 discusses *sustainable agriculture*, which includes conservation tillage and the other soil conservation practices presented in this chapter.)

> **conservation tillage** A method of cultivation in which residues from previous crops are left in the soil, partially covering it and helping to hold it in place until the newly planted seeds are established.

Visualizing

Soil Conservation

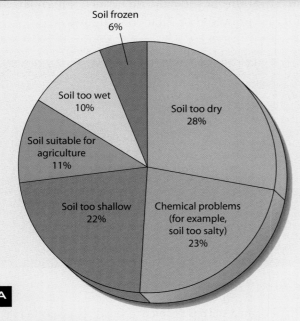

A

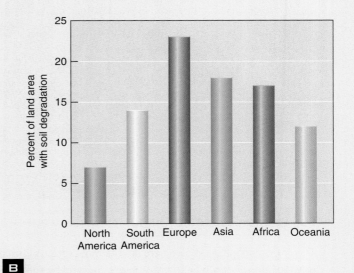

B

Soil and agriculture FIGURE 12.17

A Only 11 percent of the world's total land area has soil suitable for agriculture without taking steps to improve it. Soil that is too dry can be irrigated, whereas soil that is too wet can be drained.

B Degree of soil degradation (eroded, desertified, or salty soil) by continent. North America has the least degraded soils, and Europe, Africa, and Asia have the most degraded soils.

Farmers who practice effective soil conservation measures often use a combination of conservation tillage and **crop rotation**. When the same crop is grown over and over in one place, pests for that crop accumulate to destructive levels and the essential nutrient minerals for that crop are depleted in greater amounts. This not only makes the soil more prone to erosion but the crops less productive as well. Crop rotation is therefore effective in decreasing insect damage and disease, reducing soil erosion, and maintaining soil fertility (**FIGURE 12.19A**).

A typical crop rotation would be corn → soybeans → oats → alfalfa. Soybeans and alfalfa, both members of the legume family, increase soil fertility through their association with bacteria that fix atmospheric nitrogen into the soil. Thus, soybeans and alfalfa produce higher yields of the grain crops they alternate with in crop rotation.

> **crop rotation**
> The planting of a series of different crops in the same field over a period of years.

Contour plowing, strip cropping, and terracing

Hilly terrain must be cultivated with care because it is more prone than flatland to soil erosion. Contour plowing, strip cropping, and terracing help control erosion of farmland with variable topography. In **contour plowing**, furrows run around hills rather than in straight rows. Strip cropping, a special type of contour plowing, produces alternating strips of different crops along natural contours (Figure 3.1 on page 48). For example, alternating a row crop such as corn with a closely sown crop such as wheat reduces soil erosion. Even more effective control of soil erosion is achieved when strip cropping is done in conjunction with conservation tillage.

Farming is undesirable on steep slopes, but if it must be done, **terracing** produces level areas and thereby reduces soil erosion from gravity or water runoff (**FIGURE 12.19B**). Nutrient minerals and soil are retained on the horizontal platforms instead of being washed away.

> **contour plowing**
> Plowing that matches the natural contour of the land.

Conservation tillage FIGURE 12.18
Decaying residues from the previous year's crop (rye) surround young soybean plants in a field in Iowa. Conservation tillage reduces soil erosion as much as 70 percent because plant residues from the previous season's crops are left in the soil.

Crop rotation and terracing FIGURE 12.19
A Yellow mustard plants are planted between rows of grape vines during the season when the grape vines are dormant. This crop rotation enriches the soil for the next grape crop.
B Terracing hilly or mountainous areas, such as in the Luzon rice fields, Philippines, reduces the amount of soil erosion.

Soil reclamation   Badly eroded land can be reclaimed by: (1) preventing further erosion and (2) restoring soil fertility. To prevent further erosion, the bare ground is seeded with plants; they eventually grow to cover the soil, stabilizing it and holding it in place. The plants start to improve the quality of the soil almost immediately, as dead material decays into humus. The humus holds nutrient minerals in place, releasing them a little at a time; it also improves the water-holding capacity of the soil. One of the best ways to reduce the effects of wind erosion on soil is to plant **shelterbelts** that lessen the impact of wind (**FIGURE 12.20**).

> **shelterbelt** A row of trees planted as a windbreak to reduce soil erosion of agricultural land.

Shelterbelts surrounding kiwi orchards

FIGURE 12.20

Trees protect the delicate fruits from the wind and reduce wind erosion of farmland soil. Photographed in the North Island, New Zealand.

Restoration of soil fertility is a slow process. The land cannot be farmed or grazed until the soil has completely recovered. But the restriction of land use for an indefinite period may be difficult to accomplish. How can a government tell landowners they may not use their own land? How can land use be restricted when people's livelihoods, and maybe even their lives, depend on it?

Soil conservation policies in the United States

The Food Security Act (Farm Bill) of 1985 contained provisions for two main soil conservation programs, a conservation compliance program and the Conservation Reserve Program. The conservation compliance program requires farmers with highly erodible land to develop and adopt a 5-year conservation plan for their farms that includes erosion-control measures. If they do not comply, they lose federal agricultural subsidies such as price supports.

The **Conservation Reserve Program (CRP)** is a voluntary subsidy program that pays U.S. farmers to stop producing crops on highly erodible farmland. It requires planting native grasses or trees on such land and then "retiring" it from further use for 10 to 15 years. The CRP has benefited the environment. Annual loss of soil on CRP lands planted with grasses or trees has been reduced more than 90 percent. Because the vegetation is not disturbed once it is established, it provides biological habitat. Small and large mammals, birds of prey, and ground-nesting birds such as ducks have increased in number and kind on CRP lands. The reduction in soil erosion has improved water quality and enhanced fish populations in surrounding rivers and streams.

CONCEPT CHECK **STOP**

What is sustainable soil use?

How do human activities accelerate soil erosion?

How do contour plowing and shelterbelts contribute to soil conservation?

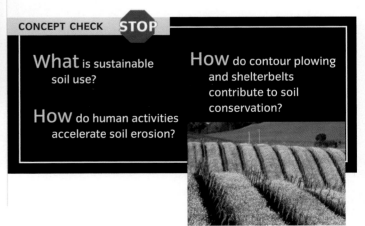

Traditional industries operate in a linear fashion: natural resources → products → wastes dumped back into the environment. However, natural resources are finite, and the environment's capacity to absorb waste is limited. The field of **industrial ecology** seeks to address these issues by using resources efficiently, regarding "wastes" as potential products, and creating **industrial ecosystems** that in many ways mimic natural ecosystems.

One pioneering industrial ecosystem in Kalundborg, Denmark (**FIGURE A**) consists of an electric power plant, an oil refinery, a pharmaceutical plant, a wallboard factory, a sulfuric acid producer, a cement manufacturer, a fish farm, greenhouses, and area homes and farms. These entities are linked in ways that resemble a food web in a natural ecosystem (**FIGURE B**).

The coal-fired electric power plant originally cooled its waste steam and released it into the local fjord. The steam is now supplied to the oil refinery and the surplus heat warms greenhouses, the fish farm, and area homes. Surplus natural gas from the oil refinery is sold to the power plant and the wallboard factory. Before selling the natural gas, the oil refinery removes excess sulfur from it (as required by law) and sells the sulfur to the sulfuric acid producer.

To meet environmental regulations, the power plant installed pollution control equipment to remove sulfur from its coal smoke. This sulfur, in the form of calcium sulfate, is sold to the wallboard plant and used as a gypsum substitute. The fly ash produced by the power plant goes to the cement manufacturer for use in road building.

A

To fertilize their fields, local farmers use sludge from the fish farm and a high-nutrient sludge generated at the pharmaceutical plant. Most pharmaceutical companies discard this sludge because it contains living microorganisms, but the Kalundborg plant heats the sludge to kill the microorganisms, converting a waste material into a commodity.

It took a decade to develop this entire industrial ecosystem. Although initiated for economic reasons, the industrial ecosystem has distinct environmental benefits, from energy conservation to a reduction of pollution.

B

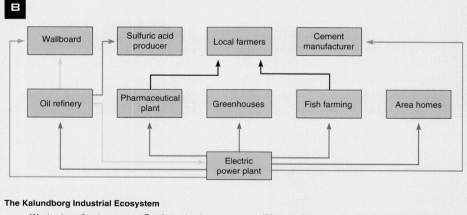

The Kalundborg Industrial Ecosystem

— Waste steam/heat   — Surplus natural gas   — Waste sulfur   — Waste calcium sulfate

— Fly ash   — Sludge

1 Plate Tectonics and Shifting Continents

1. The lithosphere, Earth's outermost rigid rock layer, is composed of plates that float on the asthenosphere, the region of the mantle where rocks become hot and soft. **Plate tectonics** is the study of the processes by which the lithospheric plates move over the asthenosphere. **Plate boundaries** are often sites of intense geologic activity: earthquakes, volcanoes, and mountain-building.

2 Economic Geology: Useful Minerals

1. **Minerals** are metallic or nonmetallic elements or compounds of elements that occur naturally in Earth's crust. An **ore** is rock with a large enough concentration of a mineral to be profitably mined and extracted. **High-grade ores** contain relatively large amounts of minerals.

2. Minerals are extracted through surface or subsurface mining. **Surface mining** removes the **overburden**: the overlying soil, subsoil, and rock strata. Strip mining, a type of surface mining, produces a **spoil bank** when the overburden from a new trench is put into an excavated trench. **Subsurface mining** extracts resources from deep underground deposits.

3. Processing minerals often involves **smelting**, melting the ore in a blast furnace to separate impurities from the metal.

3 Environmental Implications of Mineral Use

1. Surface mining destroys vegetation across large areas, increasing erosion. Open-pit mining uses huge quantities of water. Mining also affects water quality. **Acid mine drainage** is pollution caused when dissolved toxic materials wash from mines into nearby lakes and streams.

2. **Derelict lands** degraded by mining can be restored by filling in and grading the land to its natural contours, then planting vegetation to hold the soil in place.

4 Soil Properties and Processes

1. **Soil** is the uppermost layer of Earth's crust and supports terrestrial plants, animals, and microorganisms. Soil is formed from parent material: rock that is slowly fragmented into small particles by biological, chemical, and physical **weathering processes**.

2. Soil is composed of mineral particles, organic matter, water, and air. **Soil horizons** are the horizontal layers into which many soils are organized, from the surface to the underlying parent material.

3. Soil organisms provide **ecosystem services** such as maintaining soil fertility and preventing soil erosion. Soil organisms carry out **nutrient cycling**, the pathway of nutrient minerals or elements from the environment through organisms and back to the environment.

5 Soil Problems and Conservation

1. **Sustainable soil use** is the wise use of soil resources, without a reduction in the amount or fertility of soil, so it is productive for future generations. Soil used in a sustainable way renews itself by natural processes year after year.

2. Water, wind, ice, and other agents cause **soil erosion**, the wearing away or removal of soil from the land. Soil erosion reduces fertility as essential minerals and organic matter are removed. Erosion causes sediments and pesticide and fertilizer residues to pollute nearby waterways.

3. Good soil conservation practices promote sustainable soil use. In **conservation tillage**, residues from previous crops partially cover the soil to help hold it in place until newly planted seeds are established. **Crop rotation**, the planting of different crops in a field over a period of years, decreases the insect damage, disease, and mineral depletion that occur when one crop is grown continuously. **Contour plowing**, which matches the natural contour of the land, helps control erosion of land with variable topography. **Strip cropping** produces alternating strips of different crops along natural contours. **Terracing** reduces soil erosion on steep slopes. A **shelterbelt** is a row of trees planted as a windbreak to reduce soil erosion.

KEY TERMS

- plate tectonics p. 288
- minerals p. 292
- surface mining p. 294
- overburden p. 294
- spoil bank p. 295
- subsurface mining p. 295

- smelting p. 295
- acid mine drainage p. 296
- soil p. 299
- soil horizons p. 300
- nutrient cycling p. 302
- sustainable soil use p. 302

- soil erosion p. 303
- conservation tillage p. 304
- crop rotation p. 305
- contour plowing p. 305
- shelterbelt p. 306

CRITICAL AND CREATIVE THINKING QUESTIONS

1. How are plate tectonics and tsunamis related?

2. How many minerals have you come in contact with today? Which were metals; which were nonmetals?

3. What is the difference between surface and subsurface mining? Open-pit and strip mines? Shaft and slope mines? When is each most likely to be used?

4. What are the roles of weathering, organisms, and topography in soil formation?

5. Certain pests that cause plant disease reside in the plant residues left on the ground with conservation tillage. Given that these organisms are often specific for the plants they attack, how can we control them?

6. How did Copper Basin, Tennessee, become an environmental disaster? Are reclamation efforts making a difference there?

7. How does the industrial ecosystem at Kalundborg resemble a natural ecosystem? What are some environmental benefits of this industrial ecosystem?

8–9. The figure represents the flow of minerals in a low-waste society.

8. Which colors of arrows represent sustainable manufacturing, consumer reuse, and consumer recycling, respectively?

9. Would the flow of minerals be more or less complicated in a high-waste society? How would the wastes generated differ from those of the low-waste society represented here?

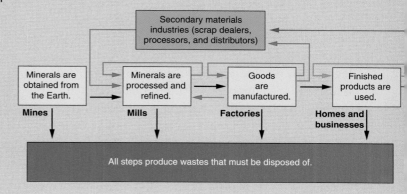

Secondary materials industries (scrap dealers, processors, and distributors)

Minerals are obtained from the Earth.	Minerals are processed and refined.	Goods are manufactured.	Finished products are used.
Mines	**Mills**	**Factories**	**Homes and businesses**

All steps produce wastes that must be disposed of.

What is happening in this picture ?

At this former coal mine site in West Virginia, special equipment carries out "hydroseeding," applying a mixture of grass seed, water, and fertilizers. What is the purpose of this effort?

Why do you think hydroseeding was chosen here rather than applying grass seed using equipment that rolls across the ground?

Hydroseeding is also used after other activities have disturbed large patches of ground; suggest one or two likely examples.

Environmental Geography: Land Resources

13

KORUP NATIONAL PARK

Korup National Park in Cameroon has the richest biological diversity in Africa *(see larger photograph)*. It is home to more than 400 tree species, 50 mammal species, and more than 320 bird species.

Korup National Park is a project of the wildlife conservation group World Wildlife Fund (WWF), in cooperation with the government of Cameroon and other organizations. The WWF helps protect, maintain, and integrate the park into regional development plans. To alleviate poverty, the WWF offers people the opportunity to train in local technical colleges. They receive education about environmental issues and the importance of conservation. Some are employed as park staff members, including game guards. This training also discourages the killing of animals—the park's most serious problem —especially of threatened or endangered species. When villagers have alternative sources of income, they are less likely to poach animals. Recently, the local people and park authorities formed an anti-poaching partnership; villagers now inform the park manager when they find poachers.

Long-term plans for Korup National Park include conservation and sustainable use of forest resources (see the inset of forest products that can be harvested sustainably). The six villages now located inside park boundaries will be voluntarily relocated to fertile land outside them. An established research program develops sustainable management practices for the park. Tourism, which depends on preservation of the area's unique biological diversity, provides income for the local economy. Korup National Park is a model of land conservation that other developing countries, in cooperation with conservation organizations and with foreign aid, might profitably emulate.

Land Use in the United States p. 312

Forests p. 314

Rangelands p. 323

National Parks and Wilderness
Areas p. 326

Conservation of Land
Resources p. 331

Case Study: The Tongass Debate
over Clear-Cutting p. 332

Land Use in the United States

Private citizens, corporations, and non-profit organizations own about 55 percent of the land in the United States and Native American tribes own about 3 percent. State and local governments own another 7 percent. The federal government owns the rest (about 35 percent).

Government-owned land encompasses all types of ecosystems, from tundra to desert, and includes land that contains important resources such as minerals and fossil fuels, land that possesses historical or cultural significance, and land that provides critical biological habitat. Most federally owned land is in Alaska and 11 western states (**FIGURE 13.1**).

It is managed primarily by four agencies, three in the U.S. Department of the Interior—the Bureau of Land Management (BLM), the Fish and Wildlife Service (FWS), and the National Park Service (NPS)—and one in the Department of Agriculture—the U.S. Forest Service (USFS) (**TABLE 13.1**).

Government-owned lands provide vital **ecosystem services** that benefit humans living far from public forests, grasslands, deserts, and wetlands. These services

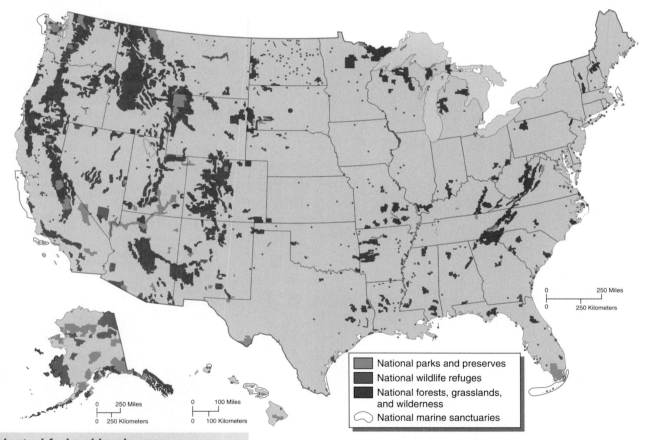

National parks and preserves

National wildlife refuges

National forests, grasslands, and wilderness

National marine sanctuaries

Selected federal lands FIGURE 13.1

Shown are national parks and preserves, national wildlife refuges, national forests and grasslands, and national marine sanctuaries in the United States. Note the preponderance of federal lands in western states and Alaska. Other federal lands, such as military installations and research facilities, aren't shown.

Administration of federal lands TABLE 13.1

Agency	Land held	Area in millions of hectares (acres)
Bureau of Land Management (Dept. of Interior)	National resource lands	109 (270)
U.S. Forest Service (Dept. of Agriculture)	National forests	77 (191)
U.S. Fish and Wildlife Service (Dept. of Interior)	National wildlife refuges	37 (92)
National Park Service (Dept. of Interior)	National Park System	34 (84)
Other—includes Department of Defense, Corps of Engineers (Dept. of the Army), and Bureau of Reclamation (Dept. of Interior)	Remaining federal lands	29 (72)
Total federal lands		**286 (709)**

include wildlife habitat, flood and erosion control, groundwater recharge, and the breakdown of pollutants.

Undisturbed public lands are ecosystems that scientists use as a benchmark, or point of reference, to determine the impact of human activity. Geologists, zoologists, botanists, ecologists, and soil scientists are some of the scientists who use government-owned lands for scientific inquiry. These areas provide perfect settings for educational experiences not only in science but also in history, because they can be used to demonstrate the condition of the land when humans originally settled here (FIGURE 13.2).

Public lands are important for their recreational value, providing places for hiking, swimming, boating, rafting, sport hunting, and fishing. Wild areas—forest-covered mountains, rolling prairies, barren deserts, and other undeveloped areas—are important to the human spirit. We can escape the tensions of the civilized world by retreating, even temporarily, to the solitude of natural areas.

Not all public lands remain undeveloped. As you will see throughout this chapter, many public lands are developed for uses ranging from logging to cattle grazing to mineral extraction.

Delta River in the Tangle Lakes area in Alaska
FIGURE 13.2

This area is one of six wild and scenic rivers managed by the U.S. Bureau of Land Management in Alaska.

CONCEPT CHECK STOP

What percentage of land in the United States is privately owned?

What percentage of public land is owned by the federal government?

Forests

LEARNING OBJECTIVES

Define sustainable forestry and explain how monocultures and wildlife corridors are related to it.

Define deforestation, including clear-cutting, and relate the main causes of tropical deforestation.

Describe national forests, stating which government agencies administer them and current issues of concern.

Forests, important ecosystems that provide many goods and services to support human society, occupy less than one-third of Earth's total land area. Timber harvested from forests is used for fuel, construction materials, and paper products. Forests supply nuts, mushrooms, fruits, and medicines. Forests provide employment for millions of people worldwide and offer recreation and spiritual sustenance to an increasingly crowded world.

Forests also provide a variety of beneficial ecosystem services, such as influencing climate condi-

Process Diagram

Role of forests in the hydrologic cycle FIGURE 13.3

Forests return most of the water that falls as precipitation to the atmosphere by transpiration. When an area is deforested, almost all precipitation is lost as runoff.

Up to 75 percent water recycled by transpiration and evaporation

25 percent or more water seeps into ground or runs off to rivers, streams, and lakes

tions. If you walk into a forest on a hot summer day, you will notice that the air is cooler and moister than it is outside the forest. This is the result of a biological cooling process called *transpiration*, in which water from the soil is absorbed by roots, transported through plants, and then evaporated from their leaves and stems. Transpiration provides moisture for clouds, eventually resulting in precipitation (see FIGURE 13.3 on facing page). Thus, forests help maintain local and regional precipitation.

Forests play an essential role in regulating global biogeochemical cycles like those for carbon and nitrogen. Photosynthesis by trees removes large quantities of heat-trapping carbon dioxide from the atmosphere and fixes it into carbon compounds, while releasing the oxygen back into the atmosphere. Forests thus act as carbon "sinks," which may help mitigate global warming and which produce the oxygen that almost all organisms require for cellular respiration.

Tree roots hold vast tracts of soil in place, reducing erosion and mudslides. Forests protect watersheds because they absorb, hold, and slowly release water; this moderation of water flow provides a more regulated flow of water downstream, even during dry periods, and helps control floods and droughts. Forest soils remove impurities from water, improving its quality. In addition, forests provide a variety of essential habitats for many organisms, such as mammals, reptiles, amphibians, fishes, insects, lichens and fungi, mosses, ferns, conifers, and numerous kinds of flowering plants.

FOREST MANAGEMENT

Management for timber production disrupts a forest's natural condition and alters its species composition and other characteristics. Specific varieties of commercially important trees are planted, and those trees not as commercially desirable are thinned out or removed. *Traditional forest management* often results in low-diversity forests. In the southeastern United States, many tree plantations of young pine grown for timber and paper production are all the same age and

Tree plantation FIGURE 13.4

This intensively managed pine plantation in the southern United States is a monoculture, with trees of uniform size and age. Such plantations supplement harvesting of trees in wild forests to provide the United States with the timber it requires.

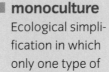

monoculture
Ecological simplification in which only one type of plant is cultivated over a large area.

are planted in rows a fixed distance apart (FIGURE 13.4). These "forests" are essentially **monocultures**—areas uniformly covered by one crop, like a field of corn. Herbicides are sprayed to kill shrubs and herbaceous plants between the rows. One of the disadvantages of monocultures is that they are more prone to damage from insect pests and disease-causing microorganisms. Also, because managed forests contain few kinds of food, they can't support the variety of organisms typically found in natural forests.

In recognition of the many ecosystem services performed by natural forests, a newer method of forest management, known as **ecologically sustainable forest management**, or simply **sustainable forestry**, is evolving.

Sustainable forestry maintains a mix of forest trees, by age and species, rather than imposing a monoculture. This broader approach seeks to conserve forests for the long-term commercial harvest of timber and nontimber forest products. Sustainable forestry also attempts to sustain biological diversity by providing habitats for a variety of species, prevent soil erosion and improve soil conditions, and preserve watersheds that produce clean water. Effective sustainable forest management involves cooperation among environmentalists; loggers; farmers; indigenous people; and local, state, and federal governments.

When loggers use sustainable forestry principles, they set aside unlogged areas and **wildlife corridors** as sanctuaries for organisms. The purpose of wildlife corridors is to provide animals with escape routes, should they be needed, and to allow them to migrate so they can interbreed. (Small, isolated, inbred populations may have a higher risk of extinction.) Wildlife corridors may also allow large animals to maintain large territories. Recent research on wildlife corridors in fragmented landscapes suggests that wildlife corridors help certain wildlife populations persist. Additional research is needed to resolve the effectiveness of wildlife corridors for all endangered species.

Methods for ecologically sustainable forest management are under development. Such practices vary from one forest ecosystem to another, in response to different environmental, cultural, and economic conditions. In Mexico, many sustainable forestry projects involve communities that are economically dependent on forests. Because trees have such long life spans, scientists and forest managers of the future will judge the results of today's efforts.

Harvesting trees   According to the U.N. Food and Agricultural Organization (FAO), 3.4 million cubic meters of wood (for fuelwood, timber, and other products) were harvested in 2004 (the latest available data). The five countries with the greatest tree harvests are the United States, Canada, Russia, Brazil, and China; these countries currently produce more than half the world's timber. About 50 percent of harvested wood is burned directly as fuelwood or used to make charcoal. Most fuelwood and charcoal are used in developing countries (see Chapter 18). Highly developed countries consume more than three-fourths of the remaining 50 percent for paper and wood products.

Loggers harvest trees in several ways—selective cutting, shelterwood cutting, seed tree cutting, and clear-cutting (see What a Scientist Sees on facing page). **Selective cutting**, in which mature trees are cut individually or in small clusters while the rest of the forest remains intact, allows the forest to regenerate naturally.

The removal of all mature trees in an area over an extended period is **shelterwood cutting**. In the first year, undesirable tree species and dead or diseased trees are removed. Subsequent harvests occur at intervals of several years, allowing time for remaining trees to grow.

In **seed tree cutting**, almost all trees are harvested from an area; a scattering of desirable trees is left behind to provide seeds for the regeneration of the forest.

Clear-cutting is harvesting timber by removing all trees from an area, then either allowing the area to reseed and regenerate itself naturally or planting the area with one or more specific varieties of trees. Timber companies prefer clear-cutting because it is the most cost-effective way to harvest trees. However, clear-cutting over wide areas is ecologically unsound. It destroys biological habitats and increases soil erosion, particularly on sloping land, sometimes degrading land so much that reforestation doesn't take place. Clear-cut areas at lower elevations are usually regenerated successfully, whereas those at high elevations are difficult to regenerate. Obviously, the recreational benefits of forests are lost when clear-cutting occurs.

■ **sustainable forestry** The use and management of forest ecosystems in an environmentally balanced and enduring way.

■ **wildlife corridor** A protected zone that connects isolated unlogged or undeveloped areas.

■ **clear-cutting** A logging practice in which all the trees in a stand of forest are cut, leaving just the stumps.

Harvesting Trees

A Aerial view of a large patch of clear-cut forest in Washington state. Clear-cutting is the most common but most controversial type of logging. The lines are roads built to haul away the logs.

www.wiley.com/
college/berg

B (1 to 3) As a forest scientist looks at the clear-cut forest, he or she may think about the various kinds of tree harvesting that are less environmentally destructive than clear-cutting (4).

(1) In **selective cutting**, the older, mature trees are selectively harvested from time to time and the forest regenerates itself naturally.

(2) In **shelterwood cutting**, less desirable and dead trees are harvested. As younger trees mature, they produce seedlings, which continue to grow as the now-mature trees are harvested.

(3) **Seed tree cutting** involves the removal of all but a few trees, which are allowed to remain, providing seeds for natural regeneration.

(4) In **clear-cutting**, all trees are removed from a particular site. Clear-cut areas may be reseeded or allowed to regenerate naturally.

The Forest Stewardship Council ecologically certifies "green" wood.

Green timber has proved profitable for its national distributors, who have seen sales increases of 20 percent to 30 percent. Often, the consumer pays no additional premium, or only slightly more, for ecologically certified wood, which has become so popular that demand threatens to exceed supply.

Many homebuilders and homeowners are interested in "green" wood for flooring and other building materials *(see photograph)*. Such wood is ecologically certified by a legitimate third party, such as the Mexico-based Forest Stewardship Council (FSC), to have come from a forest managed with environmentally sound and socially responsible practices. Although these areas remain a small percentage of world forests, by early 2005 the FSC had certified as well managed more than 48 million hectares (120 million acres) in 61 countries. Certification is based on sustainability of timber resources, socioeconomic benefits provided to local people, and forest ecosystem health, which includes such considerations as preservation of wildlife habitat and watershed stability.

Green forestry has its detractors. Traditional forestry organizations are skeptical about the reliability of FSC investigations and the economic viability of this type of forestry. Trade experts caution that government efforts to specify the purchase of certified timber could violate global free-trade agreements. Still, green timber is gaining market share, pleasing business owners and consumers alike, and offering the promise of better conservation practices in managed forests.

DEFORESTATION

The most serious problem facing the world's forests is **deforestation**. According to latest estimates by the FAO, world forests shrank by about 7.3 million hectares (18 million acres) each year, between 2000 and 2005. This estimate represents a total loss of about one percent of global forested area. Causes of this decades-long trend of deforestation include fires caused by drought and land clearing practices, expansion of agriculture, construction of roads, tree harvests, insects, disease, and mining.

Most of the world's deforestation is currently taking place in Africa and South America, according to the FAO.

> ■ **deforestation**
> The temporary or permanent clearance of large expanses of forest for agriculture or other uses.

Africa lost about 3.2 percent of its forested area from 2000 to 2005, and South America lost about 2.5 percent. Central America and the Caribbean nations lost about 3.9 percent of their forests during this period. There was a very small loss of forested area in North America, whereas Europe and Asia actually gained forested areas, either by natural regrowth or by increasing forest plantations.

Results of deforestation

Deforestation results in decreased soil fertility, as the essential mineral nutrients found in most forest soils leach away rapidly without trees to absorb them. Uncontrolled soil erosion, particularly on steep deforested slopes, affects the production of hydroelectric power as silt builds up behind dams. Increased sedimentation of waterways caused by soil erosion harms downstream fisheries. In drier areas, deforestation contributes to the formation of deserts. Regulation of water flow is disrupted when a forest is removed, so that the affected region experiences alternating periods of flood and drought.

Deforestation contributes to the extinction of many species. (See Chapter 15 for a discussion of the importance of tropical forests as repositories of biological diversity.) Many tropical species, in particular, have limited ranges within a forest, so they are especially vulnerable to habitat modification and destruction. Migratory species, including birds and butterflies, also suffer from deforestation.

Deforestation is thought to induce regional and global climate changes. Trees release substantial amounts of moisture into the air; about 97 percent of the water that roots absorb from the soil is evaporated directly into the atmosphere, then falls back to Earth in the hydrologic cycle. When a large forested area is removed, local rainfall may decline, droughts may become more common in that region, and temperatures may rise slightly. Deforestation may also contribute to an increase in global temperature by releasing carbon originally stored in the trees into the atmosphere as carbon dioxide, which enables the air to retain heat. When an old-growth forest is harvested, researchers estimate that it takes about 200 years for the replacement forest to accumulate the equivalent amount of carbon stored in the original trees.

Boreal forests and deforestation

Extensive deforestation in boreal forests due to logging began in the late 1980s. **Boreal forests** occur in Alaska, Canada, Scandinavia, and northern Russia, and are dominated by coniferous evergreen trees such as spruce, fir, cedar, and hemlock. Harvested primarily by clear-cut logging, boreal forests are the primary source of industrial wood and wood fiber. The annual loss of boreal forests is estimated to encompass an area twice as large as the rain forests of Brazil.

About 1 million hectares (2.5 million acres) of forest in Canada—currently the world's biggest timber exporter—are logged annually, and most of Canada's forests are subject to logging contracts. On the basis of current harvest quotas, logging is unsustainable in Canada (**FIGURE 13.5**).

Extensive tracts of Siberian forests in Russia are also harvested, although estimates are unavailable. Alaska's boreal forests are at risk because the U.S. government may increase logging on public lands in the future.

Logging in Canada's boreal forest
FIGURE 13.5

About 80 percent of Canada's forest products are exported to the United States.

Tropical forests and deforestation

There are two types of tropical forests: tropical rain forests and tropical dry forests. **Tropical rain forests** prevail in warm areas that receive 200 cm (at least 79 in.) or more of precipitation annually. Tropical rain forests are found in Central and South America, Africa, and Southeast Asia, but almost half of them are in just three countries: Brazil, Democratic Republic of the Congo, and Indonesia (**Figure 13.6** in Visualizing Tropical Deforestation). **Tropical dry forests** occur in other tropical areas where annual precipitation is less but is still enough to support trees. India, Kenya, Zimbabwe, Egypt, and Brazil are a few of the countries that have tropical dry forests.

Most of the remaining undisturbed tropical forests, which lie in the Amazon and Congo river basins of South America and Africa, respectively, are being cleared and burned at a rate unprecedented in human history. Tropical forests are also being destroyed at an extremely rapid rate in southern Asia, Indonesia, Central America, and the Philippines.

Why are tropical rain forests disappearing? Several studies show a strong statistical correlation between population growth and deforestation. More people need more food, and so forests are cleared for agricultural expansion.

However, tropical deforestation can't be attributed simply to population pressures because it is also affected by a variety of interacting economic, social, and governmental factors that vary from place to place. Government policies sometimes provide incentives that favor the removal of forests. The Brazilian government opened the Amazonian frontier, beginning in the late 1950s, by constructing the Belem-Brasilia Highway, which cut through the Amazon Basin. Such roads open the forest for settlement (**Figure 13.7**). An example of economic conditions encouraging deforestation is the farmer who converts forest to pasture so that he can maintain a larger herd of cattle.

Keeping in mind that the origins of tropical deforestation are complex, three agents—subsistence agriculture, commercial logging, and cattle ranching—are considered its most immediate causes. Subsistence agriculture, in which a family produces just enough food to feed itself, accounts for perhaps 60 percent of tropical deforestation. Subsistence farmers carry out **slash-and-burn agriculture** (discussed further in Chapter 14). Subsistence farmers often follow loggers' access roads until they find a suitable spot. They cut down trees and allow them to dry; then they burn the area and plant crops immediately after burning. The yield from the first crop is often quite high because the nutrients that were in the burned trees are now available in the soil. Soil productivity subsequently declines at a rapid rate, and subsequent crops are poor. In a short time, the people farming the land must move to a new part of the forest and repeat the process. Cattle ranchers often claim the abandoned land for grazing, because land not rich enough to support crops can still support livestock.

Slash-and-burn agriculture done on a small scale, with plenty of forest to shift around in so that there are periods of 20 to 100 years between cycles, is sustainable. The forest regrows rapidly after a few years of farming. But when millions of people try to obtain a living in this way, the land is not allowed to lie uncultivated long enough to recover.

Another 20 percent of tropical deforestation is the result of commercial logging. Vast tracts of tropical rain forests are harvested for export abroad. Most tropical countries allow commercial logging to proceed at a much faster rate than is sustainable. Unmanaged logging does not contribute to economic development; rather, it depletes a valuable natural resource faster than it can regenerate for sustainable use.

Approximately 12 percent of tropical deforestation provides open rangeland for cattle. Other causes of tropical rainforest destruction include the development of hydroelectric power, which inundates large areas of forest, and mining, particularly when ore smelters burn charcoal produced from rainforest trees (**Figure 13.8**).

Why are tropical dry forests disappearing? Tropical dry forests are also being destroyed at a rapid rate, primarily for fuelwood (**Figure 13.9**). About half of the wood consumed worldwide is used as heating and cooking fuel by much of the developing world. Often the wood cut for fuel is converted to charcoal, which is then used to power steel, brick, and cement factories. Charcoal production is extremely wasteful: 3.6 metric tons (4 tons) of wood produce only enough charcoal to fuel an average-sized iron smelter for 5 minutes.

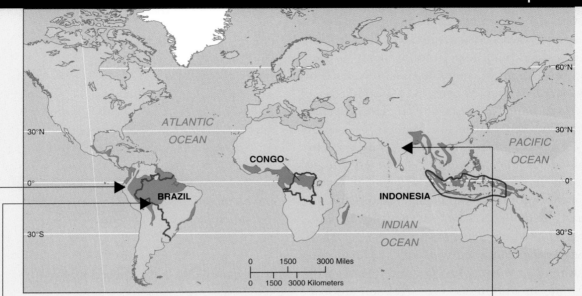

Distribution of tropical rain forests FIGURE 13.6 ▲

Rain forests (green areas) in Central and South America, Africa, and Southeast Asia. Much of the remaining forested area is highly fragmented. The three countries with the largest area of tropical rain forest are outlined.

◄
Human settlements along a road in Brazil's tropical rain forest FIGURE 13.7

This satellite photograph shows numerous smaller roads extending perpendicularly from the main roads. As farmers settle along the roads, they clear out more and more forest (dark green) for croplands and pastures (tans and pinks).

Deforestation for fuelwood ▲
FIGURE 13.9

Women gather firewood in the Ranthambore National Park buffer zone in India. Note the branches trimmed off the trees in the background. About 94 percent of wood removed from Indian forests is burned as fuel.

◄
Deforestation due to mining

FIGURE 13.8

Forest was removed for this oil exploration drilling site in Ecuador.

FORESTS IN THE UNITED STATES

In recent years, most temperate forests in the Rocky Mountains, Great Lakes region, and New England and other eastern states have been holding steady or even expanding. In Vermont, the amount of land covered by forests has increased from 35 percent in 1850 to 80 percent today. Expanding forests are the result of *secondary succession* on abandoned farms (see Figure 6.19 on page 149), the commercial planting of tree plantations on both private and public lands, and government protection. Although these second- and third-growth forests generally don't have the biological diversity of virgin stands, many organisms have successfully become reestablished in the regenerated areas.

Slightly more than one-half of U.S. forests are privately owned (**FIGURE 13.10**), and three-fourths of these private lands are in the northeast and midwest. Many private owners are under economic pressure to subdivide the land and develop tracts for housing or shopping malls, as they seek ways to recoup their high property taxes. Projected conversion of forests to agricultural, urban, and suburban lands over the next 40 years places the greatest potential impact in the South, where more than 85 percent of forest is privately owned and logging is largely unregulated.

U.S. national forests

According to the USFS, the United States has 155 national forests encompassing 77 million hectares (191 million acres) of land, mostly in Alaska and western states. The USFS manages most national forests, while the BLM oversees the remainder. National forests were established for multiple uses, including timber harvesting; mining; livestock foraging; hunting, fishing, and other forms of outdoor recreation; water resources and watershed protection; and habitat for fishes and wildlife. Recreation, which increased dramatically in national forests during the 1990s and early 2000s, ranges from camping at designated campsites to backpacking in the wilderness. Visitors to national forests swim, boat, picnic, and observe nature. With so many possible uses of national forests, conflicts inevitably arise, particularly between timber interests and those who wish to preserve the trees for other purposes.

Road building is a particularly contentious issue, in part because the USFS builds taxpayer-funded roads to

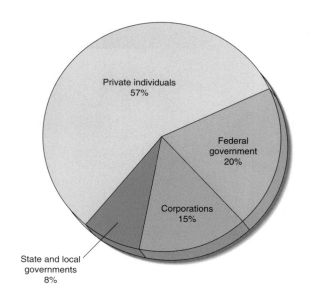

Forest ownership in the United States

FIGURE 13.10

Most forests are privately owned.

allow private logging companies access to the forest to remove timber. (See the Case Study at the end of the chapter for an example.) Road building is environmentally destructive when improper construction accelerates soil erosion and mudslides (particularly on steep terrain), and causes water pollution in streams. Biologists are concerned that so many roads fragment wildlife habitat and provide entries for disease organisms and invasive species.

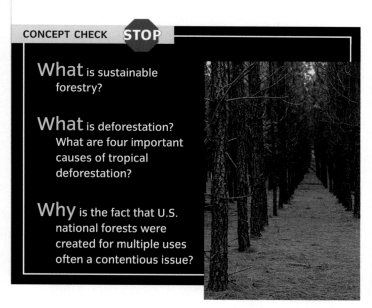

CONCEPT CHECK STOP

What is sustainable forestry?

What is deforestation? What are four important causes of tropical deforestation?

Why is the fact that U.S. national forests were created for multiple uses often a contentious issue?

Rangelands

R angelands are grasslands, in both temperate and tropical climates, that serve as important areas of food production for humans by providing fodder for livestock such as cattle, sheep, and goats (FIGURE 13.11). Rangelands may be mined for minerals and energy resources, used for recreation, and preserved for biological habitat and for soil and water resources. The predominant vegetation of rangelands includes grasses, *forbs* (small plants other than grasses), and shrubs.

rangeland
Land that isn't intensively managed and is used for grazing livestock.

RANGELAND DEGRADATION AND DESERTIFICATION

Grasses, the predominant vegetation of rangelands, have a *fibrous root system*, in which many roots form a diffuse network in the soil to anchor the plant. Plants with fibrous roots hold the soil in place quite well, thereby reducing soil erosion. Grazing animals eat the leafy shoots of the grass, and the fibrous roots continue to develop, allowing the plants to recover and regrow to their original size.

Carefully managed grazing is beneficial for grasslands. Because rangeland vegetation is naturally adapted to grazing, when grazing animals remove mature vegetation, the activity stimulates rapid regrowth. At the same time, the hooves of grazing animals disturb the soil surface enough to allow rainfall to more effectively reach the root systems of grazing plants. Several studies report that moderate levels of grazing encourage greater plant diversity.

Rangeland FIGURE 13.11

When the carrying capacity of rangeland is not exceeded, it is a renewable resource. Photographed along the Salmon River in Idaho.

The **carrying capacity** of a range-land is the maximum number of animals the natural vegetation can sustain over an indefinite period without deterioration of the ecosystem. When the carrying capacity of a rangeland is exceeded, grasses and other plants are **overgrazed**. When plants die, the ground is left barren, and the exposed soil is susceptible to erosion. Sometimes plants that do not naturally grow in a rangeland, but which can tolerate the depleted soil, invade an overgrazed area.

Most of the world's rangelands lie in semiarid areas that have natural extended droughts. Under normal conditions, native grasses in these dry lands can survive a severe drought. But when an extended drought occurs in conjunction with overgrazing, once-fertile rangeland may be converted to desert as reduced grass cover allows winds to erode the

■ **overgrazing**
When too many grazing animals consume the plants in a particular area, leaving the vegetation destroyed and unable to recover.

■ **desertification**
Degradation of once-fertile rangeland or tropical dry forest into non-productive desert.

soil. Even when the rains return, the degradation may be so extensive that the land cannot recover. **Land degradation** is a natural or human-induced process that decreases the future ability of the land to support crops or livestock. Water erosion removes the little remaining topsoil and the sand left behind forms dunes.

This progressive degradation, which induces unproductive desert-like conditions on formerly productive rangeland (or tropical dry forest), is **desertification** (FIGURE 13.12). It reduces the agricultural productivity of economically valuable land, forces many organisms out, and threatens endangered species. Worldwide, desertification seems to be on the increase. The United Nations estimates that each year since the mid-1990s, 3,560 km^2 (1,374 mi^2)—an area about the size of Rhode Island—has turned into desert.

Global Locator

Desertification in the African Sahel region, Sudan FIGURE 13.12

Note the Sahara Desert encroaching on the grasslands. There is still much that scientists don't understand about desertification and the processes related to it, such as to what extent desertification results from natural fluctuations in climate versus population pressures and human activities.

RANGELAND TRENDS IN THE UNITED STATES

Rangelands make up approximately 30 percent of the total land area in the United States, mostly in the western states. Of this, approximately one-third is publicly owned and two-thirds are privately owned. Much of the private rangeland is under increasing pressure from developers, who want to subdivide the land into lots for homes and condominiums. To preserve the open land, conservation groups often pay ranchers for **conservation easements** that prevent future owners from developing the land. An estimated 405,000 hectares (one million acres) of private rangelands are protected by conservation easements.

Excluding Alaska, there are at least 89 million hectares (220 million acres) of public rangelands in the United States. The BLM manages approximately 69 million hectares (170 million acres) of public rangelands, and the USFS manages an additional 20 million hectares (50 million acres).

Overall, the condition of public rangelands in the United States has slowly improved since the low point of the Dust Bowl in the 1930s, when the combined effects of poor agricultural practices, severe winds, and extended drought led to devastating soil erosion and dramatic declines in soil productivity. Much of this improvement is attributed to fewer livestock being permitted to graze the rangelands after the passage of the **Taylor Grazing Act** of 1934, the **Federal Land Policy and Management Act** of 1976, and the **Public Rangelands Improvement Act** of 1978. Better livestock management practices and scientific monitoring have also contributed to rangeland recovery.

But restoration is slow and costly, and more is needed. Rangeland management includes seeding in places where plant cover is sparse or absent, conducting controlled burns to suppress shrubby plants, constructing fences to allow rotational grazing, controlling invasive weeds, and protecting habitats of endangered species. Most livestock operators use public rangelands in a way that results in their overall improvement.

conservation easement A legal agreement that protects privately owned forest, rangeland, or other property from development for a specified number of years.

Issues involving public rangelands

The federal government distributes permits that allow private livestock operators to use public rangelands for grazing in exchange for a fee that is much lower than grazing costs on private land. The permits are held for many years and are not open to free-market bidding by the general public—that is, only ranchers who live in the local area are allowed to obtain grazing permits. Some environmental groups are concerned about the ecological damage caused by overgrazing of public rangelands and want to reduce the number of livestock animals allowed to graze. They want public rangelands managed for other uses, such as biological habitat, recreation, and scenic value, rather than exclusively for livestock grazing. To accomplish this goal, they would like to purchase grazing permits and set aside the land for nongrazing purposes.

Conservative economists have joined environmentalists in criticizing the management of federal rangelands. According to policy analysts at Taxpayers for Common Sense, in 2003 taxpayers contributed at least $67 million more than the grazing fees collected, in order to support grazing on public rangelands. This money is used to manage and maintain the rangelands and to repair damage caused by overgrazing. Taxpayers for Common Sense and other free-market groups want grazing fees increased to cover all costs of maintaining herds on publicly owned rangelands.

CONCEPT CHECK STOP

What are rangelands?

How can overgrazing of rangeland lead to desertification?

How do conservation easements help protect privately owned rangelands?

Which agencies manage public rangelands? What management issues do they face?

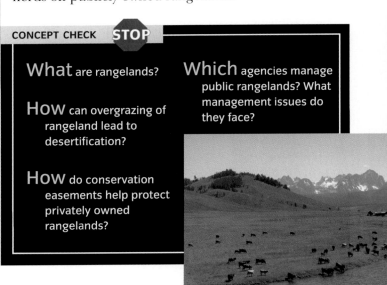

National Parks and Wilderness Areas

Many acres of federal land are set aside either as national park property or as wilderness areas. Both types of land were established to encourage the protection of the natural environment, yet both experience conflicts associated with how best to use and manage these protected areas.

NATIONAL PARKS

Created in 1916 as a federal bureau in the Department of the Interior, the **National Park System** (NPS) was originally composed of large, scenic areas in the West like Yellowstone, the Grand Canyon, and Yosemite Valley (**FIGURE 13.13** in Visualizing National Parks).

Today the NPS has more cultural and historical sites—battlefields and historically important buildings and towns—than places of scenic wilderness. The NPS currently administers 388 sites, 58 of which are national parks. The total acreage that the NPS administers is 34.1 million hectares (84.3 million acres).

Because knowledge and understanding increase enjoyment, one of the primary roles of the NPS is to teach people about the natural environment, management of natural resources, and history of a site by providing nature walks and guided tours of its parks. Exhibits along roads and trails, evening campfire programs, museum displays, and lectures are other common educational tools.

The popularity and success of U.S. national parks (**TABLE 13.2**) have encouraged many other nations to establish national parks. Today about 1,200 national parks exist in more than 100 countries (see the introduction to this chapter). As in the United States, in other countries parks usually have multiple roles, from providing biological habitat to facilitating human recreation.

Threats to U.S. parks Some national parks are overcrowded (**FIGURE 13.14**). All the problems plaguing urban areas are found in popular national parks during peak seasonal use, including crime, vandalism, litter, traffic jams, and pollution of the soil, water, and air. In addition, thousands of resource violations, from cutting live trees and collecting plants, minerals, and fossils to defacing historical structures with graffiti, are investigated in national parks each year. Park managers have had to reduce visitor access to park areas that have become degraded from overuse.

Some national parks have imbalances in wildlife populations. Populations of many mammal species are in decline, including bears, white-tailed jackrabbits, and red foxes. For example, grizzly bear populations in national parks of the western United States are threatened. Grizzlies are territorial and require large areas of wilderness as habitat, and the human presence in national parks may imperil them. Most important, the parks may be too small to support grizzlies. While grizzly bears have survived in sustainable numbers in Alaska and Canada, the populations native to the western continental United States are in considerable danger.

Yosemite National Park in California FIGURE 13.13 ▲

This winter view shows the Merced River flowing past the rock formation El Capitan.

Overcrowding in national parks FIGURE 13.14

A Summer crowd at Grand Canyon National Park in Arizona. The popularity of certain national parks threatens to overwhelm them. ▼

A

B The popularity of certain national ▲ parks threatens to overwhelm them.

The ten most popular national parks TABLE 13.2

National park	Number of recreational visitors in 2003 (in millions)
Great Smoky Mountains (North Carolina, Tennessee)	9.4
Grand Canyon (Arizona)	4.1
Yosemite (California)	3.4
Olympic (Washington)	3.2
Rocky Mountain (Colorado)	3.1
Yellowstone (Wyoming, Montana, Idaho)	3.0
Cuyahoga Valley (Ohio)	2.9
Zion (Utah)	2.5
Acadia (Maine)	2.4
Grand Teton (Wyoming)	2.4
Total visitors to the national park system	266.1

Gray wolves prey upon elk in Yellowstone National Park FIGURE 13.15

Since their reintroduction, gray wolf populations have gained a secure foothold in Yellowstone National Park. While studies of their effects on elk and deer populations are still ongoing, scientists have predicted that they may help reduce the burgeoning elk population.

Other mammal populations—notably elk—have proliferated. Elk in Yellowstone National Park's northern range have increased from a population of 3,100 in 1968 to a record high of 19,000 in 1994. Ecologists have documented that the elk have reduced the abundance of native vegetation, such as willow and aspen, and seriously eroded stream banks. The reintroduction of gray wolves to Yellowstone, which began in 1995 and 1996, may help to control the elk population (FIGURE 13.15).

National parks are increasingly becoming islands of natural habitat surrounded by human development. Development on the borders of national parks limits the areas in which wild animals may range, forcing them into isolated populations. Ecologists have found that when environmental stressors occur, several small "island" populations are more likely to become threatened than a single large population occupying a sizable range (see Chapter 15).

WILDERNESS AREAS

> ■ **wilderness** A protected area of land in which no human development is permitted.

Wilderness encompasses regions where the land and its community of organisms aren't greatly disturbed by human activities, where humans may visit but don't live permanently. The 88th Congress recognized that increased human population and expansion into wilderness areas might result in a future where no lands exist in their natural condition. Accordingly, the **Wilderness Act** of 1964 authorized the U.S. government to set aside federally owned land that retains its primeval character and lacks permanent improvements or human habitation, as part of the **National Wilderness Preservation System**. These federal lands range in size from tiny islands of uninhabited land to portions of national parks, national forests, and national wildlife refuges (FIGURE 13.16 on facing page).

Areas designated as wilderness are given the highest protection of any federal land. These areas are to remain natural and unchanged so they will be unimpaired for future generations to enjoy. The same four government agencies that regulate all publicly owned land—the NPS, USFS, FWS, and BLM—oversee 630 wilderness areas comprising 40.3 million hectares (102 million acres) of land. More than one-half of the lands in the National Wilderness Preservation System lie in Alaska, and western states contain much of the remainder.

Millions of people visit U.S. wilderness areas each year, and some areas are overwhelmed by this traffic: Eroded trails, soil and water pollution, litter and trash, and human congestion predominate over quiet, unspoiled land. Government agencies now restrict the number of people allowed into each wilderness area at one time so that human use doesn't seriously affect the wilderness.

Some of the most popular wilderness areas may require more intensive future management, such as the development of trails, outhouses, cabins, and campsites. These amenities are not encountered in true wilderness, posing a dilemma between wilderness preservation and human use and enjoyment of wild lands.

A The Blackbeard Island Wilderness in Georgia is only 1,214 hectares (3,000 acres), whereas (**B**) the Selway–Bitterroot Wilderness in Idaho is more than 530,000 hectares (1.3 million acres).

Diverse wilderness areas FIGURE 13.16

Limiting the number of human guests in a wilderness area doesn't control all the factors that threaten wilderness, however. **Invasive species** have the potential to upset the natural balance among native species. For example, white pine blister rust, a foreign (nonnative) fungus that kills white pine trees, has invaded the wilderness in the northern Rocky Mountains. Wilderness managers are concerned that declining white pine populations could harm the population of grizzly bears in the region, since pine seeds are a major part of the grizzlies' diet. The Wilderness Act specifies the avoidance of intentional ecological management. In this example, should the white pine population be scientifically manipulated to help preserve the original wilderness?

Large tracts of wilderness, most of it in Alaska, have been added to the National Wilderness Preservation System since passage of the Wilderness Act in 1964. People who view wilderness as a nonrenewable resource support the designation of additional wilderness areas. They think it is particularly important to preserve additional land in the lower 48 states, where currently less than 2 percent of the total land area is protected as wilderness. Increasing the amount of federal land in the National Wilderness Preservation System is opposed by groups who operate businesses on public lands (such as timber, mining, ranching, and energy companies) and by their political representatives.

Clear-cut forest in the Olympic National Park
FIGURE 13.17

The wise-use movement favors opening federal lands to logging and other types of economic development.

MANAGEMENT OF FEDERAL LANDS

How do we best manage the legacy of federal lands? These questions have divided many Americans into two groups, both coalitions of several hundred grassroots organizations. Those who wish to exploit resources on federal lands are known collectively as the *wise-use movement*. Those who wish to preserve the resources on federally owned lands are known collectively as the *environmental movement*.

In general, people who support the wise-use movement think that the government overregulates environmental protection and that property owners should have more flexibility to use natural resources. They believe that the primary purpose of federal lands is to enhance economic growth (**FIGURE 13.17**).

Some of the goals of the wise-use movement include the following:

1. Put all national forests under timber management, including old-growth forests.

2. Permit mining and commercial development of wilderness areas, wildlife refuges, and national parks, where appropriate.

3. Allow unrestricted development of wetlands.

4. Change the Endangered Species Act so economic factors are considered along with scientific ones (you will learn in Chapter 15 that threatened and endangered species are currently defined based on scientific information only).

5. Sell parts of resource-rich federal lands to private interests, such as mining, oil, coal, ranching, and timber groups, for resource extraction.

Many organizations that embrace the wise-use movement have environmentally friendly names. The National Wetlands Coalition, for example, consists primarily of real estate developers and energy companies who wish to drain and develop wetlands. Similarly, logging companies support the American Forest Resource Alliance.

In contrast to the wise-use movement, the environmental movement views federal lands as a legacy of U.S. citizens. They think that:

1. The primary purpose of public lands is to protect biological diversity and ecosystem integrity.

2. Those who extract resources from public lands should pay U.S. citizens compensation equal to the fair market value of the resource and not be subsidized by taxpayers.

3. Those who use public lands should be held accountable for any environmental damage they cause.

CONCEPT CHECK **STOP**

What government agency administers the National Park System? What problems does it face?

What is wilderness, and how does the U.S. National Wilderness Preservation System seek to protect it? What problems do they face?

How do the wise-use and environmental movements differ on the use of public lands?

Conservation of Land Resources

Our ancestors considered natural areas as an unlimited resource to exploit. They appreciated prairies as valuable agricultural land and forests as immediate sources of lumber and eventual farmland. This outlook was practical as long as there was more land than people needed. But as the population increased and the amount of available land decreased, it was necessary to consider land as a limited resource. Increasingly, the emphasis has shifted from exploitation to preservation of the remaining natural areas.

Although all types of ecosystems must be conserved, several are in particular need of protection (FIGURE 13.18). The USGS and the Defenders of Wildlife commissioned studies that developed a numerical ranking of the most endangered ecosystems in the United States. They used four criteria:

1. The area lost or degraded since Europeans colonized North America;

2. The number of present examples of a particular ecosystem, or the total area;

3. An estimate of the likelihood that a given ecosystem will lose a significant area or be degraded during the next 10 years;

4. The number of threatened and endangered species living in that ecosystem.

Endangered ecosytem

FIGURE 13.18

The Hitchcock Woods Natural Area in South Carolina is an example of a longleaf pine forest.

The Tongass Debate over Clear-Cutting

Despite its northern location along Alaska's southeastern coast, the Tongass National Forest is one of the world's few temperate rain forests (FIGURES A and B; also see Chapter 6 for a description of the temperate rainforest biome). It is one of the wettest places in the United States. This moisture supports old-growth forest of giant Sitka spruce, yellow cedar, and western hemlock, some of which are 700 years old. This 17-million-acre forest, the largest in the National Forest System, provides habitat for a wealth of wildlife, such as grizzly bears and bald eagles.

The Tongass is a prime logging area because a single large Sitka spruce may yield as much as 10,000 board feet of high-quality timber. The logging industry forms the basis of much of the local economy but conflicts with environmental interests seeking to protect the forests from overharvesting.

As in most national forests, it is expensive to log in the Tongass. To cover high operating costs, timber

Alaska's Tongass National Forest

This temperate rain forest (light green area) is in southeastern Alaska along the Pacific Ocean.

TABLE 13.3 lists the 15 most endangered U.S. ecosystems based on these criteria. Examples include the South Florida landscape, Southern Appalachian spruce-fir forests, and longleaf pine forests and savannas. As these ecosystems are lost and degraded, the organisms that compose them decline in number and in genetic diversity. Conservation strategies that set aside ecosystems are the best way to preserve an area's biodiversity.

As you have seen in this chapter, government agencies, private conservation groups, and private citizens have begun to set aside natural areas for permanent preservation. Such activities ensure our children and grandchildren will inherit a world with wild places and other natural ecosystems.

The fifteen most endangered ecosystems in the United States (in order of priority) TABLE 13.3

- South Florida landscape
- Southern Appalachian spruce-fir forests
- Longleaf pine forests and savannas
- Eastern grasslands, savannas, and barrens
- Northwestern grasslands and savannas
- California native grasslands
- Coastal communities in lower 48 states and Hawaii
- Southwestern riparian communities
- Southern California coastal sage scrub
- Hawaiian dry forest
- Large streams and rivers in lower 48 states and Hawaii
- Cave systems
- Tallgrass prairie
- California river- and stream-bank communities and wetlands
- Florida scrub

CONCEPT CHECK STOP

What are three U.S. ecosystems that need protection?

What are three criteria used to evaluate whether an ecosystem is endangered?

B

Aerial view of the Tongass.

C

USFS logging road in the Tongass. Roads open the forest to logging and other kinds of development.

interests such as pulp mills rely on obtaining the timber from the federal government at below-market prices. This right was granted in 1954 by a contract that expired in the 1990s. In 1990, congressional efforts to pass the Tongass Timber Reform Act, which would force timber interests to pay market prices, were bitterly opposed. The compromise agreement, reached in 1997, provided timber to the mills at market prices. As a result of this legislation, clear-cut logging continued in the Tongass, but at lower rates than in the past.

In the closing months of the Clinton administration, the USFS officially adopted the *Roadless Area Conservation Rule* to protect roadless national forests from road building and forest harvest. A federal judge blocked the roadless rule in 2001, and shortly thereafter the Bush administration opened to logging and development part of the Alaska forest formerly placed off-limits (FIGURE C).

The take-home message is that the USFS, like other government agencies, takes its lead from current presidential policies. Changes in administrations often leave the USFS and other government agencies floundering as they strive to implement established rulings that are no longer supported.

CHAPTER SUMMARY

1 Land Use in the United States

1. More than half of U.S. land is privately owned. Slightly more than a third is owned by the federal government, including many types of ecosystems and land uses. Seven percent belongs to state and local governments.

2 Forests

1. Sustainable forestry is the use and management of forest ecosystems in an environmentally balanced and enduring way. Sustainable forestry maintains a mix of forest trees, by age and species, rather than a **monoculture**, in which only one type of plant is cultivated over a large area. Adopting sustainable forestry principles requires setting aside sanctuaries and **wildlife corridors**, protected zones that connect isolated unlogged or undeveloped areas.

2. Deforestation is the temporary or permanent clearance of large expanses of forest for agriculture or other uses. **Clear-cutting** is a logging practice in which all the trees in a stand of forest are cut, leaving just the stumps; clear-cutting over a wide area is ecologically unsound. The major causes of tropical deforestation are subsistence farming, commercial logging, and cattle ranching, all accelerated by growing human populations. Increased needs for fuelwood drives deforestation of tropical dry forests.

3. Most U.S. national forests are managed by the U.S. Forest Service (USFS); the rest are overseen by the Bureau of Land Management. National forests face conflicts associated with supporting multiple uses: timber harvest; livestock forage; water resources and watershed protection; mining; hunting, fishing, and other forms of recreation; and habitat for fishes and wildlife.

3 Rangelands

1. **Rangelands** are grasslands that aren't intensively managed and are used for grazing livestock. Rangelands are also mined for mineral and energy resources, used for recreation, and preserved for biological habitat and for soil and water resources.

2. **Overgrazing** is the destruction of vegetation caused by too many grazing animals consuming the plants in a particular area, leaving them unable to recover.

Overgrazing accelerates **land degradation,** which decreases the future ability of the land to support crops or livestock. **Desertification** is the degradation of once-fertile rangeland or tropical dry forest into nonproductive desert.

3. A **conservation easement** is a legal agreement that protects privately owned forest or other property from development for a specified number of years. Conservation groups often pay for conservation easements to preserve open rangeland.

4. The BLM manages more than three-fourths of U.S. public rangelands, excluding Alaska; the USFS manages the remainder. Current issues on public rangelands include conflicts between environmental groups and ranchers over the number of livestock allowed to graze and the potential to manage the areas for uses such as biological habitat, recreation, and scenic value. Conflicts also arise over whether grazing fees paid by livestock operators on public lands should be high enough to cover all costs of maintaining herds, removing taxpayer burden.

4 National Parks and Wilderness Areas

1. The **National Park Service** administers 388 sites in the United States, including 58 national parks. The problems they encounter include overcrowding, pollution, crime, resource violations, and imbalanced wildlife populations.

2. **Wilderness** is a protected area of land in which no human development is permitted. The **National Wilderness Preservation System** consists of four U.S. government agencies—the NPS, USFS, FWS, and BLM—that oversee 630 wilderness

areas. The problems they face include overuse and overcrowding by visitors, pollution, erosion, and the introduction of invasive species.

3. Those who support the **wise-use movement** believe a primary purpose of federal lands is to enhance economic growth. They think that the government overregulates environmental protection and that property owners should have more flexibility to use natural resources. Those who support the **environmental movement** view federal lands as a legacy of U.S. citizens and thus want to preserve resources on federally owned lands.

5 Conservation of Land Resources

1. Endangered U.S. ecosystems include the southern Florida landscape, southern Appalachian spruce-fir forests, and longleaf pine forests and savannas.

2. Criteria used to evaluate whether an ecosystem is endangered and to what degree it is threatened include its history of land loss and degradation, its prospects for future loss or degradation, the area the ecosystem occupies, and the number of threatened and endangered species living in that ecosystem.

KEY TERMS

- monoculture p. 315
- sustainable forestry p. 316
- wildlife corridor p. 316

- clear-cutting p. 316
- deforestation p. 318
- rangeland p. 323
- overgrazing p. 324

- desertification p. 324
- conservation easement p. 325
- wilderness p. 328

CRITICAL AND CREATIVE THINKING QUESTIONS

1. Why is deforestation a serious global environmental problem?

2. What are the environmental effects of clear-cutting on steep mountain slopes? On tropical rainforest land?

3. Distinguish between rangeland degradation and desertification. Why is moderate grazing beneficial to rangelands, yet overgrazing leads to erosion?

4. Explain the various uses that must be considered in the management of a national park. Use Korup National Park as an example in which the needs of the human population surrounding a park have been addressed successfully.

5. Debate conflicts over logging in Tongass National Forest from two points of view, that of the wise-use movement and that of the environmental movement.

6. Do you think additional federal lands should be added to the wilderness system? Why or why not?

7. Should private landowners have control over what they wish to do to their land? How would you as a landowner handle land-use decisions that may affect the public? Present arguments for both sides of the issue.

8. Explain how economic growth and sustainable use of natural resources can be compatible goals.

9. Give at least five ecosystem services provided by forests and other non-urban lands.

10. Suppose a valley contains a small city surrounded by agricultural land. The land is encircled by a mountain wilderness. Explain why the preservation of the mountain ecosystem would support both urban and agricultural land in the valley.

11. Given the important contributions of forests in providing both timber and ecosystem services, how would you manage U.S. public forests if you were in charge?

12–14. Examine the graph of world land use that was assembled by the World Resources Institute and the U.N. Food and Agricultural Organization.

12. How much of the world's total land area is used for agriculture?

13. Approximately 29% of the world's land area consists of natural ecosystems that could potentially be developed for human use. What areas were counted in this estimate?

14. Why do you think the people who assembled these data placed roads and urban areas in the same category as rock, ice, tundra, and deserts?

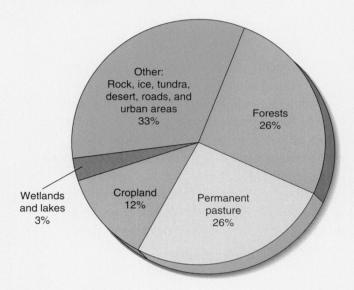

Other: Rock, ice, tundra, desert, roads, and urban areas 33%

Forests 26%

Wetlands and lakes 3%

Cropland 12%

Permanent pasture 26%

What is happening in this picture ?

Julia "Butterfly" Hill lived in this nearly-1,000-year-old, 180-foot-tall California redwood for more than two years in the late 1990s, to keep a lumber company from cutting the tree down. Would Hill's perspective on wilderness better fit the wise-use movement or the environmental movement?

Explain the likely differences in the perspectives of Hill and the lumber company, especially given that the tree is on the company's land.

Why do ecologists and environmentalists believe that the logging of old-growth trees causes such particular damage?

See if you can find out what happened to this tree after Julia left it.

Agriculture and Food Resources

MAINTAINING GRAIN STOCKPILES FOR FOOD SECURITY

W hen people have access at all times to adequate amounts and kinds of food needed
for healthy, active lives, they are said to have *food
security*. World grain carryover stocks—the amounts
of rice, wheat, corn, and other grains remaining
from previous harvests—provide a measure of world
food security *(see larger photo)*. Stockpiles of grain
have decreased since their all-time high in 1987,
when carryover stocks were large enough, if evenly
distributed, to feed every person in the world for
104 days. In contrast, the amount of grain stockpiled
in 2004 would have fed the world's people for only
80 days. According to the United Nations, world
grain carryover stocks should not fall below a minimum of 70 days' supply in a given year.

The year 2004 had the largest harvest on
record and reversed a decade-long trend of declining harvests. The bumper grain harvest in 2004
was the result of farmers planting more land in grain
crops and of good weather conditions in major
grain-growing areas.

World grain stocks are under pressure because
consumption of beef, pork, poultry, and eggs has
increased in developing countries like China *(see
inset)*, where growing affluence has led some
people to diversify their diets. This trend represents
a global pattern: In highly developed countries, animal products account for nearly half of the calories
people consume, compared to only 5 percent of
the calories people consume in developing countries. Increased consumption of meat and meat
products has prompted a surge in the amount of
grain used to feed the world's billions of livestock
animals. Mirroring this trend, in 2004 China produced more wheat than any other country, yet
was also the world's largest importer of wheat.

Challenges of Agriculture

LEARNING OBJECTIVES

Discuss recent trends in loss of U.S. agricultural land, global declines in domesticated plant and animal varieties, and efforts to increase crop and livestock yields.

Relate the benefits and problems associated with the green revolution.

Describe the environmental impacts of industrialized agriculture, including land degradation and habitat fragmentation.

The United States has more than 121 million hectares (300 million acres) of **prime farmland**, land that has the soil type, growing conditions, and available water to produce food, forage, fiber, and oilseed crops. U.S. agriculture faces a decline in prime farmland. Other challenges include coping with declining numbers of domesticated varieties, improving crop and livestock yields, and addressing environmental impacts.

LOSS OF AGRICULTURAL LAND

There is considerable concern that much of our prime agricultural land is falling victim to urbanization and sub-

urban sprawl by being converted to parking lots, housing developments, and shopping malls (**FIGURE 14.6**). In certain areas of the United States, loss of rural land is a significant problem. According to the American Farmland Trust, more than 162,000 hectares (400,000 acres) of prime U.S. farmland are lost each year.

The 1996 Farm Bill included funding for the establishment of a national *Farmland Protection Program*. This voluntary program lets farmers sell **conservation easements** that prevent their farmland from being converted to nonagricultural uses. The easements are in effect from a minimum of 30 years to forever. As with other conservation easements, the farmers retain full rights to use their property, in this case, for agricultural purposes.

Suburban spread onto agricultural land FIGURE 14.6

Virginia housing developments encroach on farmland.

GLOBAL DECLINE IN DOMESTICATED PLANT AND ANIMAL VARIETIES

A global trend is currently underway to replace the many local varieties of a particular crop or domesticated farm animal with just a few kinds. When farmers abandon traditional varieties in favor of more modern ones, which are bred for uniformity and maximum production, the traditional varieties frequently face extinction (FIGURE 14.7). This represents a great loss in genetic diversity, because each variety's characteristic combination of genes gives it distinctive nutritional value, size, color, flavor, resistance to disease, and adaptability to different climates and soil types.

> **germplasm** Any plant or animal material that may be used in breeding.

To preserve older, more diverse varieties of plants and animals, many countries, including the United States, are collecting **germplasm**: seeds, plants, and plant tissues of traditional crop varieties, and the sperm and eggs of traditional livestock breeds.

Dutch Belted cow FIGURE 14.7

The USDA recognizes Dutch Belts as a viable dairy breed that produces high-quality, flavorful milk. The breed is endangered in the United States and rare in Canada. Photographed in Wisconsin.

INCREASING CROP YIELDS

Until the 1940s, agricultural yields among various countries, whether highly developed or developing, were generally equal. Advances by research scientists since then have dramatically increased food production in highly developed countries (FIGURE 14.8). Greater knowledge of plant nutrition has resulted in fertilizers that promote high yields. The use of pesticides to control insects, weeds, and disease-causing organisms has also improved crop yields.

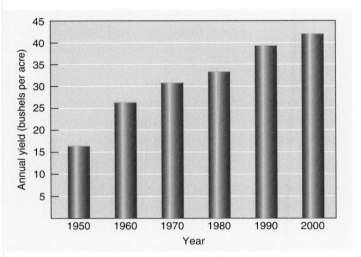

Average U.S. wheat yields, 1950 to 2000
FIGURE 14.8

Each year shown is actually an average of 3 years to minimize the effects that poor weather conditions might have in a given year. Similar increases in yield occurred in other grain crops.

The green revolution By the middle of the 20th century, serious food shortages occurred in many developing countries coping with growing populations. The development and introduction during the 1960s of high-yield varieties of wheat and rice to Asian and Latin American countries gave these nations the chance to provide their people with adequate supplies of food (FIGURE 14.9). But the high-yield varieties required intensive industrial cultivation methods, including the use of commercial inorganic fertilizers, pesticides, and mechanized machinery, to realize their potential. These agricultural technologies were passed from highly developed nations to developing nations.

Using modern cultivation methods and the high-yield varieties of certain staple crops to produce more food per acre of cropland is known as the **green revolution**. Some of the success stories of the green revolution are remarkable. During the 1920s, Mexico produced less than 700 kg (0.77 ton) of wheat per hectare annually. During the green revolution years that began in 1965, Mexico's annual wheat production rose to more than 2,400 kg (2.65 tons) per hectare. Indonesia, which formerly imported more rice than any other country in the world, today produces enough rice to feed its people and export some.

Critics of the green revolution argue that it has made developing countries dependent on imported technologies, such as agrochemicals and tractors, at the expense of traditional agriculture. The two most important problems associated with higher crop production are the high energy costs built into this type of agriculture and the environmental problems caused by the intensive use of commercial inorganic fertilizers and pesticides (discussed later in this chapter).

Increasing crop yields in the post–green revolution era
In 1999, the International Food Policy Research Institute projected that the world demand for rice, wheat, and corn will increase 40 percent between 2000 and 2020. This rise in demand will require a corresponding rise in grain production to feed the human population, as well as the livestock needed to satisfy the appetites of the increasing numbers of affluent people who can afford to buy meat.

This challenge cannot be met by increasing the amount of land under cultivation, as the best arable lands are already being cultivated. Projected freshwater shortages, rising costs of agricultural chemicals, and deteriorating soil quality caused by intensive agricultural techniques may further constrain productivity. As Figure 14.9 demonstrates, recent progress in coaxing more grain out of crops genetically improved during the green revolution has resulted in diminishing returns. Grain yields have continued to rise since the 1960s, but in recent years, the rates of increase have not been as great as they were previously.

Despite these problems, most plant geneticists think we can produce enough food in the 21st century to meet demand if countries spend more money in support of a concerted scientific effort to improve crops. Many scientists think genetic engineering is one of the keys to breeding more productive varieties. Additionally, modern agricultural methods, such as water-efficient irrigation, will have to be introduced to developing countries that do not currently have them if we are to continue increasing crop yields.

Development of high-yield rice varieties FIGURE 14.9

Tall conventional plant   Improved high-yielding plant   Low-tillering ideotype (new plant type)

Traditional rice plants on the left are taller and do not yield as much grain (clusters at top of plant) as the more modern varieties shown in the middle and on the right. The rice plant in the middle was developed during the 1960s by crossing a high-yield, disease-resistant variety with a dwarf variety to prevent the grain-heavy plants from falling over. Improvements since the green revolution have been modest, as the rice variety, developed during the 1990s (right), shows. Some researchers think rice and certain other genetically improved crops are near their physical limits of productivity.

INCREASING LIVESTOCK YIELDS

The use of hormones and antibiotics, although controversial, increases animal production. **Hormones** regulate livestock bodily functions and promote faster growth. Although U.S. and Canadian farmers use hormones, the European Union (EU) currently restricts imports of hormone-treated beef. They cite studies that suggest that these hormones or their breakdown products, both found in trace amounts in meat and meat products, could cause cancer or affect the growth of young children. In 1999 an international scientific committee organized by the FAO and WHO concluded that the trace amounts of hormones found in beef are safe because they are very low compared to the normal hormone concentrations found in the human body.

Modern agriculture has also embraced the addition of low doses of **antibiotics** to the feed for pigs, chickens, and cattle. These animals gain 4 to 5 percent more weight than untreated animals, presumably because they expend less energy fighting infections.

Several studies link the indiscriminate use of antibiotics in humans and livestock to the evolution of bacterial strains with a resistance to antibiotics. Therefore, WHO recommended in 2003 that routine use of antibiotics in livestock be eliminated. Many European countries have complied, but the United States and many other countries continue the practice.

ENVIRONMENTAL IMPACTS

Industrialized agriculture has many environmental effects (**TABLE 14.1** in Visualizing Environmental Impacts of Agriculture). The agricultural use of fossil fuels and pesticides produces air pollution. Untreated animal wastes and agricultural chemicals such as fertilizers and pesticides cause water pollution, which reduces biological diversity, harms fisheries, and leads to outbreaks of nuisance species. According to the Environmental Protection Agency, agricultural practices are the single largest cause of surface-water pollution in the United States.

Industrialized agriculture has favored the replacement of traditional family farms by large agribusiness conglomerates. In the United States, most cattle, hogs, and poultry are now grown in feedlots and livestock factories (**FIGURE 14.10**). The large concentrations of animals in livestock factories create many environmental problems, including air and water pollution.

Many insects, weeds, and disease-causing organisms have developed or are developing resistance to **pesticides**. Pesticide resistance forces farmers to apply progressively larger quantities of pesticides (**FIGURE 14.11**). Pesticide residues contaminate our food supply and reduce the number and diversity of beneficial microorganisms in the soil. Fishes and other aquatic organisms are sometimes killed by pesticide runoff into lakes, rivers, and estuaries.

Land degradation is a reduction in the potential productivity of land. Soil erosion, which is exacerbated by large-scale mechanized operations, causes a decline in soil fertility, and the eroded sediments damage water quality. Other types of degradation are compaction of soil by heavy farm machinery and waterlogging and **salinization** (salting) of soil from improper irrigation methods.

Clearing grasslands and forests and draining wetlands to grow crops result in **habitat fragmentation** that reduces biological diversity. Many species are endangered or threatened as a result of habitat loss to agriculture. The most dramatic example of habitat loss in North America is tallgrass prairie, more than 90 percent of which has been converted to agriculture.

pesticide Any toxic chemical used to kill pests.

degradation (of land) Natural or human-induced processes that decrease the future ability of the land to support crops or livestock.

habitat fragmentation The breakup of large areas of habitat into small, isolated patches.

CONCEPT CHECK STOP

What is the green revolution? What are some of its benefits and problems?

What are the major environmental problems associated with industrialized agriculture?

A researcher observes genetically modified rice plants. "Golden rice" (inset on right) has been successfully engineered to produce beta-carotene, a source of vitamin A. This variety has the potential to improve world health because about half the world's people eat rice as their staple food, and rice is a poor source of many vitamins. (Golden rice is still not available commercially.)

Genetic engineering of rice varieties FIGURE 14.14

does) or that would be rich in necessary vitamins (FIGURE 14.14). Crop plants resistant to insect pests, viral diseases, drought, heat, cold, herbicides, or salty or acidic soils are also being developed.

Genetic engineering has been used to develop more productive farm animals, including quicker-growing hogs and fishes. Perhaps the greatest potential contribution of animal genetic engineering is the production of vaccines against disease organisms that harm agricultural animals. For example, genetically engineered vaccines have been developed to protect cattle against the deadly viral disease rinderpest, which is economically devastating in parts of Asia and Africa.

Concerns about genetically modified foods

During the late 1990s and early 2000s, opposition to genetically engineered crops increased in many countries in Europe and Africa. In 1999 the EU placed a 5-year moratorium on virtually all approvals of GM crops (the moratorium is now lifted). The EU refused to buy U.S. corn because it might be genetically modified. (Currently, about 40 percent of the U.S. corn crop is genetically modified.) One concern is that the inserted genes could spread in an uncontrolled manner from GM crops to weeds or wild relatives of crop plants and possibly harm natural ecosystems in the process. Scientists recognize this concern as legitimate and must take special precautions to avoid this possibility. Critics also

worry that some consumers might develop food allergies to GM foods, although scientists routinely screen new GM crops for allergenicity.

The scientific consensus is that the risks associated with consuming food derived from GM varieties are the same as those associated with consuming food derived from varieties produced by traditional breeding techniques. A growing body of evidence, summarized in the FAO *State of Food and Agriculture 2003–2004*, concludes that current **genetically modified (GM)** crop plants are as safe for human consumption as crops grown by conventional or organic agriculture. However, more research on the environmental impact of GM crops is required. To that end, strict guidelines exist in areas of genetic engineering research in which there are unanswered questions about possible effects on the environment.

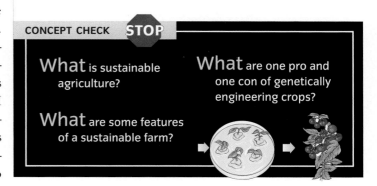

CONCEPT CHECK STOP

What is sustainable agriculture?

What are one pro and one con of genetically engineering crops?

What are some features of a sustainable farm?

Controlling Agricultural Pests

Any organism that interferes in some way with human welfare or activities is a **pest**. Some weeds, insects, rodents, bacteria, fungi, nematodes (microscopic worms), and other pest organisms compete with humans for food; other pests cause or spread disease. People try to control pests, usually by reducing the size of the pest population. *Pesticides* are the most common way of doing this, particularly in agriculture. Pesticides can be grouped by their target organisms—that is, by the pests they are supposed to eliminate. **Insecticides** kill insects, **herbicides** kill plants, **fungicides** kill fungi, and **rodenticides** kill rodents like rats and mice (**FIGURE 14.15**).

The ideal pesticide is a **narrow-spectrum pesticide** that kills only the intended organism and does not harm any other species. The perfect pesticide would readily break down, either by natural chemical decomposition or by biological organisms, into safe materials such as water, carbon dioxide, and oxygen. The ideal pesticide would stay exactly where it was put and would not move around in the environment. Unfortunately, no pesticide is perfect. Most pesticides are **broad-spectrum pesticides**. Some pesticides do not degrade readily, or they break down into compounds as dangerous as—if not more dangerous than—the original pesticide. And most pesticides move around the environment quite a bit.

> **broad-spectrum pesticide** A pesticide that kills a variety of organisms, including beneficial organisms, in addition to the target pest.

Some representative pesticides FIGURE 14.15

BENEFITS OF PESTICIDES

Pesticides can effectively control organisms, such as insects, that transmit devastating human diseases. Fleas and lice carry the microorganism that causes typhus in humans. Malaria, also caused by a microorganism, is transmitted to millions of humans each year by female *Anopheles* mosquitoes (**FIGURE 14.16**). Pesticides help control the population of mosquitoes, thereby reducing the incidence of malaria.

Pesticides also protect crops. It is widely estimated that pests eat or destroy more than one-third of the world's crops. Given our expanding population and world hunger, it is easy to see why control of agricultural pests is desirable. Pesticides reduce the amount of a crop lost through competition with weeds, consumption by insects, and diseases caused by plant **pathogens** (microorganisms, such as fungi and bacteria, that cause disease).

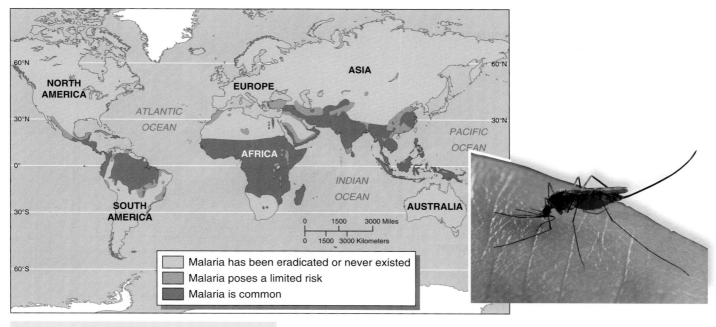

Location of malaria FIGURE 14.16

Insecticides sprayed to control mosquitoes in these locations have saved millions of lives. (inset) Mosquitos carry malaria and transfer it to other animals by biting them.

Why are agricultural pests found in such great numbers in our fields? Part of the reason is that agriculture is usually a monoculture, in which the field cultivated with a single species represents a simple ecosystem. In contrast, forests, wetlands, and other natural ecosystems are complex and contain many species, including predators and parasites that control pest populations, as well as plant species that pests do not use for food. A monoculture reduces the dangers and accidents that might befall a pest as it searches for food. In the absence of many natural predators and in the presence of plenty of food, a pest population thrives and grows, damaging more of the crop.

PROBLEMS WITH PESTICIDES

Although pesticides have their benefits, they have several problems. For one thing, the prolonged use of a particular pesticide can cause a pest population to develop **genetic resistance** to the pesticide. In the 50 years during which pesticides have been widely used, at least 520 species of insects and mites and at least 84 weed species have evolved genetic resistance to certain pesticides.

An organism exposed to a chemically stable pesticide that takes years to break down may accumulate high concentrations of the toxin, a phenomenon known as **bioaccumulation**. Organisms at higher levels on food webs tend to have greater concentrations of bioaccumulated pesticide stored in their bodies than those lower on food webs, through a process known as **biological magnification** (see Figure 4.8 on p. 81).

One of the worst problems associated with pesticide use is that pesticides affect more species than the pests for which they are intended. Beneficial insects are killed as effectively as pest insects. Pesticides do not have to kill organisms to harm them. Quite often the stress of carrying pesticides in their tissues makes organisms more vulnerable to predators, diseases, or other stressors in their environments. Because the natural enemies of pests often starve or migrate in search of food after pesticide is sprayed in an area, pesticides are indirectly responsible for a large reduction in the populations of

genetic resistance
Any inherited characteristic that decreases the effect of a given agent (like a pesticide) on an organism (like a pest).

Controlling Agricultural Pests **353**

these natural enemies. Pesticides also kill natural enemies directly, because predators may consume lethal amounts of a pesticide while consuming the pests. After a brief period, the pest population rebounds and gets larger than ever, partly because no natural predators are left to keep its numbers in check. In some instances, the use of a pesticide has resulted in a pest problem that did not exist before (see What a Scientist Sees).

Another problem associated with pesticides is that they do not stay where they are applied but tend to move through the soil, water, and air, sometimes long distances (**FIGURE 14.17**). Pesticides applied to agricultural lands wash into rivers and streams, where they can harm fishes. Pesticide mobility is also a problem for humans. About 14 million U.S. residents drink water containing traces of five widely used herbicides, and some people living where the herbicides are commonly used face a slightly elevated cancer risk because of their exposure.

Mobility of pesticides in the environment
FIGURE 14.17

A helicopter sprays pesticides on a crop—and everything else in its pathway—in California.

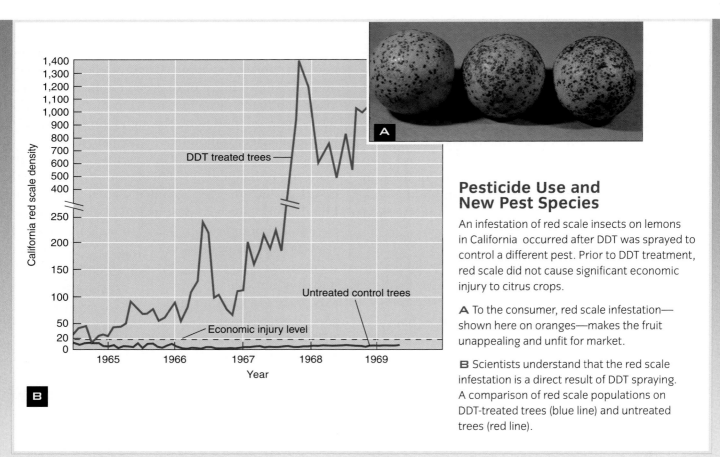

Pesticide Use and New Pest Species

An infestation of red scale insects on lemons in California occurred after DDT was sprayed to control a different pest. Prior to DDT treatment, red scale did not cause significant economic injury to citrus crops.

A To the consumer, red scale infestation—shown here on oranges—makes the fruit unappealing and unfit for market.

B Scientists understand that the red scale infestation is a direct result of DDT spraying. A comparison of red scale populations on DDT-treated trees (blue line) and untreated trees (red line).

biological control A method of pest control that involves the use of naturally occurring disease organisms, parasites, or predators to control pests.

pheromone A natural substance produced by animals to stimulate a response in other members of the same species.

ALTERNATIVES TO PESTICIDES

Given the many problems associated with pesticides, they are clearly not the final solution to pest control. Fortunately, pesticides are not the only weapons in our arsenal. Alternative ways to control pests include cultivation methods, **biological controls**, **pheromones** and hormones, reproductive controls, genetic controls, quarantine, and irradiation (TABLE 14.2 and FIGURE 14.18).

Pheromone traps FIGURE 14.18

A Japanese beetle trap uses a pheromone to lure large numbers of Japanese beetles, which fall into the bag and die.

Alternative methods of controlling agricultural pests TABLE 14.2

Pest control method	How it works	Disadvantages
Cultivation methods	Interplanting mixes different plants, as by alternating rows; strip cutting alternates crop harvest by portion—remaining portions protect natural predators, parasites of pests	No appreciable disadvantages; more care must be taken in harvest
Biological controls	Naturally occurring predators, parasites, or disease organisms are used to reduce pest populations	Organism introduced for biological control can unexpectedly affect environment or other organisms
Pheromones and hormones	Sexual attractants (pheromones) lure pest species to traps; synthetic regulatory chemicals (hormones) disrupt pests' growth and development	Might affect beneficial species
Reproductive controls	Sterilizing some members of pest population reduces population size	Expensive; must be carried out continually
Genetic controls	Selective breeding develops pest-resistant crops	Plant pathogens evolve rapidly, adapting to disease-resistant host plant; plant breeders forced to constantly develop new strains
Quarantine	Governments restrict importation of foreign pests, diseases	Not foolproof; pests are accidentally introduced
Irradiating foods	Harvested foods are exposed to ionizing radiation that kills potentially harmful microorganisms	Consumers concerned about potential radioactivity (not a true risk); irradiation forms traces of potentially carcinogenic chemicals (free radicals)

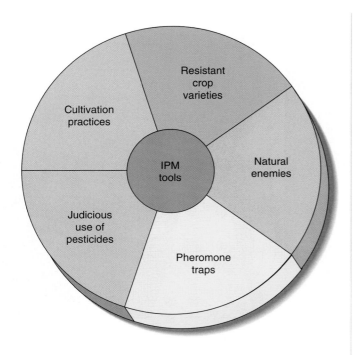

Tools of integrated pest management (IPM)
FIGURE 14.19

minimum of environmental disturbance and often at minimal cost.

To be effective, IPM requires a thorough knowledge of the life cycles and feeding habits of the pests as well as all their interactions with their hosts and other organisms. The timing of planting, cultivation, and treatments with biological controls is determined by carefully monitoring the concentration of pests. IPM is an important part of **sustainable agriculture**.

IPM is based on two fundamental premises. First, IPM is the *management* rather than the eradication of pests. Farmers who adopt IPM allow a low level of pests in their fields and accept a certain amount of economic damage from the pests. These farmers do not spray pesticides at the first sign of a pest. Instead, they periodically sample the pest population in the field to determine when the benefit of using pesticides exceeds the cost of that action.

Second, IPM requires that farmers be educated to understand what strategies will work best in their particular situations. Managing pests is more complex than trying to eradicate them. The farmer must know what pests to expect on each crop and what to do to minimize their effects.

Adoption of IPM by U.S. farmers has steadily increased since the 1960s, but the overall proportion of farmers using IPM is still small. One reason that IPM is not more widespread is that the knowledge required to use pesticides is relatively simple compared to the sophisticated knowledge needed to implement IPM.

INTEGRATED PEST MANAGEMENT

Many pests are not controlled effectively with a single technique; a combination of control methods is often more effective. **Integrated pest management (IPM)** combines the use of a variety of biological, cultivation, and pesticide controls tailored to the conditions and crops of an individual farm (**FIGURE 14.19**). Biological and genetic controls, including GM crops designed to resist pests, are used as much as possible, and conventional pesticides are used sparingly and only when other methods fail. When pesticides are required, the least toxic pesticides are applied in the lowest possible effective quantities. Thus, IPM allows the farmer to control pests with a

> ■ **integrated pest management (IPM)** A combination of pest control methods that, if used in the proper order and at the proper times, keep the size of a pest population low enough to prevent substantial economic loss.

CONCEPT CHECK **STOP**

What is a broad-spectrum pesticide?

What are two important benefits and two potential costs of pesticide use?

How can pests be controlled without pesticides?

DDT and the American Bald Eagle

Nesting pair of bald eagl[es]

The American bald eagle—the symbol of the United States and an emblem of strength—was a common sight throughout colonial North America (*photo*). More recently, its numbers dropped precipitously to only 417 nesting pairs in the lower 48 states in 1963. In danger of extinction, the bald eagle was listed as an endangered species following enactment of the *Endangered Species Act (ESA)* in 1973.

Several factors contributed to the bald eagle's decline. As European settlers pushed across North America, they cleared many thousands of square kilometers of forest near lakes and rivers, destroying the bald eagle's habitat. Eagles were hunted for sport and because it was thought they had a significant impact on commercially important fishes. In addition, eagles' numbers dwindled because they could not reproduce at high enough levels to ensure their population growth or their survival. This reproductive failure was the direct result of ingesting food contaminated with the pesticide DDT (dichlorodiphenyltrichloroethane). DDT made the eagles' eggs so thin-shelled that they cracked open before the embryos could mature and hatch. Mercury, lead, and selenium were other pollutants that harmed bald eagles.

Banning the use of DDT in the United States in 1972 started the recovery efforts for the bald eagle (see Figure 4.7 on page 80). Conservation efforts involving the U.S. Fish and Wildlife Service (FWS), other federal agencies, state and local governments, Native American tribes, conservation organizations, universities, corporations, and individuals have helped the bald eagle make a remarkable comeback.

As a result, the number of nesting pairs in the continental United States increased to more than 7,500 pairs in 2004. In 1994 the bald eagle was removed from the endangered list and transferred to the less critical threatened list. In 1999, the FWS proposed that it be removed from the threatened list, an action that has not yet occurred because of concerns about managing the eagle once it is delisted.

CHAPTER SUMMARY

1 World Food Problems

1. **Malnutrition** is the impairment of health due to consuming too few or too many calories. **Undernutrition** is a serious underconsumption of calories that leaves the body weakened and susceptible to disease. **Overnutrition** is a serious overconsumption of calories that leaves the body susceptible to disease.

2. **Food insecurity**, a condition in which people live with chronic hunger and malnutrition, is exacerbated by population growth, environmental problems, and poverty.

2 The Principal Types of Agriculture

1. **Industrialized agriculture** uses modern methods requiring large capital input and less land and labor than traditional methods. **Subsistence agriculture** uses traditional methods dependent on labor and a large amount of land to produce enough food to feed a family. There are three types of subsistence agriculture. In **slash-and-burn agriculture**, small patches of tropical forests are cleared to plant crops. **Nomadic herding**, carried out on arid land, requires herders to move livestock continually to find food for them. **Intercropping** involves growing a variety of plants simultaneously on the same field.

CHAPTER SUMMARY

3 Challenges of Agriculture

1. **Prime farmland** in the United States is being lost to urbanization and urban sprawl. Global declines in plant and animal varieties have led many countries to collect **germplasm**, any plant or animal material that may be used in breeding. Farmers and ranchers strive to increase yields in many ways, including administering **hormones** and antibiotics to livestock.

2. The **green revolution** introduced modern cultivation methods and high-yield crop varieties to Asia and Latin America. These methods require developing nations to import energy-intensive technologies and to face environmental problems caused by inorganic fertilizers and pesticides.

3. Environmental problems caused by industrialized agriculture include air pollution from the use of fossil fuels and pesticides, water pollution from untreated animal wastes and agricultural chemicals, pesticide-contaminated foods and soils, and increased resistance of pests to pesticides. **Land degradation** decreases the future ability of the land to support crops or livestock. Clearing grasslands and forests and draining wetlands to grow crops have resulted in **habitat fragmentation**, the breakup of large areas of habitat into small, isolated patches.

4 Solutions to Agricultural Problems

1. **Sustainable agriculture** uses methods that maintain soil productivity and a healthy ecological balance while minimizing long-term impacts. Unlike industrialized agriculture, sustainable agriculture relies on beneficial biological processes and environmentally friendly chemicals.

5 Controlling Agricultural Pests

1. A **pesticide** is any toxic chemical used to kill pests. A **narrow-spectrum pesticide** kills only the intended organism and does not harm other species. Most pesticides are **broad-spectrum pesticides**, which kill a variety of organisms, including beneficial ones, in addition to the target pest.

2. Pesticides can effectively control disease-carrying organisms and crop pests. The abundance of pests in agriculture is partly due to the common practice of **monoculture**, the cultivation of only one type of plant over a large area.

3. Pesticide use leads to several problems: pests evolve **genetic resistance**, an inherited characteristic that decreases the effect of a given agent (like a pesticide) on an organism; ecosystem imbalances occur when pesticides affect species other than the intended pests; and pesticides exhibit persistence, degrading

2. **Genetic engineering** is the manipulation of genes to produce a particular trait. Genetic engineering produces more productive livestock varieties, more nutritious crops, or crop plants resistant to pests, diseases, or drought. Concerns about genetic engineering include its potential to produce harmful organisms and to trigger food allergies.

very slowly. **Bioaccumulation** is the buildup of a persistent pesticide or other toxins in an organism's body. **Biological magnification** is the increased concentration of toxins, such as certain pesticides, in the tissues of organisms at higher levels in food webs. Pesticides also show mobility, moving to places other than where they were applied.

4. Safer alternates to pesticides include **biological controls**, which use naturally occurring disease organisms, parasites, or predators to control pests. **Pheromones**, natural substances produced by animals to stimulate a response in other members of the same species, are used to attract and trap pest species. **Integrated pest management** is a combination of pest control methods that, if used in the proper order and at the proper times, keep a pest population low enough to prevent substantial economic loss.

KEY TERMS

- undernutrition p. 338
- malnutrition p. 338
- overnutrition p. 338
- food insecurity p. 338
- economic development p. 340
- industrialized agriculture p. 341

- subsistence agriculture p. 342
- germplasm p. 344
- pesticide p. 346
- degradation (of land) p. 346
- habitat fragmentation p. 346
- sustainable agriculture p. 348

- genetic engineering p. 350
- broad-spectrum pesticide p. 352
- genetic resistance p. 353
- biological control p. 355
- pheromone p. 355
- integrated pest management (IPM) p. 356

CRITICAL AND CREATIVE THINKING QUESTIONS

1. Compare undernutrition, malnutrition, and overnutrition. Describe the types of people and the regions of the world most likely to be affected by each.

2. How and why do (a) population growth and (b) the rising consumption of meat affect world grain carryover stocks and world food security?

3. Distinguish between shifting cultivation and slash-and-burn agriculture.

4. What was the green revolution? Describe a few of its successes and shortcomings.

5. Name two environmental problems associated with industrialized agriculture, and give at least three examples of ways that industrialized agriculture could be made more sustainable.

6. Overall, do you think the benefits of pesticide use outweigh its disadvantages? Give at least two reasons for your answer.

7. Is a major goal of integrated pest management (IPM) to eradicate the pest species? Explain your answer.

8-9. Use the following graph to answer questions 8 and 9.

School attainment and undernourishment by region, 2000

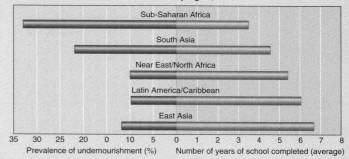

8. Based on the graph, what conclusion can you make about how the level of education people attain in a region is related to that region's prevalence of undernutrition? Give examples of data from specific regions to support your conclusion.

9. How do these results fit in with our understanding of the role that poverty plays in the world's hunger problems?

What is happening in this picture ?

- A pest controller drops insects, young lacewings, over crops. The lacewings attack aphids, a very destructive pest.

- What type of pest control is this?

- Would this practice typically be a part of industrialized agriculture or of sustainable agriculture?

- Can you think of any environmental costs associated with this approach?

Biological Resources

DISAPPEARING FROGS

Amphibians have existed as a group for more than 350 million years. Despite their evolutionary resilience, frogs, toads, and salamanders are remarkably sensitive environmental indicators in both aquatic and terrestrial ecosystems. Most frogs lay gelatinous and unprotected eggs in ponds and other pools of standing water, where the tadpoles metamorphose into adults. As adults, frogs breathe primarily through their moist, permeable skin, which makes them susceptible to environmental contaminants *(see larger photo)*. Amphibians are bellwether species. *Bellwether species,* also called *sentinel species*, provide early warnings of environmental damage that may affect other species.

Since the 1970s, many of the world's frog populations have dwindled or disappeared. As much as 38 percent of the native U.S. amphibian species are declining. Worldwide, 32 amphibian species have gone extinct in the last few decades, and about 200 species are in decline, even in remote, pristine locations. Biologists think multiple potential causes may be involved, including pollutants, increased UV radiation, infectious diseases, and global climate warming.

In 1995 Minnesota schoolchildren found almost half the leopard frogs they caught at a local pond were deformed—with extra legs or toes, eyes located on the shoulder or back, deformed jaws, bent spines, and missing legs, toes, and eyes *(see inset)*. Since then, most states have reported abnormally large numbers of amphibian deformities. Frog deformities have now been found on four continents. Scientists have demonstrated that several factors produce amphibian deformities, including pesticides, which affect development in frog embryos; parasite infestations; and multiple environmental stressors like habitat loss, disease, and air and water pollution.

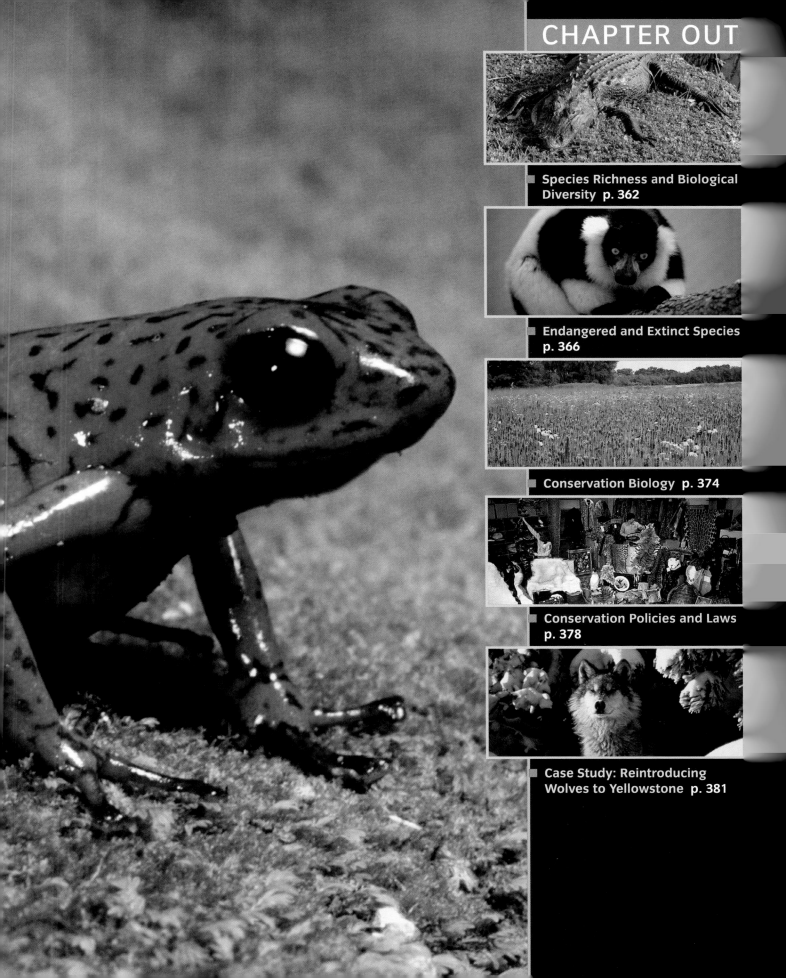

CHAPTER OUT

Species Richness and Biological Diversity p. 362

Endangered and Extinct Species p. 366

Conservation Biology p. 374

Conservation Policies and Laws p. 378

Case Study: Reintroducing Wolves to Yellowstone p. 381

Species Richness and Biological Diversity

A **species** is a group of distinct organisms capable of interbreeding with one another in the wild but which do not interbreed with organisms outside of their group. We do not know exactly how many species exist. In fact, biologists now realize how little we know about Earth's diverse organisms.

species richness
The number of different species in a community.

HOW MANY SPECIES ARE THERE?

Scientists estimate there may be as few as 5 million or as many as 100 million different species inhabiting Earth. To date, about 1.8 million species have been scientifically named and described, including about 270,000 plant species, 45,000 vertebrate animal species, and some 950,000 insect species. About 10,000 new species are identified each year.

Species richness varies greatly from one community to another. It is related to the abundance of potential ecological niches. A complex community, like the tropical rain forest or a coral reef, offers a greater variety of potential ecological niches than does a simple community like mountain chaparral (**FIGURE 15.1**).

Effect of community complexity on species richness
FIGURE 15.1

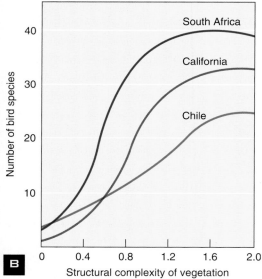

A Chilean dry scrub (chaparral) has low species richness and supports few bird species.

B The structural complexity of vegetation (*x*-axis) is based on its height and density. Note that species richness in birds increases as vegetation becomes more structurally complex. Chile has fewer bird species than either California or South Africa because mountains and deserts isolate Chile from the rest of South America. The chaparral habitats of California and South Africa are not isolated from the rest of their continents.

Species richness is inversely related to the geographic isolation of a community. Isolated island communities are much less diverse than communities in similar environments found on continents, for two reasons. Many species have difficulty reaching and successfully colonizing the island, and locally extinct species are not readily replaced in isolated environments such as islands or mountaintops.

Species richness is also inversely related to environmental stress. Only those species capable of tolerating extreme environmental conditions can live in an environmentally stressed community, such as a polluted stream or a polar region exposed to a harsh climate. Species richness is also reduced when one species enjoys a decided dominance within a community, because it may appropriate a disproportionate share of resources, thus crowding out other species.

Species richness is usually greater at the edges of adjacent communities than in their centers. This is because an **ecotone**—a transitional zone where communities meet—contains all or most of the ecological niches of the adjacent communities as well as some niches unique to the ecotone. The change in species composition produced at ecotones is the **edge effect**.

Geologic history greatly affects species richness. Tropical rain forests are probably old, stable communities that have undergone few climate changes over time, which allowed a large number of species to evolve. In contrast, glaciers have repeatedly altered temperate and arctic regions during Earth's history. An area recently vacated by glaciers will have a low species richness because few species will as yet have had a chance to enter it and become established.

WHY WE NEED BIODIVERSITY

The variation among organisms is referred to as **biological diversity** or **biodiversity**. Biological diversity occurs at all levels of biological organization, from populations to ecosystems (Chapter 5). It takes into account three components: species richness; **genetic diversity**, the genetic variety *within* all populations of that species (FIGURE 1 5.2); and **ecosystem diversity**, the variety of interactions among organisms in natural communities. For example, a forest community with its trees, shrubs, vines, herbs, insects, worms, vertebrate animals, fungi, bacteria, and other microorganisms has greater ecosystem diversity than a cornfield.

Humans depend on the contributions of thousands of species for their survival. For example, insects are instrumental in several ecological and agricultural processes, including pollination of crops, weed control, and insect pest control. Bacteria and fungi provide us with foods, antibiotics and other medicines, and biological processes such as nitrogen fixation (see Chapter 5). However, relatively few species have been evaluated for their potential usefulness to humans. There are approximately 270,000 known plant species, but as many as 250,000 of them have yet to be assessed for industrial, medicinal, or agricultural potential. The same is true for most of the millions of microorganisms, fungi, and animals.

> ■ **biological diversity**
> The number and variety of Earth's organisms; consists of three components: genetic diversity, species richness, and ecosystem diversity.

Genetic diversity in corn FIGURE 1 5.2

The variation in corn kernels and ears is evidence of the genetic diversity in the species *Zea mays*.

Ecosystem services and species richness

The living world functions much like a complex machine. Each ecosystem is composed of many parts that are organized and integrated to maintain the ecosystem's overall performance. The activities of all organisms are interrelated; we depend on one another and on the physical environment, often in subtle ways (FIGURE 15.3). When one species declines, other species linked to it may either decline or increase in number.

Ecosystems supply human societies with many environmental benefits, or **ecosystem services** (TABLE 15.1). Forests are not just a source of lumber; they provide watersheds from which we obtain fresh water, limit the number and severity of local floods, and reduce soil erosion. Many flowering plant species depend on insects to transfer pollen for reproduction. Soil dwellers develop and maintain soil fertility for plants. Bacteria and fungi perform the crucial task of decomposition, which allows nutrients to cycle in the ecosystem. Conservationists maintain that ecosystems with greater species richness supply ecosystem services better than ecosystems with lower species richness.

You might think that the loss of some species from an ecosystem might not endanger the rest of the organisms. However, if enough species are removed, the entire ecosystem will change. Species richness within an ecosystem provides the ecosystem with resilience, the ability to recover from environmental changes or disasters.

Alligators in the environment FIGURE 15.3

The American alligator plays an integral role in its natural ecosystem. Alligators help maintain populations of smaller fishes by eating the gar, a fish that preys on them. They dig underwater holes that other aquatic organisms use during droughts when the water level is low. Their nest mounds eventually form small islands colonized by trees and other plants. The trees on these islands support heron and egret populations. The alligator habitat is maintained in part by underwater "gator trails," which help clear out aquatic vegetation that might eventually form a marsh.

Ecosystem services TABLE 15.1	
Ecosystem	**Services provided**
Forests	Purify air and water
	Produce and maintain soil
	Absorb carbon dioxide (carbon storage)
	Provide wildlife habitat
	Provide humans with wood and recreation
Freshwater systems (rivers, lakes, and groundwater)	Moderate water flow and mitigate floods
	Dilute and remove pollutants
	Provide wildlife habitat
	Provide humans with drinking and irrigation water, food, transportation corridors, electricity, and recreation
Grasslands	Purify air and water
	Produce and maintain soil
	Absorb carbon dioxide (carbon storage)
	Provide wildlife habitat
	Provide humans with livestock and recreation
Coasts	Provide a buffer against storms
	Dilute and remove pollutants
	Provide wildlife habitat
	Provide humans with food, harbors, transportation routes, and recreation

IMPORTANCE OF GENETIC DIVERSITY

The maintenance of a broad genetic base is critical for each species' long-term health and survival. Consider economically important crop plants. During the 20th century, plant scientists developed genetically uniform, high-yielding varieties of important food crops like wheat. However, genetic uniformity resulted in increased susceptibility to pests and disease. By crossing the "super strains" with more genetically diverse relatives, disease and pest resistance can be reintroduced into such plants.

Genetic engineering, the incorporation of genes from one organism into a different species (see Chapter 14), makes it possible to use organisms' genetic resources on a wide scale. Genetic engineering has provided new vaccines, more productive farm animals, and disease-resistant agricultural plants.

Evolution has taken hundreds of millions of years to produce the genetic diversity found on our planet today. This diversity may hold solutions to today's problems and to future problems we have not begun to imagine. It would be unwise to allow such an important part of our heritage to disappear.

Medicinal, agricultural, and industrial importance of organisms

The genetic resources of organisms are vitally important to the pharmaceutical industry, which incorporates hundreds of chemicals derived from plants and other organisms into its medicines (**FIGURE 15.4**). Many of the natural products taken directly from marine organisms are promising anticancer or antiviral drugs. The AIDS (acquired immune deficiency syndrome) drug AZT (azidothymidine), for example, is a synthetic derivative of a compound from a sponge. The 20 best-selling prescription drugs in the United States are either natural products, natural products that are slightly modified chemically, or synthetic drugs whose chemical structures were obtained from organisms.

The agricultural importance of plants and animals is indisputable, because we must eat to survive. However, the number of different kinds of foods we eat is limited compared to the total number of edible species available in any given region. Many species are probably nutritionally superior to our common foods.

Medicinal value of the rosy periwinkle
FIGURE 15.4

The rosy periwinkle produces chemicals effective against certain cancers. Drugs from the rosy periwinkle have increased the chance of surviving childhood leukemia from about 5 percent to more than 95 percent.

Modern industrial technology depends on a broad range of products from organisms. Plants supply oils and lubricants, perfumes and fragrances, dyes, paper, lumber, waxes, rubber and other elastic latexes, resins, poisons, cork, and fibers. Animals provide wool, silk, fur, leather, lubricants, waxes, and transportation, and they are important in medical research. The armadillo, for example, is used for research in Hansen's disease (leprosy) because it is one of only two species known to be susceptible to that disease (the other species is humans). Certain beetles produce steroids with birth-control potential, and fireflies produce a compound that may be useful in treating viral infections.

Aesthetic, ethical, and spiritual value of organisms

Organisms not only contribute to human survival and physical comfort, they provide recreation, inspiration, and spiritual solace. Our natural world is a thing of beauty largely because of the diversity of living forms found in it. Artists have attempted to capture this beauty in drawings, paintings, sculpture, and photography; and poets, writers, architects, and musicians have created works reflecting and celebrating the natural world.

Traditionally, many human cultures have viewed themselves as superior beings, subduing and exploiting other forms of life for human benefit. An alternative view is that organisms have intrinsic value and that as stewards of the life forms on Earth, humans should protect their existence (see Chapter 2 for a discussion of *environmental ethics*).

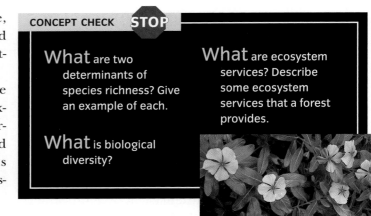

CONCEPT CHECK **STOP**

What are two determinants of species richness? Give an example of each.

What is biological diversity?

What are ecosystem services? Describe some ecosystem services that a forest provides.

Endangered and Extinct Species

LEARNING OBJECTIVES

Define extinction and distinguish between background extinction and mass extinction.

Contrast threatened and endangered species.

Describe four human causes of species endangerment and extinction.

Extinction, the death of a life form, occurs when the last member of a species dies. Once a species is extinct, it will never reappear. Biological extinction is the fate of all species, much as death is the fate of all individuals. Biologists estimate that for every 2,000 species that have ever lived, 1,999 of them are extinct today.

During the time in which organisms have occupied Earth, a continuous, low-level extinction of species, or **background extinction**, has occurred. Perhaps five or six times, a second kind of extinction, **mass extinction**, has occurred, in which a large number of species disappear during a relatively short period of geologic time.

The causes of past mass extinctions are not well understood, but biological and environmental factors were probably involved. Major climate change could have triggered the mass extinction or a catastrophe such as a collision between a large asteroid or comet and Earth.

extinction The elimination of a species from Earth.

Although extinction is a natural biological process, it is greatly accelerated by human activities. The burgeoning human population has spread into almost all areas of Earth. Whenever humans invade an area, the habitats of many organisms are disrupted or destroyed, which contributes to their extinction.

Currently, Earth's biological diversity is disappearing at an unprecedented rate (**FIGURE 15.5**). Conservation biologists estimate that species are now becoming extinct at a rate of 100 to 1,000 times the natural rate of background extinctions. More than 34,000 plant species are currently threatened with extinction.

ENDANGERED AND THREATENED SPECIES

The Endangered Species Act legally defines an **endangered species** as a species in imminent danger of extinction throughout all or a significant portion of its **range**. (The area in which a particular species is found is its range.) A species is endangered when its numbers are so severely reduced that it is in danger of becoming extinct without human intervention. A species is legally defined as **threat-**

endangered species A species that faces threats that may cause it to become extinct within a short period.

www.wiley.com/
college/berg

San Joaquin kit fox	Dusky seaside sparrow	Monterey manzanita	Atitlan giant pied-billed grebe	Black rhinoceros	Pitcher plant
Iguana	Mountain gorilla	Trumpeter swan	Giant weta	Golden lion tamarin	Whooping crane
Green turtle	Bladderpod	Red wolf	Abingdon tortoise	Partula snail	Cheetah
Arizona century plant	Dodo	Tiger	Coelacanth	White rhinoceros	Kiwi
Snow leopard	Cyanea	Saddle-backed tamarin	Black mamo	Grizzly bear	Great auk
Hawaii 'o'o	Ivory-billed woodpecker	Kemp's ridley sea turtle	Texas snowball	Black-footed ferret	Horned guan

Representative endangered or extinct species FIGURE 15.5

Officials at the U.S. Fish and Wildlife Service estimate more than 500 U.S. species have gone extinct during the past 200 years. Of these, roughly half have gone extinct since 1980.

ened when extinction is less imminent but its population is quite low and the species is likely to become endangered in the foreseeable future.

Endangered and threatened species represent a decline in biological diversity because as their numbers decrease, their

threatened species A species whose population has declined to the point that it may be at risk of extinction.

genetic variability is severely diminished. Long-term survival and evolution depend on genetic diversity, so a decline in genetic diversity heightens the risk of extinction for endangered and threatened species, as compared to species that have greater genetic variability.

Endangered and Extinct Species   367

AREAS OF DECLINING BIOLOGICAL DIVERSITY

Declining biological diversity is a concern throughout the United States, but is most serious in the states of Hawaii (where 63 percent of species are at risk) and California (where about 29 percent of species are at risk). At least two-thirds of Hawaii's native forests are gone.

As serious as declining biological diversity is in the United States, it is even more serious abroad, particularly in tropical rain forests. Tropical rain forests are being destroyed faster than almost all other ecosystems; approximately 1 percent of these ecosystems are being cleared or severely degraded each year. The forests are making way for human settlements, banana plantations, oil and mineral explorations, and other human activities. (For further discussion of tropical rain forests see Chapters 6 and 13.)

Tropical rain forests are home to thousands or even millions of species. Many species in tropical rain forests are **endemic** (that is, they are not found anywhere else in the world), and the clearing of tropical rain forests contributes to their extinction.

Perhaps the most unsettling outcome of tropical **deforestation** is its disruptive effect on evolution. In Earth's past, mass extinctions were followed over millions of years by the evolution of new species as replacements for those that died out. In the past, tropical rain forests may have supplied ancestral organisms from which other organisms evolved. Destroying tropical rain forests may be reducing nature's ability to replace its species.

■ **endemic species**
Organisms that are native to or confined to a particular region.

EnviroDiscovery   Is Your Coffee Bird Friendly®?

Many species of migratory songbirds, favorites among North American bird lovers, are in decline, and Americans' coffee habits may play a role. In the tropics, high-yield farms cultivating coffee in full sunlight—known as *sun plantations*—are rapidly replacing traditional *shade plantations*. Shade plantations grow coffee plants in the shade of tropical rainforest trees. These trees support a vast diversity of songbird species that winter in the tropics (one study counted 150 species in 5 hectares), as well as large numbers of other vertebrates and insects. In contrast, sun plantations provide poor bird habitat. Sun-grown varieties of coffee, treated with large inputs of chemical pesticides and fertilizers, out-produce the shade-grown varieties but lack the diverse products that come from the shade trees. About half of the region's shade plantations have been converted to sun plantations since the 1970s.

Songbird populations have declined alarmingly during this period. Various conservation organizations and development agencies like the Smithsonian Migratory Bird Center and USAID have initiated programs to certify coffee as "shade-grown," which allows consumers the chance to support the preservation of tropical rain forest. Shade-grown coffee typically costs more than sun-grown coffee because it is hand picked, involves more care in selecting only ripe beans, and is often certified organic.

The SMBC allows certified shade-grown and organic coffee farmers to use this label on their product

Zebra mussels FIGURE 15.9

A clump of zebra mussels, which have caused billions of dollars in damage, in addition to displacing native clams and mussels.

The zebra mussel's strong appetite for algae and zooplankton reduces the food supply of native fishes, mussels, and clams, threatening their survival. The U.S. Coast Guard estimates economic losses and control efforts associated with the zebra mussel cost the United States about $5 billion each year.

More than 200 foreign species have been introduced in the United States since 1980, and 25 percent of these have caused significant enough damage to be considered invasive. Worldwide, most regions are estimated to contain 10 percent to 30 percent foreign species.

Overexploitation

Sometimes species become endangered or extinct as a result of deliberate efforts to eradicate or control their numbers. Ranchers, hunters, and government agents have reduced populations of large predators such as the wolf and grizzly bear. Some animals are killed because their lifestyles cause problems for humans. The Carolina parakeet, a beautiful green, red, and yellow bird endemic to the southern United States, was exterminated as a pest by farmers because it ate fruit and grain crops. It was extinct by 1920.

Prairie dogs and pocket gophers were poisoned and trapped so extensively by ranchers and farmers that between 1900 and 1960 they disappeared from most of their original geographic range. As a result of sharply decreased numbers of prairie dogs, the black-footed ferret, the natural predator of these animals, became endangered. A successful captive-breeding program has allowed black-footed ferrets to be reintroduced into the wild and reproduce successfully, though some populations have been decimated by disease.

Unregulated hunting, or overhunting, was a factor contributing to the extinction of certain species in the past but is now strictly controlled in most countries. The passenger pigeon was one of the most common birds in North America in the early 1800s, but a century of overhunting resulted in its extinction in the early 1900s. Unregulated hunting was one of several factors that caused the near extinction of the American bison.

Illegal commercial hunting, or **poaching**, endangers many larger animals, such as the tiger, cheetah, and snow leopard, whose beautiful furs are quite valuable. Rhinoceroses are slaughtered primarily for their horns, used for ceremonial dagger handles in the Middle East, and for purported medicinal purposes in Asian medicine. Bears are killed for their gallbladders, used in Asian medicine to treat ailments from indigestion to heart ailments. Caimans (reptiles similar to crocodiles) are killed for their skins and made into shoes and handbags. Although these animals are legally protected, the demand for their products on the black market has caused them to be hunted illegally.

In West Africa, poaching has contributed to the decline in lowland gorilla and chimpanzee populations. The meat (called *bushmeat*) of these rare primates and other protected species, such as anteaters, elephants, and mandrill baboons, provides an important source of protein for indigenous people. Bushmeat is also sold to urban restaurants. This demand for a meat source increases the incidence of poaching.

Illegal animal trade FIGURE 15.10

These hyacinth macaws were seized in French Guiana in South America as part of the illegal animal trade there.

Live organisms collected through **commercial harvest** end up in zoos, aquaria, biomedical research laboratories, circuses, and pet stores. Several million birds are commercially harvested each year for the pet trade, but unfortunately many of them die in transit, and many more die from improper treatment after they are in their owners' homes. At least 40 parrot species

are now threatened or endangered, in part because of unregulated commercial trade (**FIGURE 15.10**).

Animals are not the only organisms threatened by excessive commercial harvest. Many unique and rare plants have been collected from nature to the point that they are endangered. These include carnivorous plants, wildflowers, grasses, and ferns, certain cacti, and orchids.

CONCEPT CHECK **STOP**

What is background extinction? Mass extinction?

What is the difference between a threatened species and an endangered species?

How do human activities cause species to become endangered or extinct?

Conservation Biology

LEARNING OBJECTIVES

Define conservation biology and compare in situ and ex situ conservation.

Describe restoration ecology.

Studies in the field of **conservation biology** cover everything from the processes that influence biological diversity to the protection and restoration of endangered species, to the preservation of entire ecosystems and landscapes.

Conservation biology includes two problem-solving techniques to save organisms from extinction: in

situ and ex situ conservation. **In situ conservation**, which includes the establishment of parks and reserves, concentrates on preserving biological diversity in nature. With increasing demands on land, in situ conservation cannot guarantee the preservation of all types of biological diversity. Sometimes only ex situ conservation can save a species. **Ex situ conservation** involves conserving biological diversity in human-controlled settings. The breeding of captive species in zoos and the seed storage of genetically diverse plant crops are examples of ex situ conservation.

conservation biology The scientific study of how humans impact organisms and of the development of ways to protect biological diversity.

PROTECTING HABITATS

Protecting animal and plant habitats—that is, conserving and managing the ecosystem as a whole—is the single best way to preserve biological diversity. Because human activities adversely affect the sustainability of many ecosystems, direct conservation management of protected areas is often required (**FIGURE 15.11**).

Currently, more than 3,000 national parks, sanctuaries, refuges, forests, and other protected areas exist worldwide. Protected areas are not always effective in preserving biological diversity. Many existing protected areas are too small or too isolated from other protected areas to efficiently conserve species. In developing countries where biological diversity is greatest, there is little money or expertise to manage them. Finally, many of the world's protected areas are in lightly populated mountain areas, tundra, and the driest deserts, places that often have spectacular scenery but relatively few species. In contrast, ecosystems in which biological diversity is greatest often receive little protection. Protected areas are urgently needed in tropical rain forests, deserts, the tropical grasslands and savannas of Brazil and Australia, many islands and temperate river basins, and dry forests all over the world.

Wildlife refuges

The **National Wildlife Refuge System**, established in 1903 by President Theodore Roosevelt, is the most extensive network of lands and waters committed to wildlife habitat in the world. The National Wildlife Refuge System contains more than 535 refuges, with at least one in each of the 50 states, and encompasses 38.4 million hectares (95 million acres) of land (see Figure 13.1, page 312). The refuges represent all major U.S. ecosystems, from tundra to temperate rain forest to desert, and are home to some of North America's most endangered species, such as the whooping crane. The mission of the National Wildlife Refuge System, which the U.S. Fish and Wildlife Service (FWS) administers, is to preserve lands and waters for the conservation of fishes, wildlife, and plants of the United States.

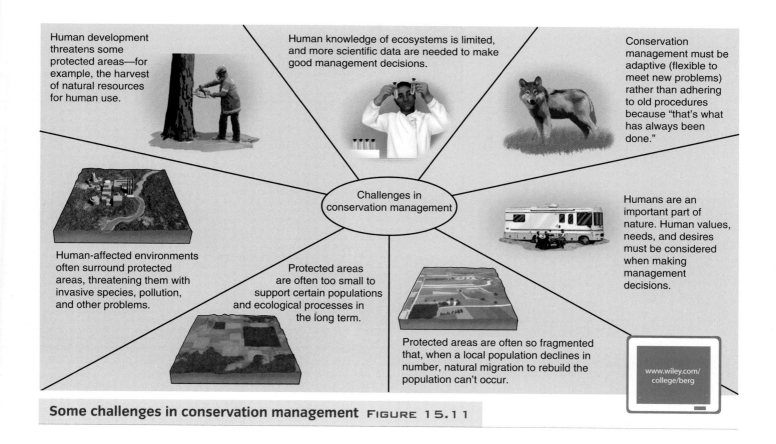

Human development threatens some protected areas—for example, the harvest of natural resources for human use.

Human knowledge of ecosystems is limited, and more scientific data are needed to make good management decisions.

Conservation management must be adaptive (flexible to meet new problems) rather than adhering to old procedures because "that's what has always been done."

Human-affected environments often surround protected areas, threatening them with invasive species, pollution, and other problems.

Challenges in conservation management

Protected areas are often too small to support certain populations and ecological processes in the long term.

Humans are an important part of nature. Human values, needs, and desires must be considered when making management decisions.

Protected areas are often so fragmented that, when a local population declines in number, natural migration to rebuild the population can't occur.

www.wiley.com/college/berg

Some challenges in conservation management FIGURE 15.11

RESTORING DAMAGED OR DESTROYED HABITATS

When preserving habitats is not possible, scientists can reclaim disturbed lands and convert them into areas with high biological diversity. In **restoration ecology**, ecological principles are used to help return a degraded environment to a more functional and sustainable one (FIGURE 15.12).

Restoration of disturbed lands creates biological habitats and provides additional benefits, such as the regeneration of soil damaged by agriculture or mining. The disadvantages of restoration include the expense and the time required to restore an area. Even so, restoration is an important aspect of in situ conservation, as restoration may reduce extinction.

restoration ecology The study of the historical condition of a human-damaged ecosystem, with the goal of returning it as close as possible to its former state.

CONSERVING SPECIES

Zoos, aquaria, botanical gardens, and other conservation organizations often play a critical role in saving individual species on the brink of extinction. Eggs or seeds may be collected from nature, or the few remaining wild animals may be captured and bred in research environments (FIGURE 15.13 in Visualizing Efforts to Conserve Species). But attempting to save a species on the brink of extinction is expensive, and only a small proportion of endangered species can be saved.

Conservation organizations are an essential part of the effort to maintain biological diversity through species and habitat conservation. These groups help educate policy makers and the public about the importance of biological diversity. In certain instances, they galvanize public support for important biodiversity preservation efforts. They provide financial support for conservation projects, from basic research to the purchase of land that is a critical habitat for a particular species or group of species (FIGURE 15.14).

Prairie restoration FIGURE 15.12

The University of Wisconsin–Madison Arboretum pioneered restoration ecology.

A The restoration of the prairie was at an early stage in November 1935.

B The prairie as it looks today. This picture was taken at approximately the same location as the 1935 photograph.

Reintroducing endangered species to nature

The ultimate goal of the captive-breeding programs practiced by zoos, aquaria, and other conservation organizations is to produce offspring in captivity and then release them into nature so wild populations are restored. However, only one of every ten reintroductions using animals raised in captivity is successful. Before attempting a reintroduction, conservation biologists make a feasibility study. This includes determining (1) what factors originally caused the species to become extinct in nature, (2) whether these factors still exist, and (3) whether any suitable habitat still remains. Captive-breeding programs are sometimes unsuccessful because it is impossible to teach critical survival skills to animals raised in captivity.

Seed banks

More than 100 seed collections, called **seed banks**, or *gene banks*, exist around the world and collectively hold more than 3 million samples at low temperatures (**FIGURE 15.15**). They offer the advantage of storing a large amount of plant genetic material in a small space. Seeds stored in seed banks are safe from habitat destruction, climate warming, and general neglect. There have even been some instances of using seeds from seed banks to reintroduce to nature a plant species that had become extinct.

Visualizing

Efforts to Conserve Species

Captive breeding FIGURE 15.13

Lucky (right) is the first whooping crane born to parents that were raised in captivity and then released, and the first in almost 70 years to be born in the wild in the United States. Here Lucky stands with one of his parents at their central Florida home.

Minimum critical size of ecosystems project FIGURE 15.14

When a protected area is set aside, it is important to know what the minimum size of that area must be so it is not affected by encroaching species from surrounding areas. Shown are 1-hectare and 10-hectare plots of a long-term study on the effects of habitat fragmentation on Amazonian rain forest by the World Wildlife Fund and Brazil's National Institute for Amazon Research. Preliminary data indicate that the smaller forest fragments do not maintain their ecological integrity.

Seeds from a seed bank ▲
FIGURE 15.15

Shown are small vials and packets of seeds from the seed bank in Svalbard, Norway.

Some disadvantages to seed banks exist. Seeds of many types of plants, such as avocados and coconuts, cannot be stored because they do not tolerate being dried out, a necessary step in freezing the seeds. Seeds do not remain alive indefinitely and must be germinated periodically so new seeds can be collected. Also, growing, harvesting, and returning seeds to storage is expensive. Perhaps the most important disadvantage of seed banks is that plants stored in this manner remain stagnant in an evolutionary sense. Thus, they may be less fit for survival when they are reintroduced into nature. Despite their shortcomings, seed banks are increasingly viewed as an important method of safeguarding seeds for future generations.

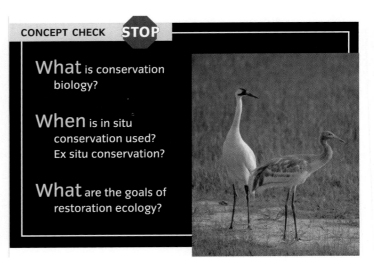

CONCEPT CHECK STOP

What is conservation biology?

When is in situ conservation used? Ex situ conservation?

What are the goals of restoration ecology?

Conservation Policies and Laws

LEARNING OBJECTIVES

Briefly describe the benefits and shortcomings of the U.S. Endangered Species Act.

Relate the purpose of the World Conservation Strategy.

In 1973 the **Endangered Species Act (ESA)** was passed in the United States, authorizing the FWS to protect endangered and threatened species in the United States and abroad. Many other countries now have similar legislation. International laws and policies also seek to conserve Earth's biological resources.

THE ENDANGERED SPECIES ACT

Currently, nearly 1,300 species in the United States are listed as endangered or threatened (**TABLE 15.2, FIGURE 15.16**). The ESA provides legal protection to listed species so that their danger of extinction is reduced. The ESA requires the FWS to select critical habitats and design a detailed recovery plan for each species listed. The recovery plan includes an estimate of the current population size, an analysis of the factors contributing to its endangerment, and a list of activities to help the population recover.

The ESA was updated in 1982, 1985, and 1988. It is considered one of the strongest pieces of U.S. environmental legislation, in part because species are designated as endangered or threatened entirely on biological grounds. Currently, economic considerations cannot influence the designation of endangered or threatened species. Biologists generally agree that fewer species have become extinct than would have if the ESA had not been passed.

The ESA is also one of the most controversial pieces of environmental legislation. The ESA does not provide compensation for private property owners who suffer financial losses because they cannot develop their land if a threatened or endangered species lives there. The ESA has also interfered with some federally funded development projects.

The ESA was scheduled for congressional reauthorization in 1992 but has been entangled since then in political wrangling between conservation advocates and supporters of private property rights. Conservation advocates think the ESA does not do enough to save endangered species, whereas those who own the land on which rare species live think the law goes too far and infringes on property rights. Another contentious issue is over the financial cost of the law.

Some critics view the ESA as an impediment to economic progress, as when the timber industry was blocked from logging old-growth forests in certain parts of the Pacific Northwest to protect the habitat of the northern spotted owl (see Chapter 3).

Those who defend the ESA point out that of 34,000 past cases of endangered species versus development, only 21 cases were not resolved by some sort of a compromise. When the black-footed ferret was reintroduced on the Wyoming prairie, for example, it was classified as an "experimental, nonessential species" so that its reintroduction would not block ranching and mining in the area. Thus, the ferret release program obtained the support of local landowners, support that was deemed necessary to the ferrets' survival in nature.

Endangered species FIGURE 15.16

The Florida panther is an endangered subspecies of cougar that exists in small pockets of isolated habitat in southern Florida.

This type of compromise is crucial to the success of saving endangered species, because, according to the U.S. General Accounting Office, more than 90% of endangered species live on at least some privately owned lands. Some critics of the ESA think the law should be changed so that private landowners are given economic incentives to help save endangered species living on their lands. For example, tax cuts for property owners who are good land stewards could make the presence of endangered species on their properties an asset instead of a liability.

Defenders of the ESA agree that it is not perfect. Few endangered species have recovered enough to be delisted—that is, removed from protection of the ESA. However, the FWS says that hundreds of listed species are stable or improving; they expect as many as several dozen additional species to be delisted in the next decade or so.

Conservationists would like the ESA strengthened in such a way as to manage whole ecosystems and maintain complete biological diversity rather than attempt to save endangered species as isolated entities. This approach offers collective protection to many declining species rather than to single species.

U.S. organisms listed as endangered or threatened, 2005 TABLE 15.2		
Type of organism	Number of endangered species	Number of threatened species
Mammals	69	9
Birds	77	13
Reptiles	14	22
Amphibians	11	10
Fishes	71	43
Snails	21	11
Clams	62	8
Crustaceans	18	3
Insects	35	9
Spiders	12	0
Flowering plants	571	144
Conifers	2	1
Ferns and other plants	24	2
Lichens	2	0
TOTAL	989	275

INTERNATIONAL CONSERVATION POLICIES AND LAWS

The **World Conservation Strategy**, a plan designed to conserve biological diversity worldwide, was formulated in 1980 by the World Conservation Union (IUCN), the World Wildlife Fund, and the U.N. Environment Program. In addition to conserving biological diversity, the World Conservation Strategy seeks to preserve the vital ecosystem services on which all life depends for survival and to develop sustainable uses of organisms and the ecosystems they comprise.

The Convention on Biological Diversity produced by the 1992 Earth Summit requires that each signatory nation must inventory its own biodiversity and develop a **national conservation strategy**, a detailed plan for managing and preserving the biological diversity of that specific country. Currently, 188 nations participate in the Convention on Biological Diversity.

The exploitation of endangered species is somewhat controlled through legislation. At the international level, 160 countries participate in the **Convention on International Trade in Endangered Species of Wild Flora and Fauna (CITES)**, which went into effect in 1975. Originally drawn up to protect endangered animals and plants considered valuable in the highly lucrative international wildlife trade, CITES bans hunting, capturing, and selling of endangered or threatened species and regulates trade of organisms listed as potentially threatened (**FIGURE 15.17**). Unfortunately, enforcement of this treaty varies from country to country. Even where enforcement exists, the penalties are not severe. As a result, illegal trade continues in rare, commercially valuable species.

The goals of CITES often stir up controversy over such issues as who actually owns the world's wildlife and whether global conservation concerns take precedence over competing local interests. These conflicts often highlight socioeconomic differences between wealthy consumers of CITES products and poor people who trade the endangered organisms.

The case of the African elephant, discussed earlier in the chapter, bears out these controversies. Listed as an endangered species since 1989 to halt the slaughter of elephants driven by the ivory trade, the species seems to have recovered in southern Africa. Organiza-

Illegal trade in products made from endangered species FIGURE 15.17

tions such as the Humane Society in the United States are developing a birth control vaccine to reduce the number of elephant births. However, the African people living near the elephants want to cull the herd periodically and sell elephant meat, hides, and ivory for profit. In the late 1990s and early 2000s, CITES transferred elephant populations in Namibia, Botswana, and South Africa to a less restrictive listing to allow a one-time trade of legally obtained stockpiled ivory (from animals that died of natural causes). The money earned from the sale of ivory is funding elephant conservation programs and community development projects for people living near the elephants.

CONCEPT CHECK STOP

What are the goals of the Endangered Species Act?

Why is the ESA considered controversial?

What is the World Conservation Strategy?

Reintroducing Wolves to Yellowstone

Gray wolves once ranged across North America from northern Mexico to Greenland, but they were trapped, poisoned, snared, and hunted to extinction in most areas by 1960. Under the provisions of the Endangered Species Act, gray wolf populations in the northern Rocky Mountains were listed as endangered in 1974. Many scientists recommended reintroducing wolves as a way to restore their populations, but the controversial proposal was not acted on for more than two decades. Beginning in 1995, the U.S. Fish and Wildlife Service captured a small number of gray wolves in Canada and released them into Yellowstone National Park in Wyoming (see photograph). The population thrived and has increased to an estimated 300 individuals in the Yellowstone area.

The wolves in Yellowstone prey on elk, mule deer, moose, and bison. In some areas, intensive hunting by wolf packs has helped reduce the park's elk population, which was at an all-time high before the wolves returned (see Figure 13.15 on page 328). When elk are not managed properly, they overgraze their habitat, and thousands starve during hard winters. The reduction and redistribution of Yellowstone's elk population has relieved heavy grazing pressure on various plant species. As a result of a more lush and varied plant composition, herbivores such as beavers and snow hares have increased in number, which in turn supports small predators such as foxes, badgers, and martens.

Wolf packs have severely reduced some coyote populations, allowing populations of the coyotes' prey, like ground squirrels, chipmunks, and pronghorns, to increase. Scavengers like ravens, magpies, bald eagles, wolverines, and bears benefit from dining on scraps from wolf kills.

The reintroduction of wolves did not occur without a fight. Ranchers and farmers who live in the area were against the reintroduction be-cause their livelihood depends on livestock being safe from predators. To address this concern, ranchers are allowed to kill Yellowstone wolves that attack their cattle and sheep, and federal officers can remove any wolf that threatens humans or livestock. Also, ranchers are reimbursed the full market value for cattle, sheep, and other livestock lost to wolves.

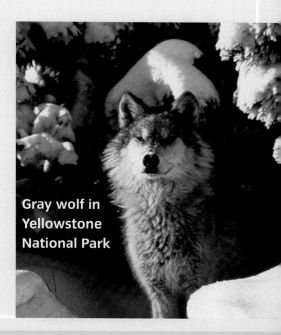

Gray wolf in Yellowstone National Park

CHAPTER SUMMARY

1 Species Richness and Biological Diversity

1. **Species richness** is the number of different species in a community. High species richness is associated with communities that are ecologically complex, not isolated, geologically old and stable, and not subject to environmental stress.

Species richness is also higher when no one species dominates the community.

2. **Biological diversity** is the number and variety of Earth's organisms; it consists of three components: genetic diversity, species richness, and ecosystem diversity. **Genetic diversity** is the genetic variety within all populations of a given species; **Ecosystem diversity** is the variety of in-teractions among organisms in natural communities.

3. Ecosystems with greater species richness are better able to supply **ecosystem services**: environmental benefits, such as clean air to breathe, clean water to drink, and fertile soil in which to grow crops.

2 Endangered and Extinct Species

1. **Extinction** is the elimination of a species from Earth. **Background extinction**, a continuous, low-level extinction of species, has occurred throughout the time in which organisms have occupied Earth. **Mass extinction**, in which many species disappear during a relatively short period of geologic time, has occurred at certain times in Earth's history.

2. An **endangered species** is a species that faces threats that may cause it to become extinct within a short period. A species is defined as **threatened** when extinction is less imminent but its population is quite low.

3. Humans cause species endangerment through habitat destruction, fragmentation, and degradation; pollution; the spread of invasive species; and the overexploitation of biological resources. **En-**demic species are organisms that are native to, or whose range is limited to, a specific place. **Biodiversity hotspots** are areas that contain particularly high numbers of endemic species. **Invasive species** are foreign species, usually introduced by humans, that spread rapidly in a new area where they are free of predators, parasites, or resource limitations that may have controlled their population in their native habitat.

3 Conservation Biology

1. **Conservation biology** is the scientific study of how humans impact organisms and of the development of ways to protect biological diversity. Conservation biology involves two problem-solving tools: **in situ conservation** includes the establishment of parks and reserves, concentrating on preserving biological diversity in nature; **ex situ conservation** involves conserving biological diversity in human-controlled settings such as zoos and seed banks.

2. **Restoration ecology** is the study of the historical condition of a human-damaged ecosystem, with the goal of returning it as close as possible to its former state.

4 Conservation Policies and Laws

1. The **Endangered Species Act (ESA)** authorizes the U.S. Fish and Wildlife Service (FWS) to protect endangered and threatened species in the United States and abroad. The ESA requires the FWS to select critical habitats and design a detailed recovery plan for each species listed. The act is controversial because it does not compensate private property owners who suffer financial losses when they cannot develop land if it hosts a threatened or endangered species. Species are designated as endangered or threatened entirely on biological grounds, not economic factors.

2. The **World Conservation Strategy** was formulated by the World Conservation Union (IUCN), the World Wildlife Fund, and the U.N. Environment Program. The program seeks to conserve biological diversity worldwide, to preserve vital ecosystem services, and to develop sustainable uses of organisms and their ecosystems.

KEY TERMS

- **species richness** p. 362
- **biological diversity** p. 363
- **ecosystem services** p. 364
- **extinction** p. 366
- **endangered species** p. 366
- **threatened species** p. 367
- **endemic species** p. 368
- **biodiversity hotspots** p. 369
- **invasive species** p. 372
- **conservation biology** p. 374
- **restoration ecology** p. 376

CRITICAL AND CREATIVE THINKING QUESTIONS

1. Is biological diversity a renewable or nonrenewable resource? Why could it be seen both ways?

2. Give at least five important ecosystem services provided by living organisms.

3. If we preserve species solely on the basis of their potential economic value—such as a source of a novel drug—do they lose their "value" after we have capitalized on a newly discovered chemical? Why or why not?

4. What are the four main causes of species endangerment and extinction? Which cause do biologists consider most important?

5. What are invasive species?

6. Why are frogs and other amphibians considered bellwether species?

7. If you had the assets and authority to take any measure to protect and preserve biological diversity, but could take only one, what would it be?

8. The most recent version of the World Conservation Strategy includes stabilizing the human population. How would stabilizing the human population affect biological diversity?

9. In *A Sand County Almanac and Sketches Here and There*, Aldo Leopold wrote, "To keep every cog and wheel is the first precaution of intelligent tinkering." How does his statement relate to this chapter?

10–12. The Nature Conservancy evaluated the extent of human-caused habitat disturbance in the world's various biomes. Use these data (graph on right) to answer the following questions.

10. On which global region have humans had the greatest impact—polar, temperate, or tropical? Suggest why human impact has been greatest in this region.

11. What percent of tropical rain forests has been disturbed by human activities? What percent of temperate deciduous forests? Given these two values, why do you think people are more concerned about destruction in tropical rain forests than in temperate deciduous forests?

12. Which biome has had the lowest percent of habitat disturbance? Suggest two possible reasons why human impacts on this biome may increase greatly in the future.

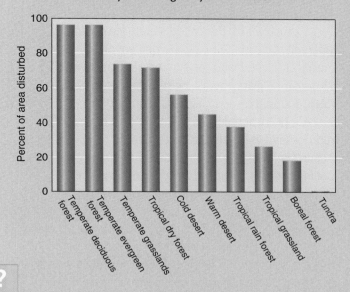

What is happening in this picture ?

- At Zoo Atlanta in the United States, an actor playing an African park guard and a volunteer teach children about ivory poaching.

- What are the goals of this effort?

- Which conservation law bans hunting elephants for their ivory?

- Do you think displaying elephants in zoos helps or hurts these species? Why?

Solid and Hazardous Waste: An Unrecognized Resource

16

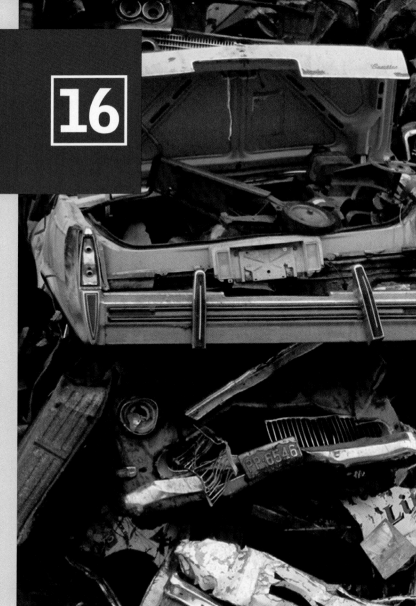

REUSING AND RECYCLING OLD AUTOMOBILES

About 11 million cars and trucks are discarded each year in the United States *(see photo of a Virginia salvage yard)*. Although 75 percent of a car can be reused as secondhand parts or recycled, the remaining 25 percent—glass, metals, plastics, fabrics, rubber, foam, and leather—usually ends up in landfills.

About 37 percent of the iron and steel scrap reprocessed in the United States comes from old cars. According to the Environmental Protection Agency, recycling scrap iron and steel produces 86 percent less air pollution and 76 percent less water pollution than mining and refining an equivalent amount of iron ore.

Recycling plastic, which automakers use because it is lightweight and improves fuel efficiency, is one of the biggest challenges in auto recycling. No industry standards for plastic parts currently exist, so the kinds and amounts used in cars vary a great deal.

Auto manufacturers worldwide have begun to address the challenge of recycling old cars. Toyota developed a way to recycle urethane foam and other shredded materials. Chrysler is developing a concept vehicle with completely recyclable body sections. Honda, Mercedes-Benz, Peugeot, Toyota, Volkswagen, Volvo, and others have started to design cars with completely recyclable or reconditionable parts (see inset of recyclable plastic and composite parts on a Toyota vehicle).

The European Union has mandated that by 2015, 95 percent of each discarded car must be recoverable. U.S. legislators may one day make the same requirement. When they design new models, European auto manufacturers now take into account the car's entire life cycle. This kind of *product stewardship* encourages optimal reuse and recycling when products are returned to manufacturers.

384

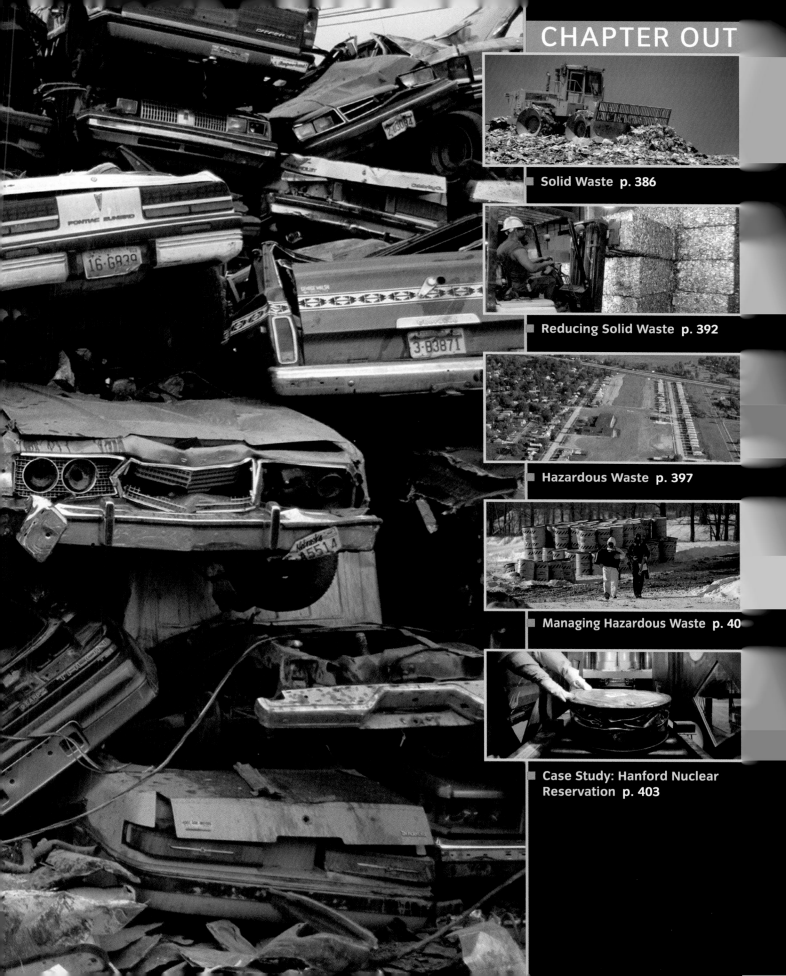

CHAPTER OUT

Solid Waste p. 386

Reducing Solid Waste p. 392

Hazardous Waste p. 397

Managing Hazardous Waste p. 40

Case Study: Hanford Nuclear Reservation p. 403

Solid Waste

The United States generates more solid waste per capita than any other country. (Canada is a close second.) Each person in the United States produces an average of 2 kg (4.4 lb) of solid waste per day. This amount corresponded to a total of 214 million metric tons (236 million tons) in 2003. The problem worsens each year as the U.S. population increases.

Waste generation is an unavoidable consequence of the prosperous, high-technology, industrial economies of the United States and other highly developed nations. Many products that would be repaired, reused, or recycled in less affluent nations are simply thrown away. Nobody likes to think about solid waste, but it is certainly a concern of modern society—we keep producing it, and places to dispose of it safely are dwindling in number (**FIGURE 16.1**).

TYPES OF SOLID WASTE

Municipal solid waste consists of the combined residential and commercial waste produced in a municipal area. Municipal solid waste is a heterogeneous mixture composed primarily of paper and paperboard; yard waste; food waste; plastics; metals; rubber, leather, and textiles; glass; and wood (**FIGURE 16.2**). The proportions of the major types of solid waste in this mixture change over time. Today's solid waste contains more paper and plastics, but less glass and steel, than it did in the past.

> **municipal solid waste**
> Solid materials discarded by homes, offices, stores, restaurants, schools, hospitals, prisons, libraries, and other facilities.

Recycling drop-off center
FIGURE 16.1

These people are dropping off plastic at a recycling center.

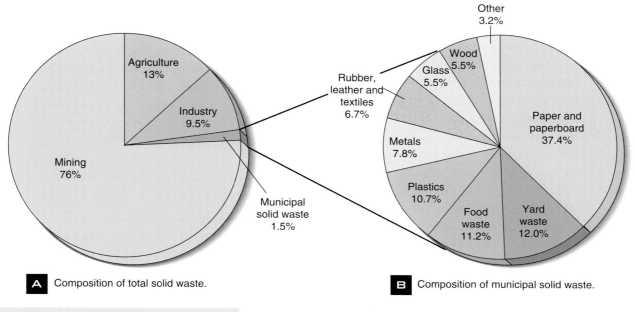

A   Composition of total solid waste.

B   Composition of municipal solid waste.

Municipal solid waste FIGURE 16.2

nonmunicipal solid waste Solid waste generated by industry, agriculture, and mining.

Municipal solid waste is actually only a small proportion—less than 2 percent—of the total solid waste produced each day. **Nonmunicipal solid waste**, which includes mining, agricultural, and industrial wastes, is produced in substantially larger amounts. Most solid waste generated in the United States is from nonmunicipal sources.

DISPOSAL OF SOLID WASTE

Solid waste has been traditionally regarded as material that is no longer useful and should be disposed of. We can get rid of solid waste in four ways: dump it, bury it, burn it, or recycle and compost it (FIGURE 16.3).

Open dumps   The old method of solid waste disposal was dumping. Open dumps, which are now illegal, were unsanitary, malodorous places where disease-carrying vermin like rats and flies proliferated. Methane gas was released into the surrounding air as

microorganisms decomposed the solid waste, and fires polluted the air with acrid smoke. Liquid oozed and seeped through the heaps of solid waste, often leaching hazardous materials that then contaminated soil, surface water, and groundwater.

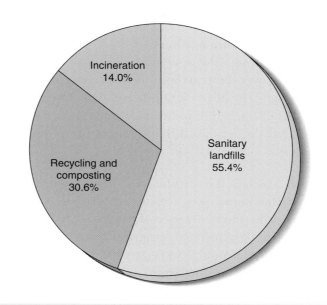

U.S. disposal of municipal solid waste in 2003 FIGURE 16.3

Sanitary landfills Open dumps have been replaced by **sanitary landfills**, which receive about 55 percent of the municipal solid waste generated in the United States today. Sanitary landfills differ from open dumps in that the solid waste is placed in a hole, compacted, and covered with a thin layer of soil every day (see What a Scientist Sees figure). This process reduces the number of rats and other vermin usually associated with solid waste, lessens the danger of fires, and decreases the amount of odor. If a sanitary landfill is operated in accordance with solid waste management–approved guidelines, it does not pollute local surface and groundwater. Safety is ensured by layers of compacted clay and plastic sheets at the bottom of the landfill, which prevent liquid waste from seeping into groundwater.

Newer landfills possess a double liner system (plastic, clay, plastic, clay) and use sophisticated systems to collect **leachate** (liquid that seeps through the solid waste) and gases that form during decomposition.

The location of an "ideal" sanitary landfill is based on a variety of factors, including the geology of the area, soil drainage properties, and the proximity of nearby bodies of water and wetlands. The landfill should be far enough away from centers of dense population so it is inoffensive but close enough so as not to require high transportation costs.

Although the operation of sanitary landfills has improved over the years with the passage of stricter and stricter guidelines, few landfills are ideal. Most sanitary landfills in operation today do not meet current legal standards for new landfills and encounter a variety of problems:

- Methane gas, produced by microorganisms that decompose organic material anaerobically (in the absence of oxygen), may seep through the solid waste and accumulate in underground pockets, creating the possibility of an explosion—even in basements of nearby homes. Landfill operators typically collect and burn off methane, but a growing number use the methane for gas-to-energy projects. The 115 U.S. landfills participating in these projects

> **■ sanitary landfill**
> The most common method of disposal of solid waste, by compacting it and burying it under a shallow layer of soil.

collectively produce enough electricity for about 500,000 homes.

- Leachate that seeps from unlined landfills or through cracks in the lining of lined landfills can potentially contaminate surface water and groundwater. Because even household trash contains toxic chemicals like heavy metals, pesticides, and organic compounds, the leachate must be collected and treated to neutralize its negative effects.

- Landfills, by their nature, fill up. They are not a long-term remedy for waste disposal. From 1988 to 2002 the number of U.S. landfills in operation decreased from 7,924 to 1,767; many reached their capacity, and others did not meet state or federal environmental standards. Fewer new sanitary landfills are being opened; many desirable sites are already taken, and people are usually adamantly opposed to the construction of a landfill near their homes.

The Special Problem of Plastic The amount of plastic in our solid waste, more than half of it from packaging, is growing faster than any other component of municipal solid waste. Most plastics are chemically stable and do not readily decompose. This characteristic, although essential in the packaging of products like food and medicine, causes long-term problems: Most plastic debris disposed of in sanitary landfills will probably last for centuries. In response to concerns about the volume of plastic waste, some countries, like Taiwan, have banned or started taxing the use of certain plastic items, especially plastic bags and eating utensils.

Special plastics that degrade or disintegrate have been developed. Some of these are **photodegradable**—that is, they break down after being exposed to sunlight—which means they will not break down buried in a sanitary landfill. Other plastics are **biodegradable**—they are decomposed by microorganisms like bacteria. Whether biodegradable plastics actually break down under the conditions found in a sanitary landfill is not yet clear, although preliminary studies indicate that they probably do not. (Other waste management options for plastic are discussed later in this chapter.)

Sanitary landfill

A A bulldozer compacts trash at a saniitary landfill in California. To the average person, a sanitary landfill is just a "dump."

B Environmental engineers know that sanitary landfills constructed today have protective liners of compacted clay and high-density plastic and sophisticated leachate collection systems that minimize environmental problems like groundwater contamination. Solid waste is spread in a thin layer, compacted into small sections called "cells," and covered with soil.

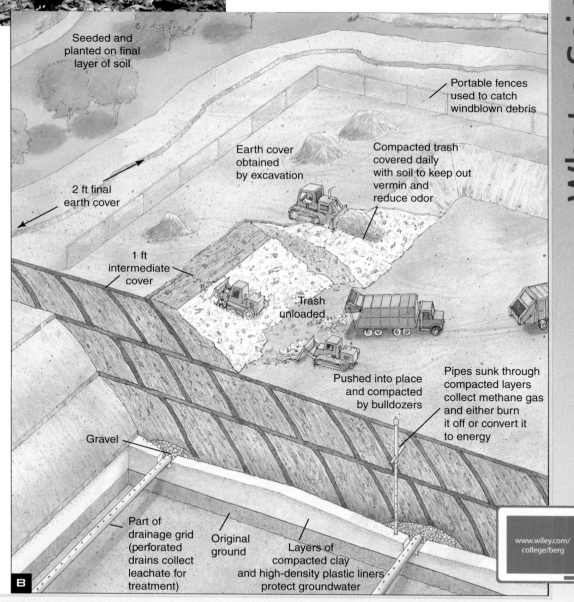

Seeded and planted on final layer of soil

2 ft final earth cover

1 ft intermediate cover

Portable fences used to catch windblown debris

Earth cover obtained by excavation

Compacted trash covered daily with soil to keep out vermin and reduce odor

Trash unloaded

Pushed into place and compacted by bulldozers

Pipes sunk through compacted layers collect methane gas and either burn it off or convert it to energy

Gravel

Part of drainage grid (perforated drains collect leachate for treatment)

Original ground

Layers of compacted clay and high-density plastic liners protect groundwater

www.wiley.com/college/berg

Incineration When solid waste is incinerated, two positive things are accomplished. First, the volume of solid waste is reduced by up to 90 percent: ash is more compact than unburned solid waste. Second, incineration produces heat that can make steam to warm buildings or generate electricity. In 2004 the United States had 89 waste-to-energy incinerators, which produce substantially less carbon dioxide emissions than power plants that burn fossil fuels (**FIGURE 16.4**). (Recall from Chapter 9 that carbon dioxide is a potent greenhouse gas.)

The best materials for incineration are paper, plastics, and rubber, all of which produce a lot of heat. Paper burns readily, and one kilogram of plastic waste yields almost as much heat as a kilogram of fuel oil.

Tires produce as much heat as coal and often generate less pollution. Some electric utilities in the United States and Canada burn tires instead of or in addition to coal (**FIGURE 16.5**). In 2001, 42 percent of all discarded tires were incinerated.

Most problems associated with incineration arise from the potential for environmental contamination:

Tires that will be burned to generate electricity
FIGURE 16.5

This mountain in Westley, California, contains 4 to 6 million old tires. The power plant that burns them supplies electricity to 3,500 homes. (The person wearing red gives a sense of scale.)

- Incinerators pollute the air with carbon monoxide, particulates, heavy metals like mercury, and other toxic materials, unless expensive air pollution control devices are used.

- Incinerators produce large quantities of ash, which must be disposed of properly. **Bottom ash**, or slag, is the ash left at the bottom of the incinerator when combustion is completed. **Fly ash** is the ash from the flue (chimney) that is trapped by air pollution control devices. Fly ash usually contains more toxic materials, including heavy metals and possibly dioxins, than bottom ash. Both types of incinerator ash are best disposed of in specially licensed hazardous waste landfills (discussed later in this chapter).

- Like sanitary landfills, site selection for incinerators is controversial. People may recognize the need for an incinerator, but they do not want it near their homes.

- Incinerators are expensive to run. Prices have escalated because costly pollution control devices are now required. Economic factors have also restricted construction of new plants.

The three types of incinerators are mass burn, modular, and refuse-derived fuel. Most **mass burn incinerators** are large and designed to recover the energy produced from combustion (**FIGURE 16.6**). **Modular incinerators** are smaller incinerators that burn

Carbon dioxide emissions per kilowatt-hour of electricity generated FIGURE 16.4

Waste-to-energy incinerators release less carbon dioxide into the atmosphere than do equivalent power plants that burn fossil fuels.

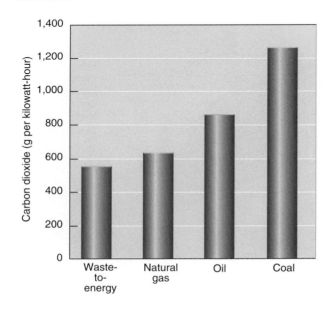

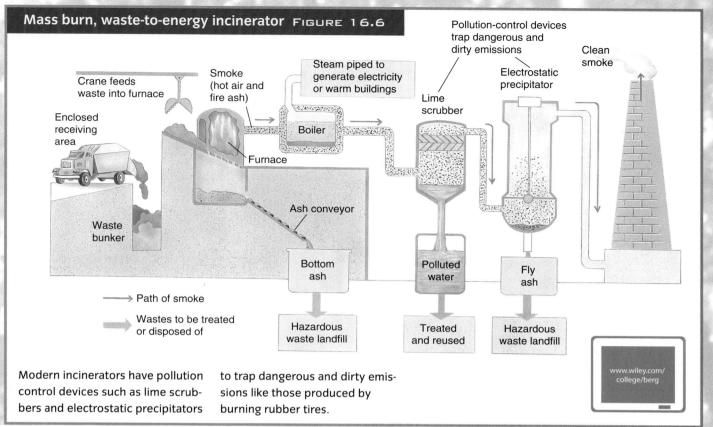

Mass burn, waste-to-energy incinerator FIGURE 16.6

Crane feeds waste into furnace

Enclosed receiving area

Smoke (hot air and fire ash)

Steam piped to generate electricity or warm buildings

Pollution-control devices trap dangerous and dirty emissions

Electrostatic precipitator

Clean smoke

Boiler

Lime scrubber

Furnace

Waste bunker

Ash conveyor

Bottom ash

Polluted water

Fly ash

→ Path of smoke

⇒ Wastes to be treated or disposed of

Hazardous waste landfill

Treated and reused

Hazardous waste landfill

www.wiley.com/college/berg

Modern incinerators have pollution control devices such as lime scrubbers and electrostatic precipitators to trap dangerous and dirty emissions like those produced by burning rubber tires.

mass burn incinerator A large furnace that burns all solid waste except for unburnable items such as refrigerators.

all solid waste. Assembled at factories, they are less expensive to build. **Refuse-derived fuel incinerators** burn the combustible portion of solid waste. First, noncombustible waste such as glass and metals are removed by machine or by hand. The remaining solid waste, including plastic and paper, is shredded or shaped into pellets and burned.

Composting Yard waste, such as grass clippings, branches, and leaves, is a substantial component of municipal solid waste (see Figure 16.2). As space in sanitary landfills becomes more limited, other ways to dispose of yard waste are being developed and implemented.

One of the best recovery methods for yard waste is to convert it into soil conditioners like compost or mulch (FIGURE 16.7). Food scraps, sewage sludge, and agricultural manure are other forms of solid waste

Home composting of household waste
FIGURE 16.7

The drum of this composting bin can be rotated on its base to mix the decomposing materials, thereby speeding up the microbial decomposition process.

New York City's Rikers Island, site of 10 jails and home to 10,000 inmates, serves as the city's surprising frontrunner in innovative recycling, composting, and gardening. Prisoners compost 6 tons of food scraps each day, in a facility with a roof made of photovoltaic cells. The composting produces high-grade fertilizer used in the prison's garden and farm programs. At the city's largest community garden, inmates raise vegetables for prison meals.

Staten Island's Fresh Kills landfill closed in 2002, and New York City faces strong opposition to its efforts to export more of its trash to other states. If the program at Rikers Island proves economical, additional composting centers could be established elsewhere in the city, making New York potentially capable of one day recycling all its food wastes (about one-fourth of the city's total garbage).

that can be used to make compost. Compost and mulch are used for landscaping in public parks and playgrounds or as part of the daily soil cover at sanitary landfills. Compost and mulch are also sold to gardeners.

Composting as a means of managing solid waste first became popular in Europe. Many municipalities in the United States have composting facilities as part of their comprehensive solid waste management plans, and more than 20 states have banned yard waste from sanitary landfills. This trend is likely to continue, making composting even more desirable. The United States currently has 3,260 municipal composting programs that recycle about 57 percent of yard wastes.

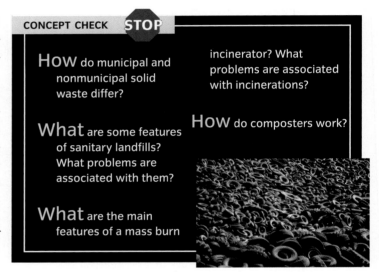

CONCEPT CHECK STOP

How do municipal and nonmunicipal solid waste differ?

What are some features of sanitary landfills? What problems are associated with them?

What are the main features of a mass burn incinerator? What problems are associated with incinerations?

How do composters work?

Reducing Solid Waste

LEARNING OBJECTIVES

Define source reduction.

Summarize how source reduction, reuse, and recycling help reduce the volume of solid waste.

Define integrated waste management.

Given the problems associated with sanitary landfills and incinerators, it makes sense to do whatever we can to reduce the wastes we generate. The three goals of waste prevention, in order of priority, are: (1) reduce the amount of waste as much as possible; (2) reuse products as much as possible; and (3) recycle materials as much as possible.

Reducing the amount of waste includes purchasing products that have less packaging and that last longer or are repairable (**FIGURE 16.8**). Consumers can also decrease their consumption of products to reduce waste. Before deciding to purchase a product, a consumer should ask, "Do I really *need* this product, or do I merely *want* it?"

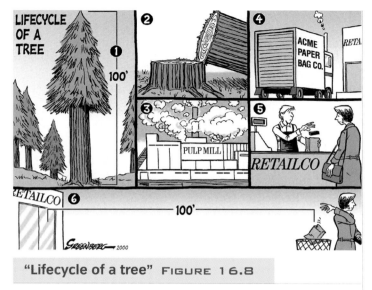

"Lifecycle of a tree" FIGURE 16.8

This editorial cartoon illustrates the six steps of wasteful packaging.

SOURCE REDUCTION

The most underutilized aspect of waste management is **source reduction**. Source reduction is accomplished in a variety of ways, including the use of raw materials that introduce less waste during manufacturing, and reusing and recycling wastes at the plants where they are generated. Innovation and product modifications play a key role. Consider aluminum cans: they are 35 percent lighter now than in the 1970s, because less material is introduced into their manufacture. Dry-cell batteries are another example: they contain much less mercury today than they did in the early 1980s.

source reduction An aspect of waste management in which products are designed and manufactured in ways that decrease the amount of solid and hazardous waste in the solid waste stream.

Dematerialization, the progressive decrease in the size and weight of a product as a result of technological improvements, is an example of source reduction only if the new product is at least as durable as the one it replaces. If the smaller, lighter product has a shorter life span and must be replaced more often, source reduction is not achieved.

REUSING PRODUCTS

One example of reuse is refillable glass beverage bottles. Years ago, refillable beverage bottles were used a great deal in the United States. Today their use is rare, although about 11 states still have them. For a glass bottle to be reused, it must be considerably thicker (and heavier) than one-use bottles. Because of the increased weight, transportation costs are higher. The centralization of bottling facilities makes it economically difficult to go back to the days of refillable bottles. Several other countries still reuse glass to a large extent, including Japan, Ecuador, Denmark, Finland, Germany, the Netherlands, Norway, Sweden, Switzerland, and parts of Canada.

RECYCLING MATERIALS

Many materials found in solid waste can be collected and reprocessed into new products. Recycling is preferred over landfill disposal because it conserves our natural resources and is more environmentally benign. Every ton of recycled paper saves 17 trees, 7,000 gallons of water, 4,100 kilowatt-hours of energy, and 3 cubic yards of landfill space. Recycling also has a positive effect on the economy by generating jobs and revenues from selling the recycled materials. Recycling does have environmental costs: like all human activities, it uses energy and generates pollution. For example, the de-inking process in paper recycling requires energy and produces a toxic sludge that contains heavy metals.

The many different materials in municipal solid waste must be separated before recycling. The separation of materials in items with complex compositions is difficult. Some food containers are composed of thin layers of metal foil, plastic, and paper, and trying to separate these layers is a daunting prospect.

The number of communities with recycling programs increased remarkably during the 1990s but leveled off in the early 2000s. In 2002 there were almost 8,900 curbside collection programs in the United States. The average U.S. family recycles aluminum and steel cans, plastic bottles, glass containers, newspapers, and cardboard (see Figure 2.14 on page 39). Many

INTEGRATED WASTE MANAGEMENT

<table>
<tr>
<td>

integrated waste management

A combination of the best waste management techniques into a consolidated program to deal effectively with solid waste.

</td>
<td>

The most effective way to deal with solid waste is with a combination of techniques. In **integrated waste management**, a variety of waste minimization methods, including the three R's of waste prevention (reduce, reuse, and recycle), are incorporated into an overall waste management plan (**FIGURE 16.12**). Even on a large scale, recycling and source reduction can substan-

</td>
</tr>
</table>

tially reduce, but not entirely eliminate, the need for disposal facilities such as incinerators and landfills.

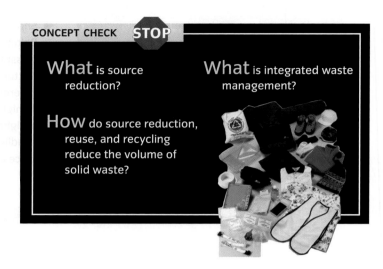

CONCEPT CHECK **STOP**

What is source reduction?

How do source reduction, reuse, and recycling reduce the volume of solid waste?

What is integrated waste management?

Process Diagram

Integrated waste management FIGURE 16.12

Source reduction, reuse, recycling, and composting are part of integrated waste management, in addition to incineration and disposal in landfills.

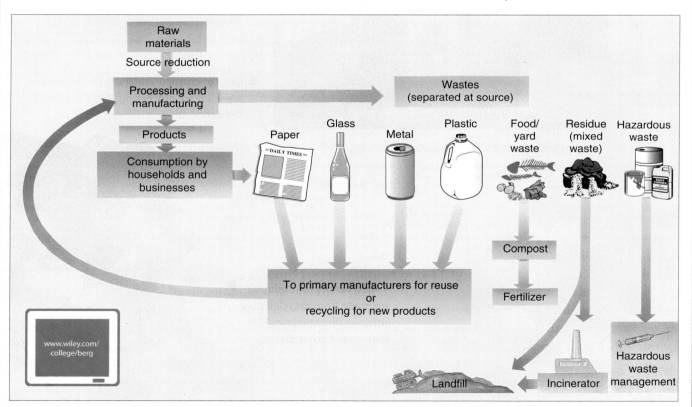

Hazardous Waste

Define hazardous waste.

Briefly characterize two types of hazardous waste: dioxins and PCBs.

Hazardous waste (also called **toxic waste**) accounts for about 1 percent of the solid waste stream in the United States. Hazardous waste includes dangerously reactive, corrosive, explosive, or toxic chemicals. The chemicals may be solids, liquids, or gases. More than 700,000 different chemicals are known to exist. How many are hazardous is unknown, because most have never been tested for toxicity, but without a doubt, there are thousands.

> **hazardous waste** Any discarded chemical that threatens human health or the environment.

Hazardous waste has held national attention since 1977, when it was discovered that toxic waste from an abandoned chemical dump had contaminated homes and possibly people in **Love Canal**, a small neighborhood on the edge of Niagara Falls, New York. Love Canal became synonymous with chemical pollution caused by negligent hazardous waste management. In 1978, it became the first location ever declared a national emergency disaster area because of toxic waste; more than 700 families were evacuated (**FIGURE 16.13**).

From 1942 to 1953, a local industry, Hooker Chemical Company, disposed of about 20,000 metric tons (22,000 tons) of toxic chemical waste in the 914-m-long (3,000-ft-long) Love Canal. When the site was filled, Hooker added topsoil and donated the land to the local board of education. A school and houses were built on the site, which began oozing toxic waste several years later. Over 300 chemicals, many of them carcinogenic, have been identified in Love Canal's toxic waste.

In 1990, after almost 10 years of cleanup, the EPA and the New York Department of Health declared the area safe for resettlement. Today, the canal is a 40-acre mound covered by clay and surrounded by a chain-link fence and warning signs. The Love Canal episode resulted in the passage of the federal Superfund law, which held polluters accountable for the cost of cleanups (discussed shortly) and generated immediate and ongoing concern about hazardous wastes.

Global Locator

Aerial view of Love Canal toxic waste site in the early 1980s FIGURE 16.13

All the homes shown in this photograph were evacuated and demolished.

TYPES OF HAZARDOUS WASTE

Hazardous chemicals include a variety of acids, dioxins, abandoned explosives, heavy metals, infectious waste, nerve gas, organic solvents, polychlorinated biphenyls (PCBs), pesticides, and radioactive substances (TABLE 16.1). Many of these chemicals are discussed in other chapters; see Chapters 4, 9, 10, 14, and 17, which examine endocrine disrupters, radioactive waste, air pollution, water pollution, and pesticides. Here we discuss dioxins and PCBs; the Case Study at the end of the chapter describes the difficulties of radioactive waste disposal.

Dioxins   **Dioxins** are a group of 75 similar chemical compounds formed as byproducts during the combustion of chlorine compounds. Incineration of medical and municipal wastes accounts for 70 to 95 percent of known human emissions of dioxins. Some other known sources of dioxins are iron ore mills, copper smelters, cement kilns, metal recycling, coal combustion, pulp and paper plants that use chlorine for bleaching (FIG-URE 16.14), and chemical accidents. Motor vehicles, barbecues, and cigarette smoke emit minor amounts of dioxins. Forest fires and volcanic eruptions are natural sources of dioxins. Dioxins also form during the production of some pesticides.

Dioxins are emitted in smoke and then settle on plants, the soil, and bodies of water; from there they are incorporated into the food web. When humans and other animals ingest dioxins—primarily in contaminated meat, dairy products, and fish—they store and accumulate the dioxins in their fatty tissues (see the bioaccumulation and biomagnification discussion in Chapter 4). Because dioxins are so widely distributed in the environment, virtually everyone has dioxins in their body fat.

Dioxins cause several kinds of cancer in laboratory animals, but the data conflict on their cancer-causing ability in humans. A 2001 EPA report suggests that dioxins probably cause several kinds of cancer in humans and likely affect the human reproductive, immune, and nervous systems. Because human milk contains dioxins, nursing infants are considered particularly at risk.

Examples of hazardous waste TABLE 16.1

Hazardous material	Some possible sources
Acids	Ash from power plants and incinerators; petroleum products
CFCs (chlorofluorocarbons)	Coolant in air conditioners and refrigerators
Cyanides	Metal refining; fumigants in ships, railway cars, and warehouses
Dioxins	Emissions from incinerators and pulp and paper plants
Explosives	Old military installations
Heavy metals	Paints, pigments, batteries, ash from incinerators, sewage sludge with industrial waste, improper disposal in landfills
Arsenic	Industrial processes, pesticides, additives to glass, paints
Cadmium	Rechargeable batteries, incineration, paints, plastics
Lead	Lead-acid storage batteries, stains and paints; TV picture tubes and electronics discarded in landfills
Mercury	Coal-burning power plants; paints, household cleaners (disinfectants), industrial processes, medicines, seed fungicides
Infectious waste	Hospitals, research labs
Nerve gas	Old military installations
Organic solvents	Industrial processes; household cleaners, leather, plastics, pet maintenance (soaps), adhesives, cosmetics
PCBs (polychlorinated biphenyls)	Older appliances (built before 1980); electrical transformers and capacitors
Pesticides	Household products
Radioactive waste	Nuclear power plants, nuclear medicine facilities, weapons factories

Air pollution from a paper mill
FIGURE 16.14

Dioxins may be in these emissions.

Studying bacteria that break down PCBs in contaminated soil FIGURE 16.15

A microbiologist adds soil to a "bioreactor" to test the ability of certain bacteria to treat contaminated soil. Note the three bioreactors in the foreground.

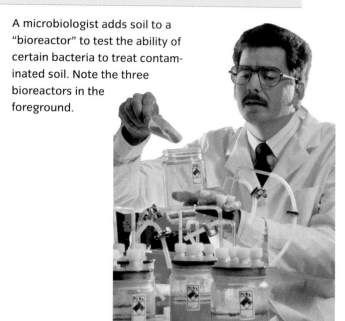

PCBs   **Polychlorinated biphenyls (PCBs)** are a group of 209 industrial chemicals composed of carbon, hydrogen, and chlorine. PCBs were manufactured in the United States between 1929 and 1979 for a wide variety of uses: as cooling fluids in electrical transformers, electrical capacitors, vacuum pumps, and gas-transmission turbines; and in hydraulic fluids, fire retardants, adhesives, lubricants, pesticide extenders, inks, and other materials. Prior to the EPA ban in the 1970s, PCBs were dumped in large quantities into landfills, sewers, and fields. Such improper disposal is one of the reasons PCBs are still a threat today.

The dangers of PCBs first became evident in Japan in 1968, where hundreds of people ate rice bran oil accidentally contaminated with PCBs and consequently experienced serious health problems, including liver and kidney damage. A similar mass poisoning tied to PCBs occurred in Taiwan in 1979. Since then, toxicity tests conducted on animals indicate that PCBs harm the skin, eyes, reproductive organs, and gastrointestinal system. PCBs are endocrine disrupters: they interfere with hormones released by the thyroid gland. Several studies have demonstrated that in utero exposure to PCBs can lead to certain intellectual impairments in children. PCBs may be carcinogenic; they are known to cause liver cancer in rats, and studies in Sweden and the United States have shown a correlation between high PCB concentrations in the body and incidences of certain cancers.

Although high-temperature incineration is one of the most effective ways to destroy PCBs in most solid waste, it is too costly to be used for the removal of PCBs that have leached into soil and water. One way to remove PCBs from soil and water is to extract them with solvents. This method is undesirable because the solvents themselves are hazardous chemicals, and these extraction methods are also costly.

More recently, researchers have discovered several bacteria that degrade PCBs at a fraction of the cost of incineration. Additional research is needed to make the biological degradation of PCBs practical (**FIGURE 16.15**).

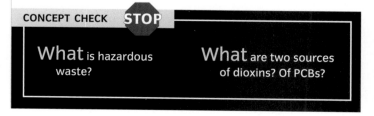

CONCEPT CHECK **STOP**

What is hazardous waste?

What are two sources of dioxins? Of PCBs?

Managing Hazardous Waste

W e have the technology to manage toxic waste in an environmentally responsible way, but it is extremely expensive. Although great strides have been made in educating the public about the problems of hazardous waste, we have only begun to address the issues of hazardous waste disposal. No country currently has an effective hazardous waste management program, but several European countries have led the way by producing smaller amounts of hazardous waste and by using fewer hazardous substances.

CHEMICAL ACCIDENTS

When a chemical accident occurs in the United States, whether at a factory or during the transport of hazardous chemicals, the National Response Center (NRC) is notified. Most chemical accidents reported to NRC involve oil, gasoline, or other petroleum spills. The remaining accidents involve some 1,032 other hazardous chemicals, such as PCBs, ammonia, sulfuric acid, and chlorine. In 2005, the NRC database indicated that toxic chemicals were released into the environment during 35,714 separate reported accidents. The state with the greatest number of toxic chemical accidents in 2005 was Texas, which had 5 significant environmental incidents.

Chemical safety programs have traditionally stressed accident mitigation and adding safety systems to existing procedures. More recently, industry and government agencies have stressed accident prevention through the **principle of inherent safety**, in which industrial processes are redesigned to involve less toxic materials so dangerous accidents are less likely to occur in the first place.

PUBLIC POLICY AND TOXIC WASTE CLEANUP

Currently, two federal laws dictate how hazardous waste should be managed: (1) the Resource Conservation and Recovery Act, which is concerned with managing hazardous waste being produced now, and (2) the Superfund Act, which provides for the cleanup of abandoned and inactive hazardous waste sites.

The **Resource Conservation and Recovery Act (RCRA)** was passed in 1976 and amended in 1984. Among other things, RCRA instructs the EPA to identify which waste is hazardous and to provide guidelines and standards to states for hazardous waste management programs. RCRA bans hazardous waste from land disposal unless it is treated to meet EPA's standards of reduced toxicity. In 1992 the EPA initiated a major reform of RCRA to expedite cleanups and streamline the permit system to encourage hazardous waste recycling.

In 1980 the **Comprehensive Environmental Response, Compensation, and Liability Act (CERCLA)**, commonly known as the **Superfund Act**, established a program to tackle the huge challenge of cleaning up abandoned and illegal toxic waste sites across the United States. At many of these sites, hazardous chemicals have migrated deep into the soil and have polluted groundwater. The greatest threat to human health from toxic waste sites comes from drinking water laced with such contaminants.

Cleaning up existing toxic waste: the Superfund Program
The federal government estimates that the United States has more than 400,000 hazardous waste sites with leaking chemical storage tanks and drums (both above and below ground), pesticide dumps, and piles of mining waste. This estimate does not include the hundreds or thousands of toxic waste sites at military bases and nuclear weapons facilities.

By 2005, 12,247 sites were in the CERCLA inventory, which means the EPA has identified them as qualifying for cleanup (FIGURE 16.16). (The count does not include more than 34,000 sites that were cleaned up and removed from the CERCLA inventory since 1980.)

The sites posing the greatest threat to public health and the environment are placed on the **Superfund National Priorities List**, which means that the federal government will assist in their cleanup. As of 2005, 1,609 sites were on the National Priorities List. The five states that lead the list are New Jersey (138 sites in 2005), Pennsylvania (122 sites), New York (110 sites), California (106 sites), and Michigan (84 sites). The average cost of cleaning up a site is $20 million.

The Superfund Act expired in 1994, and the tax on oil and chemical companies that funded the program expired in 1995. Meanwhile, the pace of cleanup has slowed to about 40 sites a year. Because the federal government cannot clean up every old dump in the United States, the current landowner, prior owners, and anyone who has dumped waste on, or transported waste to, a particular site may be liable for cleanup costs.

Although critics decry the slow pace and high cost of cleaning up Superfund sites, the very existence of CERCLA is a deterrent to further polluting. Companies that produce hazardous waste are now fully aware of the costs of liability and cleanup and are more likely to dispose properly of their hazardous wastes.

MANAGING TOXIC WASTE PRODUCTION

The Superfund deals only with hazardous waste produced in the past, not the large amount of toxic waste produced today. There are three ways to manage hazardous waste: (1) source reduction, (2) conversion to less hazardous materials, and (3) long-term storage.

As with municipal solid waste, the most effective approach is source reduction—that is, using less hazardous or nonhazardous materials in industrial processes. Source reduction relies on the increasingly important field of **environmental chemistry**, or *green chemistry*.

For example, chlorinated solvents are widely used in electronics, dry cleaning, foam insulation, and industrial cleaning. To accomplish source reduction, it is sometimes possible to substitute a less hazardous water-based solvent for the toxic chlorinated one. Substantial source reduction of chlorinated solvents can also be procured by reducing solvent emissions. Installing solvent-saving devices benefits the environment and also saves money, because smaller amounts of chlorinated solvents must be purchased. No matter how efficient source reduction becomes, however, it will never entirely eliminate hazardous waste.

The second-best way to deal with hazardous waste is to reduce its toxicity by chemical, physical, or biological means, depending on the nature of the waste. High-temperature incineration, for example, reduces dangerous compounds like pesticides, PCBs, and organic solvents to safe products such as water and carbon dioxide. The

environmental chemistry A subdiscipline of chemistry in which commercially important chemical processes are redesigned to reduce environmental harm.

Cleaning up hazardous waste FIGURE 16.16

A Toxic waste in deteriorating drums at a site near Washington, D.C. The metal drums in which much of the waste is stored have corroded and started to leak. Old toxic waste dumps are commonplace around the United States.

B Cleanup of a hazardous waste site near Minneapolis, Minnesota. Removal and destruction of the wastes are complicated by the fact that usually nobody knows what chemicals are present.

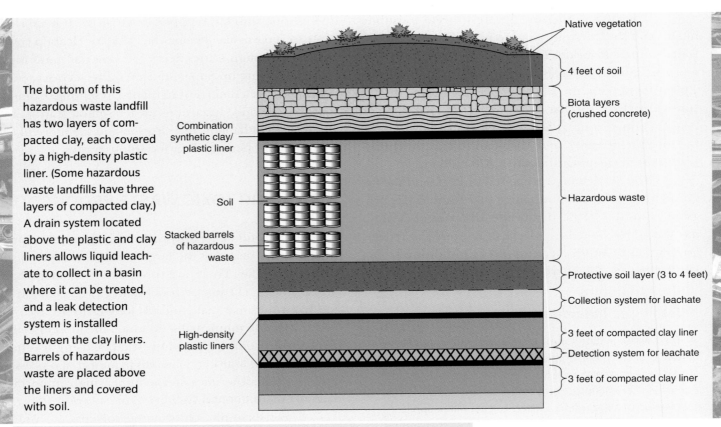

The bottom of this hazardous waste landfill has two layers of compacted clay, each covered by a high-density plastic liner. (Some hazardous waste landfills have three layers of compacted clay.) A drain system located above the plastic and clay liners allows liquid leachate to collect in a basin where it can be treated, and a leak detection system is installed between the clay liners. Barrels of hazardous waste are placed above the liners and covered with soil.

Native vegetation

4 feet of soil

Biota layers (crushed concrete)

Combination synthetic clay/plastic liner

Soil

Stacked barrels of hazardous waste

Hazardous waste

Protective soil layer (3 to 4 feet)

Collection system for leachate

High-density plastic liners

3 feet of compacted clay liner

Detection system for leachate

3 feet of compacted clay liner

Cutaway view through a hazardous waste landfill FIGURE 16.17

resulting ash is hazardous and must be disposed of in a landfill designed for hazardous materials. Incineration using a *plasma torch* produces such high temperatures (up to 10,000°C, five times higher than conventional incinerators) that hazardous waste is almost completely converted to harmless gases.

Hazardous waste that is not completely detoxified must be placed in long-term storage. Hazardous waste landfills are subject to strict environmental criteria and design features. They are located as far as possible from aquifers, streams, wetlands, and residences. These landfills include several layers of compacted clay and high-density plastic liners at the bottom of the landfill to prevent leaching of hazardous substances into surface water and groundwater (FIGURE 16.17). Leachate (liquid that percolates through the landfill) is collected and treated to remove contaminants. The entire facility and nearby groundwater deposits are carefully monitored to make sure there is no leakage.

Few facilities are certified to handle toxic waste. In 2005 the United States had only 21 commercial hazardous waste landfills, although many larger companies were licensed to treat their hazardous waste on-site. As a result, much of our hazardous waste is still dumped in sanitary landfills, burned in incinerators that lack the required pollution control devices, or discharged into sewers.

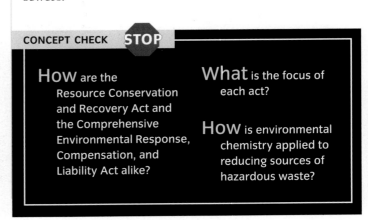

CONCEPT CHECK STOP

HOW are the Resource Conservation and Recovery Act and the Comprehensive Environmental Response, Compensation, and Liability Act alike?

What is the focus of each act?

HOW is environmental chemistry applied to reducing sources of hazardous waste?

Hanford Nuclear Reservation

While the U.S. no longer actively manufactures nuclear weapons, the production facilities still present us with an enormous challenge—reducing and managing the radioactive and toxic wastes that have accumulated since the 1940s. The main facility for the production of the plutonium used in nuclear weapons, Hanford Nuclear Reservation in Washington State, is the largest, most seriously contaminated site in the U.S. nuclear weapons infrastructure (see FIGURE A).

The immensity of the cleanup task at Hanford is daunting. Tons of highly radioactive solid and liquid wastes were stored or dumped into trenches, pits, tanks, ponds, and underground cribs—a total of 1,700 waste sites within 1,450 sq km (560 sq mi). Two concrete pools of water store more than 100,000 spent fuel rods, which release highly radioactive uranium, plutonium, cesium, and strontium into the water as they corrode. Because these pools are leaking, soil and groundwater have been contaminated, and the nearby Columbia River is in danger. The Columbia is also threatened by millions of gallons of toxic chemical and radioactive liquid wastes stored in 177 large underground tanks, some of which are potentially explosive. Many of these aging tanks are now leaking toxic chemicals into the ground.

The task of cleanup is not only immense but extremely complicated: the extent of the pollution and kinds of toxins present are not well known. As a result, scientists and engineers must assess the damage, prioritize the cleanup process, and determine how best to proceed for each unknown type of contamination (see FIGURE B).

Most experts believe the cleanup, which is under the direction of the U.S. Department of Energy (DOE), will take at least three decades to complete and cost hundreds of billions of tax dollars. Cleaning up the mess may even be a more dangerous occupation than working at Hanford when it was producing the nation's nuclear arms stockpile. After the cleanup is finished, Hanford will remain hazardous for hundreds or even thousands of years, or at least until we develop technologies to address the widespread soil contamination.

A Location of Hanford along the Columbia River in Washington State.

B A worker at the Waste Receiving and Processing Facility crushes drums of low-level nuclear waste. The crushed drums will then be packed into larger drums for permanent burial in the hazardous waste landfill at Hanford.

CHAPTER SUMMARY

1 Solid Waste

1. **Municipal solid waste** consists of solid materials discarded by homes, office buildings, stores, restaurants, schools, hospitals, prisons, libraries, and other facilities. **Nonmunicipal solid waste** consists of solid waste generated by industry, agriculture, and mining.

2. The **sanitary landfill** is the most common method of solid waste disposal, by compacting and burying it under a shallow layer of soil. Layers of compacted clay and plastic sheets prevent **leachate** (liquid waste) from seeping into groundwater. Newer landfills possess a double liner system and collect leachate and gases that form during decomposition. Problems include the potential for methane gas to seep out and cause explosions, the accidental leaking of toxic leachate, a lack of existing landfill space, and resistance to new landfills near homes and businesses.

3. A **mass burn incinerator** is a large furnace that burns all solid waste except for unburnable items such as refrigerators. Problems associated with incineration of solid waste include the potential for air pollution, difficulties in disposing of the toxic ash produced, the high costs of the process, and the difficulties in choosing incinerator sites.

4. In composting, yard waste, food scraps, and other organic wastes are transformed by microbial action into a material that, when added to soil, improves its condition.

2 Reducing Solid Waste

1. In **source reduction**, products are designed and manufactured in ways that decrease the volume of solid waste and the amount of hazardous waste in the solid waste stream.

2. The volume of solid waste produced can be reduced through source reduction, reusing products, and recycling materials. Recycling conserves natural resources and is more environmentally benign than landfill disposal but requires a market for the recycled goods.

3. **Integrated waste management** is a combination of the best waste management techniques into a consolidated program to deal effectively with solid waste.

3 Hazardous Waste

1. **Hazardous waste** is any discarded chemical that threatens human health or the environment. The chemicals may be solids, liquids, or gases and include a variety of acids, dioxins, abandoned explosives, heavy metals, infectious waste, nerve gas, organic solvents, polychlorinated biphenyls (PCBs), pesticides, and radioactive substances.

2. **Dioxins** are hazardous chemicals formed as unwanted byproducts during the combustion of many chlorine compounds. **Polychlorinated biphenyls (PCBs)** are hazardous, oily, industrial chemicals composed of carbon, hydrogen, and chlorine.

4 Managing Hazardous Waste

1. The **Resource Conservation and Recovery Act (RCRA)** instructs the EPA to identify hazardous waste and to provide guidelines and standards for states' hazardous waste management programs. The **Comprehensive Environmental Response, Compensation, and Liability Act (CERCLA)** or **Superfund Act** established a program whose goal is to clean up abandoned and illegal toxic waste sites across the United States.

2. The most effective approach to managing hazardous waste is source reduction, reducing the amount and toxicity of hazardous materials used in industrial processes. Source reduction relies on **environmental chemistry**, a subdiscipline of chemistry in which commercially important chemical processes are redesigned to reduce environmental harm.

KEY TERMS

- municipal solid waste p. 386
- nonmunicipal solid waste p. 387
- sanitary landfill p. 388
- mass burn incinerator p. 391
- source reduction p. 393
- integrated waste management p. 396
- hazardous waste p. 397
- environmental chemistry p. 401

CRITICAL AND CREATIVE THINKING QUESTIONS

1. Compare the advantages and disadvantages of disposing of waste in sanitary landfills and by incineration.

2. How could source reduction efforts reduce the volume of waste that arises from abandoned automobiles?

3. List what you think are the best ways to treat each of the following types of solid waste, and explain the benefits of the processes you recommend: paper, plastic, glass, metals, food waste, and yard waste.

4. How do industries such as Goodwill, which accepts donations of clothing, appliances, and furniture for resale, affect the volume of solid waste?

5. What are dioxins, and how are they produced? What harm do they cause?

6. Suppose hazardous chemicals were suspected to be leaking from an old dump near your home. Outline the steps you would take to (1) have the site evaluated to determine if there is a danger and (2) mobilize the local community to get the site cleaned up.

7. What are the goals, strengths, and weaknesses of the Superfund program?

8. What is integrated waste management? Why must a sanitary landfill always be included in any integrated waste management plan?

9–11. In an effort to reduce municipal solid waste, many communities have required customers to pay for garbage collection according to the amount of garbage they generate, an approach termed unit pricing or "pay as you throw." The figure to the right illustrates the effects of unit pricing in San Jose, California, on garbage sent to landfills, and on wastes diverted through recycling and through separation of yard wastes.

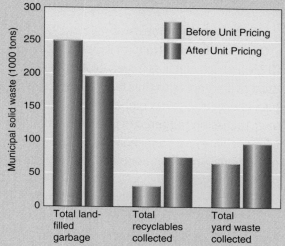

Municipal solid waste, San Jose, California

9. How did the implementation of unit pricing in San Jose affect the amount of garbage sent to landfills?

10. How did the implementation of unit pricing affect the quantity of materials recycled or of yard wastes collected?

11. Which component of municipal solid wastes pictured in the graph was most affected by the implementation of unit pricing? Why do you suppose this was the result?

What is happening in this picture ?

- This papermaker is converting denim waste into paper the color of faded jeans.

- How would this process reduce solid waste?

- What type of industrial cooperation would be needed for this type of papermaking?

- What are some other industries that might use these recycled materials?

Nonrenewable Energy Resources

OIL PRICES AND THE WORLD OIL MARKET

At the start of the Israel-Arab War in 1973, the Organization of Petroleum Exporting Countries (OPEC), which consists largely of Arab nations, restricted oil shipments to the United States, precipitating an energy crisis. The embargo resulted in an economic recession triggered in part by escalating prices for gasoline and home heating oil. Long lines at filling stations were commonplace in many states (*see larger photograph*).

Supported by presidents Nixon, Ford, and Carter, the United States embarked on a massive campaign to reduce U.S. dependence on oil and to conserve energy. Americans adjusted their thermostats, began carpooling to work, and installed additional insulation in homes and older buildings. Technological advances led to more efficient automobiles, lighting fixtures, appliances, and buildings.

By the 1980s, the embargo of the 1970s was largely forgotten as oil prices declined. Gasoline was so cheap and abundant that its consumption increased. Oil imports increased, and domestic oil production in the United States declined. The United States now relies more than ever on foreign oil. Currently, more than half the oil used in the United States is imported, up from 35 percent in 1973. The Persian Gulf region provides the United States with more than 12 percent of that—almost three times the 1973 level.

The world's oil supplies will not remain abundant indefinitely; prices will reflect the shortages. U.S. consumers felt the crunch as gas prices rose dramatically, beginning in 2005 after hurricanes Katrina and Rita disrupted domestic oil production and oil distribution from the Gulf of Mexico (*see inset*). As global supplies continue to tighten, experts predict the United States will have to scramble desperately to return to energy conservation measures.

CHAPTER OUT

Energy Consumption p. 408

Coal p. 409

Oil and Natural Gas p. 412

Nuclear Energy p. 416

Case Study: The Arctic National
Wildlife Refuge p. 425

Energy Consumption

LEARNING OBJECTIVE

Compare per-capita energy consumption in highly developed and developing countries.

Human society depends on energy. We use it to warm our homes in winter and cool them in summer; to grow, store, and cook our food; to light our homes; to extract and process natural resources for manufacturing items we use daily; and to power various forms of transportation. Many of the conveniences of modern living depend on a ready supply of energy.

A conspicuous difference in per-capita energy consumption exists between highly developed and developing nations (**FIGURE 17.1**). As you might expect, highly developed nations consume much more energy per person than developing nations—approximately eight times as much. In the United States, industry uses 42 percent of the nation's total energy, buildings like homes and offices consume 33 percent, and transportation 25 percent.

World energy consumption has increased every year since 1982, with most of the increase occurring in developing countries. From 2003 to 2004, for example, energy consumption increased worldwide by about 2.5 percent, most of it in China and India. One of the goals of developing countries is to improve their standards of living. One way to achieve this is through economic development, a process usually accompanied by a rise in per-capita energy consumption. Furthermore, the world's energy requirements will continue to increase during the 21st century, as the human population becomes larger, particularly in developing countries.

In contrast, the population in highly developed nations is more stable, and many energy experts think that their per-capita energy consumption may be at or near saturation. Additional energy demands in highly developed nations may be met by increasing the energy efficiency of things like appliances, automobiles, and home insulation (see Chapter 18).

Energy consumption FIGURE 17.1

A Annual per-capita commercial energy consumption in selected countries.

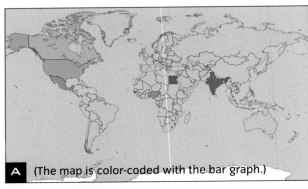

A (The map is color-coded with the bar graph.)

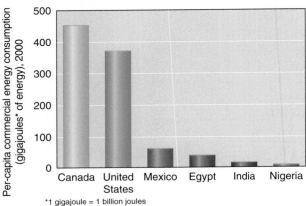

*1 gigajoule = 1 billion joules

B Projected total energy consumption, to 2030.

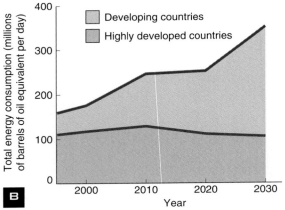

CONCEPT CHECK STOP

HOW does per-capita energy consumption compare in highly developed and developing countries?

Coal

LEARNING OBJECTIVES

Distinguish between surface mining and subsurface mining.

Summarize the environmental problems associated with using coal.

Describe two technologies that make coal a cleaner fuel.

Coal, the most abundant fossil fuel in the world, is found primarily in the Northern Hemisphere. The largest coal deposits are in the United States, Russia, China, Australia, India, Germany, and South Africa. The United States, in its massive deposits, has 25 percent of the world's coal supply. According to the World Resources Institute, known world coal reserves could last for more than 200 years at the present rate of consumption. Coal resources currently too expensive to develop have the potential to provide enough coal to last for 1,000 or more years at current consumption rates.

Utility companies use coal to produce electricity, and heavy industries use coal for steel production. Coal consumption has surged in recent years, particularly in the rapidly growing economies of China and India (FIGURE 17.2).

COAL MINING

The two basic types of coal mines are surface and subsurface (underground) mines. If the coal bed is within 30 m (100 ft) or so of the surface, **surface mining** is usually done. In **strip mining**, one type of surface mining, a trench is dug to extract the coal, which is scraped out of the ground and loaded into railroad cars or trucks. Surface mining is used to obtain approximately 60 percent of the coal mined in the United States.

surface mining The extraction of mineral and energy resources near Earth's surface by first removing the soil, subsoil, and overlying rock strata.

When the coal is deeper in the ground or runs deep into the ground from an outcrop on a hillside, it is mined underground. **Subsurface mining** accounts for approximately 40 percent of the coal mined in the United States.

Surface mining has several advantages over subsurface mining: It is usually less expensive, safer for miners, and generally allows a more complete removal of coal from the ground. However, surface mining disrupts the land much more extensively than subsurface mining and has the potential to cause serious environmental problems.

subsurface mining The extraction of mineral and energy resources from deep underground deposits.

Chinese energy consumption is on the rise FIGURE 17.2

Trucks in China deliver coal straight from the mine to homes.

ENVIRONMENTAL IMPACTS OF COAL

Coal mining, especially surface mining, has substantial effects on the environment (FIGURE 17.3). Prior to the 1977 **Surface Mining Control and Reclamation Act (SMCRA)**, abandoned surface coal mines were usually left as large open pits or trenches. Acid and toxic mineral drainage from such mines, along with the removal of topsoil, which was buried or washed away by erosion, prevented most plants from naturally recolonizing the land. Streams were polluted with sediment and **acid mine drainage**, which is produced when rainwater seeps through iron sulfide minerals exposed in mine wastes (see Chapter 12). Dangerous landslides occurred on hills unstable from the lack of vegetation.

> ■ **acid mine drainage**
> Pollution caused when sulfuric acid and dangerous dissolved materials, such as lead, arsenic, and cadmium, wash from coal and metal mines into nearby lakes and streams.

One of the most land-destructive types of surface mining is **mountaintop removal**. According to Environmental Media Services, mountaintop removal has leveled between 15 and 25 percent of the mountaintops in southern West Virginia. The valleys and streams between the mountains are gone as well, filled with mine tailings and debris. At the current rate, half the peaks in that area will be gone by 2020. Mountaintop removal is also occurring in Kentucky, Pennsylvania, Tennessee, and Virginia.

Coal burning generally contributes more of the common air pollutants than burning either oil or natural gas. In the United States, coal-burning electric power plants currently produce one-third of all airborne mercury emissions. Some coal contains sulfur and nitrogen that, when burned, are released into the atmosphere as sulfur oxides (SO_2 and SO_3) and nitrogen oxides (NO, NO_2, and N_2O), many of which form acids when they react with water. These reactions result in **acid deposition**, which is particularly prevalent downwind from coal-burning electric power plants (FIGURE 17.4). Acid precipitation and forest decline are discussed in greater detail in Chapter 9.

Surface coal mine near Douglas, Wyoming
FIGURE 17.3

The overlying vegetation, soil, and rock are stripped away, and then the coal is extracted out of the ground.

Dead trees enveloped in acid fog on Mt. Mitchell, North Carolina FIGURE 17.4

Forest decline was first documented in Germany and Eastern Europe. More recently, it has been observed in eastern North America, particularly at higher elevations. Acid deposition contributes to forest decline.

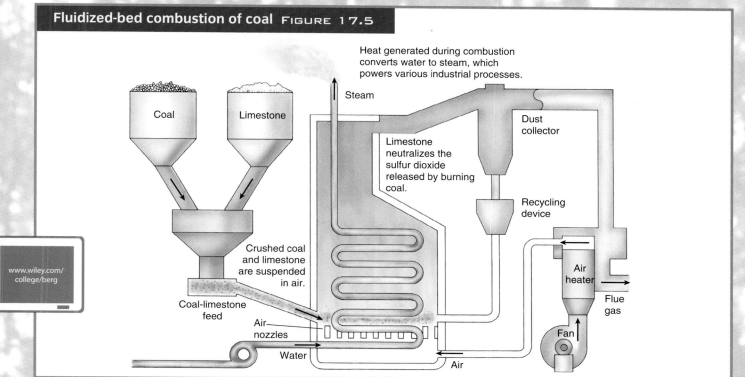

Fluidized-bed combustion of coal FIGURE 17.5

Heat generated during combustion converts water to steam, which powers various industrial processes.

Steam

Coal

Limestone

Limestone neutralizes the sulfur dioxide released by burning coal.

Dust collector

Recycling device

Crushed coal and limestone are suspended in air.

Air heater

Flue gas

Coal-limestone feed

Air nozzles

Fan

Water

Air

www.wiley.com/college/berg

Burning any fossil fuel releases carbon dioxide (CO_2), a greenhouse gas, into the atmosphere. Currently we are releasing so much CO_2 into the atmosphere that global temperature may be affected (because the increasing concentration of greenhouse gases prevents heat from escaping from the planet). Burning coal causes a more severe CO_2 problem than burning other fossil fuels, because coal releases more CO_2 per unit of heat energy produced.

MAKING COAL CLEANER

Sulfur emissions associated with the combustion of coal can be reduced by **scrubbers**, or desulfurization systems, that clean the power plants' exhaust. As the polluted air passes through a scrubber, chemicals in the scrubber react with the pollution and cause it to precipitate, or settle out.

Clean coal technologies are methods of burning coal that reduce air pollution. **Fluidized-bed combustion** mixes crushed coal with limestone particles in a strong air current during combustion (FIGURE 17.5). This clean coal technology produces fewer nitrogen ox-

ides and removes sulfur from the coal. It produces more heat from a given amount of coal, thereby reducing CO_2 emissions per unit of electricity produced.

In the United States several large power plants are testing fluidized-bed combustion, and a few small plants are already using this technology. **The Clean Air Act Amendments of 1990** provide incentives for utility companies to convert to clean coal technologies.

> **fluidized-bed combustion**
> A clean-coal technology in which crushed coal is mixed with limestone to neutralize acidic compounds produced during combustion.

CONCEPT CHECK STOP

Which type of coal mining—surface or subsurface mining—is more land-intensive?

What are the environmental impacts of mining and burning coal?

Oil and Natural Gas

LEARNING OBJECTIVES

Describe existing reserves of oil and natural gas.

Discuss the environmental problems caused by using oil and natural gas.

Oil and natural gas supply approximately 63 percent of the energy used in the United States. In comparison, other U.S. energy sources include coal (23 percent), nuclear power (8.2 percent), and hydropower (3.1 percent). Globally, oil and natural gas provide 61.5 percent of the world's energy (**FIGURE 17.6**).

Petroleum, or **crude oil**, is a liquid composed of hundreds of hydrocarbon compounds. During petroleum refining, the compounds are separated into different products—like gases, jet fuel, heating oil, diesel, and asphalt—based on their different boiling points (**FIGURE 17.7**). Oil is also used to produce **petrochemicals**, compounds used to make products like fertilizers, plastics, paints, pesticides, medicines, and synthetic fibers.

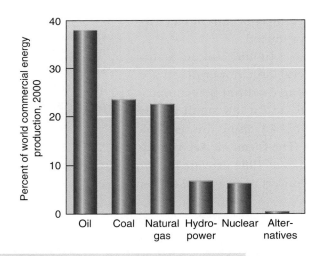

World commercial energy sources

FIGURE 17.6

Note the overwhelming importance of oil, coal, and natural gas as commercial energy sources. "Alternatives" include geothermal, solar, wind, and wood electric power.

Petroleum refining FIGURE 17.7

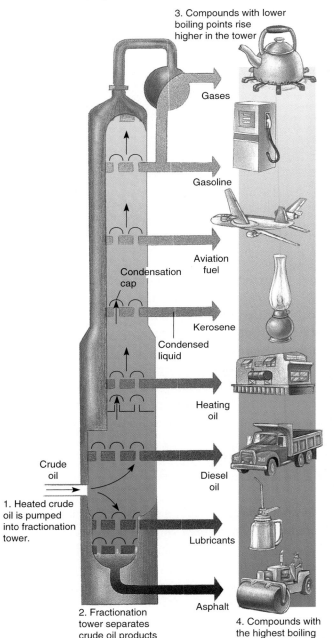

Fractionation tower (may be 30.5m [100 ft.] tall)

Petroleum products

3. Compounds with lower boiling points rise higher in the tower

Gases

Gasoline

Aviation fuel

Condensation cap

Kerosene

Condensed liquid

Heating oil

Crude oil

Diesel oil

1. Heated crude oil is pumped into fractionation tower.

Lubricants

2. Fractionation tower separates crude oil products based on boiling point.

Asphalt

4. Compounds with the highest boiling points are produced lowest in the tower.

Compared to petroleum, natural gas contains only a few hydrocarbons: methane and smaller amounts of ethane, propane, and butane. Propane and butane are separated from the natural gas, stored in pressurized tanks as a liquid called **liquefied petroleum gas**, and used primarily in rural areas as fuel for heating and cooking. Methane is used to heat residential and commercial buildings, to generate electricity in power plants, and for a variety of purposes in the organic chemistry industry.

Natural gas use is increasing in three main areas: electricity generation, transportation, and commercial cooling. An increasingly popular use of natural gas is **cogeneration**, a clean and efficient process in which natural gas is used to produce both electricity and steam; the heat of the exhaust gases provides energy to make steam for water and space heating (see Chapter 18). Cogeneration systems that use natural gas provide electricity cleanly and efficiently.

As a fuel for trucks, buses, and cars, natural gas offers significant environmental advantages over gasoline or diesel: Natural gas vehicles emit 80 percent to 93 percent fewer hydrocarbons, 90 percent less carbon monoxide, 90 percent fewer toxic emissions, and almost no soot. As of 2000, the United States had about 100,000 vehicles running on compressed natural gas, most of them fleet vehicles like city buses. The city of Los Angeles has the largest fleet of gas-powered transit buses in North America.

Natural gas efficiently fuels residential and commercial air-cooling systems. One example is the use of natural gas in a desiccant-based (air-drying) cooling system, which is ideal for restaurants and supermarkets, where humidity control is as important as temperature control.

The main disadvantage of natural gas is that deposits are often located far from where the energy is used. Because it is a gas and less dense than a liquid, natural gas costs four times more to transport through pipelines than crude oil. To transport natural gas over long distances, it must first be compressed to form liquefied natural gas (LNG) and then carried on specially constructed refrigerated ships. After LNG arrives at its destination, it must be returned to the gaseous state at regasification plants before being piped to where it will be used. Currently, the United States has only four such plants, which severely restricts the importation of natural gas from other countries. American energy companies claim the United States needs at least 40 regasification plants to keep costs down for natural gas and to meet increasing demands.

RESERVES OF OIL AND NATURAL GAS

Oil and natural gas deposits exist on every continent, but their distribution is uneven. More than half of the world's total estimated reserves are situated in the Persian Gulf region, which includes Iran, Iraq, Kuwait, Oman, Qatar, Saudi Arabia, Syria, United Arab Emirates, and Yemen (**FIGURE 17.8**). Major oil fields also exist in Venezuela, Mexico, Russia, Kazakhstan, Libya, and the United States (in Alaska and the Gulf of Mexico).

Almost half of the world's proved recoverable reserves of natural gas are located in two countries, Russia and Iran. The United States has more deposits of natural gas than Western Europe.

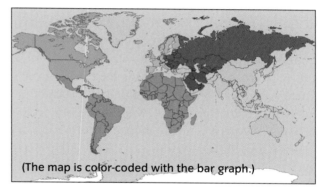

(The map is color-coded with the bar graph.)

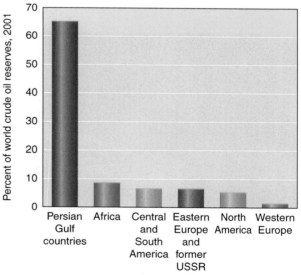

Distribution of oil deposits FIGURE 17.8

The Persian Gulf region contains huge oil deposits in a relatively small area, whereas other regions have few.

Large oil deposits probably exist under the **continental shelves**, the relatively flat underwater areas that surround continents, and in deepwater areas adjacent to the continental shelves. Despite problems like storms at sea and the potential for oil spills, many countries engage in offshore drilling. As many as 18 billion barrels (756 billion gallons) of oil and natural gas may exist in the deep water of the Gulf of Mexico, just off the continental shelf from Texas to Alabama (**Figure 17.9**). Continental shelves off the coasts of western Africa and Brazil are also promising potential sources of oil. Environmentalists and coastal industries like fishing generally oppose opening the continental shelves for oil and natural gas exploration because of the threat of a major oil spill.

How long will oil and natural gas supplies last?

We cannot predict how many reserves will be discovered, whether technological breakthroughs will allow us to extract more fuel from each deposit, or whether world consumption of oil and natural gas will increase, remain the same, or decrease. Even with technological advances, most industry analysts predict that global oil production will peak between 2050 and 2100. Natural gas is more plentiful than oil. Experts estimate that, at current rates of consumption, readily recoverable reserves of natural gas will keep production rising for at least 10 years after conventional supplies of petroleum have begun to decline.

ENVIRONMENTAL IMPACTS OF OIL AND NATURAL GAS

The environmental problems associated with oil and natural gas result from burning and transporting them. As with coal, the burning of oil and natural gas produces CO_2 that can contribute to global climate warming. Every gallon of gasoline your car or truck burns releases about 9 kg (20 lb) of CO_2 into the atmosphere. Burning oil also leads to acid deposition and the formation of photochemical smog. Oil combustion produces almost no sulfur oxides, but it does produce nitrogen oxides. In fact, gasoline combustion contributes about half the nitrogen oxides released into the atmosphere by human activities. (Coal combustion contributes the rest.)

Natural gas, on the other hand, is a relatively clean, efficient source of energy that contains almost no sulfur and releases far less CO_2, fewer hydrocarbons, and almost no particulate matter, as compared to oil and coal.

One risk of oil and natural gas production relates to their transport, often long distances by pipelines or ocean tankers. A serious spill along the route creates an environmental crisis, particularly in aquatic ecosystems, where the oil slick can travel great distances.

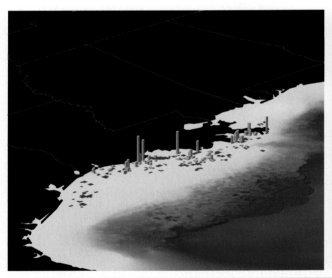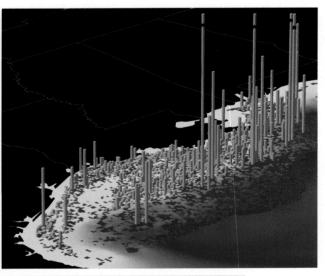

Rapid expansion of oil exploration in the Gulf of Mexico, 1961–2001 Figure 17.9

Active oil leases give drilling companies the right to obtain oil in a given location.

The largest oil spill in the United States   In 1989 the supertanker *Exxon Valdez* hit Bligh Reef and spilled 260,000 barrels (10.9 million gallons) of crude oil into Prince William Sound along the coast of Alaska, creating the largest oil spill in U.S. history (FIGURE 17.10 in Visualizing the *Exxon Valdez* Oil Spill). According to the U.S. Fish and Wildlife Service and the Alaska Department of Environmental Conservation, more than 30,000 birds (sea ducks, loons, cormorants, bald eagles, and other species) and between 3,500 and 5,500 sea otters died as a result of the spill. The area's killer whale and harbor seal populations declined, salmon migration was disrupted, and the fishing season in the area was halted that year.

Although Exxon declared the cleanup "complete" in late 1989, it left behind contaminated shorelines; continued damage to some species of birds, fishes, and mammals (TABLE 17.1); and a reduced commercial salmon catch, among other problems.

Visualizing
The *Exxon Valdez* Oil Spill

Alaskan oil spill, 1989 FIGURE 17.10

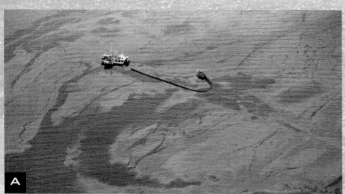

A An aerial view of the massive oil slick at the southwest end of Prince William Sound.

B Workers try to clean the rocky shoreline of Eleanor Island, Alaska, several months after the spill.

C The extent of the spill (black arrows). Water currents caused it to spread rapidly for hundreds of kilometers.

Nearshore coastal wildlife status, Prince William Sound, Alaska, 2005 TABLE 17.1

Populations that are recovering	Populations that are not yet recovering
Clams	Common loons
Intertidal communities	Cormorants (3 species)
Killer whales	Harbor seals
Marbled murrelets	Harlequin ducks
Mussels	Pacific herring
Sea otters	Pigeon guillemots

One positive outcome of the disaster was passage of the **Oil Pollution Act** of 1990. This legislation establishes liability for damages to natural resources resulting from a catastrophic oil spill, including a trust fund that pays to clean up spills when the responsible party cannot, and requires by 2015 double hulls on all oil tankers that enter U.S. waters.

The largest global oil spill   The world's most massive oil spill occurred in 1991 during the Persian Gulf War, when about 6 million barrels (250 million gallons) of crude oil—more than 20 times the amount of the *Exxon Valdez* spill—were deliberately dumped into the Persian Gulf. Many oil wells were set on fire, and lakes of oil spilled into the desert around the burning oil wells. In 2001 Kuwait began a massive remediation project to clean up its oil-contaminated desert. Progress is slow, and it may take a century or more for the area to completely recover.

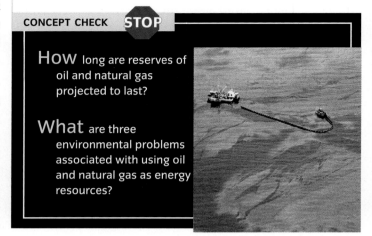

CONCEPT CHECK STOP

How long are reserves of oil and natural gas projected to last?

What are three environmental problems associated with using oil and natural gas as energy resources?

Nuclear Energy

LEARNING OBJECTIVES

Define nuclear energy and describe a typical nuclear power reactor.

Discuss the pros and cons of electric power produced by nuclear energy versus coal.

Describe safety issues associated with nuclear power plants and risks associated with the storage of radioactive wastes.

A
ll atoms are composed of positively charged protons, negatively charged electrons, and electrically neutral neutrons (**FIGURE 17.11**). Protons and neutrons, which have approximately the same mass, are clustered in the center of the atom, making up its nucleus. Electrons, which possess little mass in comparison with protons and neutrons, orbit the nucleus in distinct regions.

As a way to obtain energy, nuclear processes are fundamentally different from the combustion that produces energy from fossil fuels. Combustion is a chemical reaction. In chemical reactions, atoms of one element do not change into atoms of another element, nor does any of their mass (matter) change into energy. The energy re-

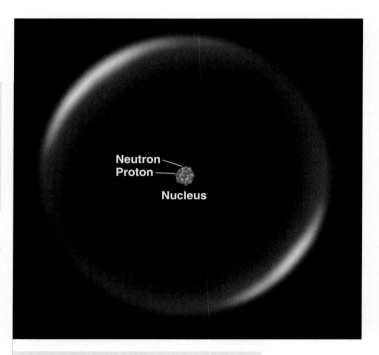

Neutron
Proton
Nucleus

Atomic structure FIGURE 17.11

Atoms contain a nucleus made of protons and neutrons. Circling the nucleus is a "cloud" of electrons.

leased in chemical reactions comes from changes in the chemical bonds that hold the atoms together. Chemical bonds are associations between electrons, and chemical reactions involve the rearrangement of electrons.

nuclear energy
The energy released by nuclear fission or fusion.

fission The splitting of an atomic nucleus into two smaller fragments, accompanied by the release of a large amount of energy.

In contrast, **nuclear energy** involves changes in the nuclei of atoms; small amounts of matter from the nucleus are converted into large amounts of energy. There are two different nuclear reactions that release energy: fission and fusion. In nuclear **fission**, the process nuclear power plants use, energy is released when a single neutron crashes into a large atom of one element, like uranium, and splits it into two smaller atoms of different elements (**FIGURE 17.12**). In **fusion**, the process that powers the sun and other stars, two small atoms are combined, forming one larger atom of a different element.

CONVENTIONAL NUCLEAR FISSION

Uranium ore, the mineral fuel used in conventional nuclear power plants, is a nonrenewable resource present in limited amounts in sedimentary rock in Earth's crust. Uranium ore contains three isotopes, U-238 (which makes up 99.28 percent of uranium), U-235 (0.71 percent), and U-234 (less than 0.01 percent). Because U-235, the isotope used in conventional fission reactions, is such a minor part of uranium ore, uranium ore must be refined after mining to increase the concentration of U-235 to about 3 percent. This refining process, called **enrichment**, requires a great deal of energy.

fusion The joining of two lightweight atomic nuclei into a single, heavier nucleus, accompanied by the release of a large amount of energy.

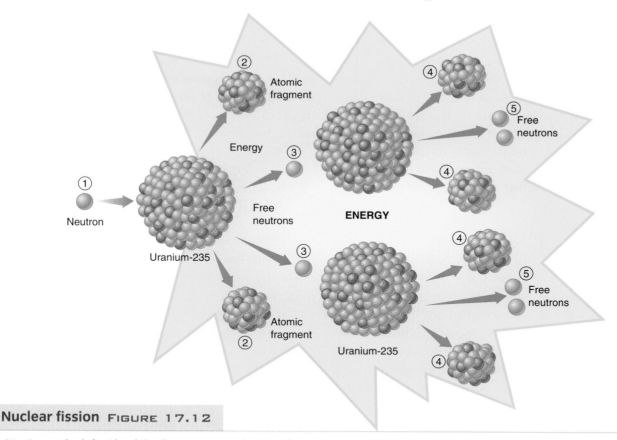

Nuclear fission FIGURE 17.12

Starting at the left side of the figure, neutron bombardment (1) of a uranium-235 (U-235) nucleus causes it to split into two smaller radioactive fragments (2) and several free neutrons (3). The free neutrons bombard nearby U-235 nuclei, causing them to split (4) and release still more free neutrons (5) in a chain reaction. Many different pairs of radioactive atomic fragments are produced during the fission of U-235.

Uranium fuel FIGURE 17.13

A Uranium dioxide pellets, held in a gloved hand, contain about 3 percent uranium-235, the fission fuel in a nuclear reactor. Each pellet contains the energy equivalent of one ton of coal.

B The uranium pellets are loaded into long fuel rods, which are grouped into square fuel assemblies (shown).

After enrichment, the uranium is processed into small pellets of uranium dioxide; each pellet contains the energy equivalent to a ton of coal (**FIGURE 17.13A**). The pellets are then placed in closed pipes, often as long as 3.7 m (12 ft), called **fuel rods**. The fuel rods are then grouped into square **fuel assemblies**, generally made up of 200 rods each (**FIGURE 17.13B**). A typical **nuclear reactor** contains about 250 fuel assemblies.

The fission of U-235 releases an enormous amount of heat, which is used to transform water into steam. The steam, in turn, is used to generate electricity. The production of electricity is possible because the fission reaction is controlled. Operators of a nuclear power plant can start or stop and increase or decrease the fission reactions in the reactor to produce the desired amount of heat energy.

A typical nuclear power plant has four main parts: the reactor core, the steam generator, the turbine, and the condenser (**FIGURE 17.14**). Fission occurs in the **reactor core**, and the heat produced by nuclear fission is used to produce steam from liquid water in the **steam generator**. The **turbine** uses the steam to generate electricity, and the **condenser** cools the steam, converting it back to a liquid.

enrichment The process by which uranium ore is refined after mining to increase the concentration of fissionable U-235.

IS NUCLEAR ENERGY CLEANER THAN COAL?

One of the reasons supporters of nuclear energy argue for its widespread adoption is that nuclear energy impacts the environment less than fossil fuels like coal (**TABLE 17.2**). The combustion of coal to generate electricity is responsible for more than one-third of the air pollution in the United States and contributes to acid precipitation and global warming. In comparison, nuclear energy emits few pollutants into the atmosphere. Nuclear energy also provides power without producing climate-altering CO_2.

nuclear reactor A device that initiates and maintains a controlled nuclear fission chain reaction to produce energy for electricity.

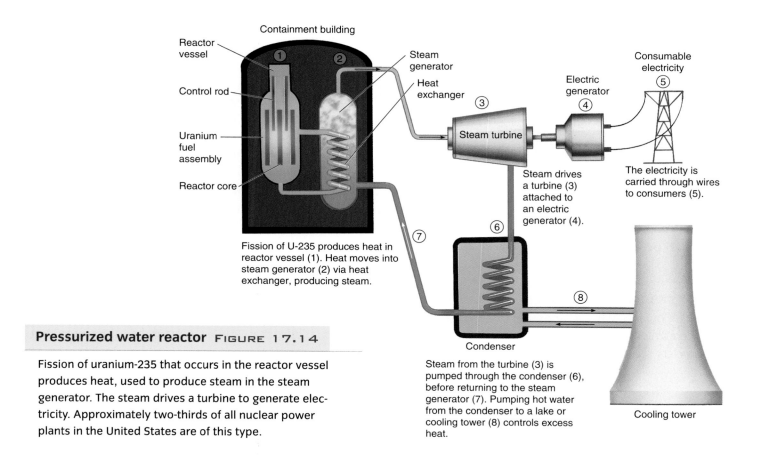

Pressurized water reactor FIGURE 17.14

Fission of uranium-235 that occurs in the reactor vessel produces heat, used to produce steam in the steam generator. The steam drives a turbine to generate electricity. Approximately two-thirds of all nuclear power plants in the United States are of this type.

Fission of U-235 produces heat in reactor vessel (1). Heat moves into steam generator (2) via heat exchanger, producing steam.

Steam drives a turbine (3) attached to an electric generator (4).

The electricity is carried through wires to consumers (5).

Steam from the turbine (3) is pumped through the condenser (6), before returning to the steam generator (7). Pumping hot water from the condenser to a lake or cooling tower (8) controls excess heat.

Cooling tower

Comparison of environmental impacts of 1000-MWe coal and conventional nuclear power plants*
TABLE 17.2

Impact	Coal	Nuclear (conventional fission)
Land use	17,000 acres	1900 acres
Daily fuel requirement	9000 tons (of coal)/day	3 kg (of enriched uranium)/day
Availability of fuel, based on present economics	A few hundred years	100 years, maybe longer
Air pollution	Moderate to severe, depending on pollution controls	Low
Climate change risk (from CO_2 emissions)	Severe	No risk
Radioactive emissions, routine	1 curie	28,000 curies
Water pollution	Often severe	Potentially severe at nuclear waste disposal sites
Risk from catastrophic accidents	Short-term local risk	Long-term risk over large areas
Link to nuclear weapons	No	Yes
Annual occupational deaths	0.5 to 5	0.1 to 1

* Impacts include extraction, processing, transportation, and conversion. Assumes coal is strip-mined. (A 1000-MWe utility, at a 60% load factor, produces enough electricity for a city of 1 million people.)

However, nuclear energy generates radioactive waste in the form of **spent fuel**. Nuclear power plants also produce radioactive coolant fluids and gases in the reactor. These radioactive wastes are extremely dangerous, and the hazards of their health and environmental impacts mean that special measures must be taken to ensure their safe storage and disposal.

Opponents of nuclear energy contend that the replacement of coal-burning power plants with nuclear power plants does not significantly lessen the threat of global warming, because only 15 percent of greenhouse gases come from power plants. Auto emissions and industrial processes produce most greenhouse gases, and neither is affected by nuclear power. Also, uranium mining, reactor construction, and disposal of nuclear wastes require the combustion of fossil fuels. This means that nuclear power indirectly contributes a small amount to the greenhouse effect—about 2 to 6 grams of carbon per kilowatt hour. This is two orders of magnitude lower than fossil fuels.

> ■ **spent fuel** Used fuel elements that were irradiated in a nuclear reactor.

CAN NUCLEAR ENERGY DECREASE OUR RELIANCE ON FOREIGN OIL?

International crises, such as the oil embargo of the early 1970s (see the chapter introduction), the Gulf War of the early 1990s, and the Iraq War in the 2000s, occasionally threaten the supply of oil to the United States. Some supporters of nuclear energy assert that our dependence on foreign oil would lessen if all oil-burning power plants were converted to nuclear plants.

However, oil is responsible for generating only about 3 percent of the electricity in the United States. Replacing electricity generated by oil with electricity generated by nuclear power would do little to lessen our dependence on foreign oil, because we would still need oil for heating buildings and for gasoline. Technological advances could change nuclear power's potential contribution in the future.

If electric heat pumps and electric motor vehicles become more common, however, nuclear power plants have the potential to heat buildings and power automobiles, thus decreasing our reliance on foreign oil.

SAFETY AND ACCIDENTS IN NUCLEAR POWER PLANTS

Although conventional nuclear power plants cannot explode like atomic bombs, accidents do happen in which dangerous levels of radiation are released into the environment and result in human casualties. At high temperatures, the metal encasing uranium fuel can melt, releasing radiation; this is called a **meltdown**. Also, the water used in a nuclear reactor to transfer heat can boil away during an accident, contaminating the atmosphere with radioactivity.

Although the nuclear industry considers the probability of a major accident as low, the consequences of such accidents are drastic and life threatening, both immediately and long after the accident has occurred.

Three Mile Island   The most serious nuclear reactor accident in the United States occurred in 1979 at the Three Mile Island power plant in Pennsylvania as the result of human error after the cooling system failed. A partial meltdown of the reactor core took place. Had there been a complete meltdown of the fuel assembly, dangerous radioactivity would have been emitted into the surrounding countryside. Fortunately, the containment building kept almost all the radioactivity released by the core material from escaping. Although a small amount of radiation entered the environment, there were no substantial environmental damages and no immediate human casualties. A study conducted within a 10-mile radius around the plant 10 years after the accident concluded that cancer rates were in the normal range and found no association between cancer rates and radiation from the accident. Numerous other studies have failed to link abnormal health problems (other than increased stress) to the accident.

In the aftermath of the accident, public wariness prompted construction delays and cancellations of several new nuclear power plants across the United States. The accident at Three Mile Island reduced the complacency that was commonplace in the nuclear industry. New safety regulations were put in place, including more frequent safety inspections, new risk assessments, and improved

Chernobyl, Ukraine FIGURE 17.15

The arrow indicates the site of the explosion. The upper part of the reactor was completely destroyed.

emergency and evacuation plans for nuclear power plants and surrounding communities. The Institute of Nuclear Power Operations was created to promote safety and improve the training of nuclear operators.

Chernobyl   The world's worst nuclear power plant accident took place in 1986 at the Chernobyl plant, located in the former Soviet republic of the Ukraine. One or possibly two explosions ripped apart a nuclear reactor and expelled large quantities of radioactive material into the atmosphere (FIGURE 17.15). The effects of this accident were not confined to the area immediately surrounding the power plant: significant amounts of radioisotopes quickly spread across large portions of Europe. The Chernobyl accident affected and will continue to affect many nations.

Although cleanup in the immediate vicinity of Chernobyl is finished, the people in Ukraine face many long-term problems. Ultimately, more than 170,000 people had to permanently abandon their homes. Much of the farmland and forests are so contaminated that they cannot be used for more than a century. Inhabitants of many areas of Ukraine cannot drink the water or con-

sume locally produced milk, meat, fish, fruits, or vegetables. Mothers do not nurse their babies because their milk is contaminated by radioactivity. The frequency of birth defects and mental retardation in newborns has increased in affected areas, and children exposed to the Chernobyl fallout experience higher incidences of leukemia, thyroid cancer, abnormalities of the immune system, and other birth defects (FIGURE 17.16).

THE LINK BETWEEN NUCLEAR ENERGY AND NUCLEAR WEAPONS

Fission is involved in both the production of electricity by nuclear energy and the destructive power of nuclear weapons. Countries that own nuclear power plants have access to the fuel needed for nuclear weapons (by reprocessing spent fuel to make plutonium). Responsible world leaders are concerned about the consequences of terrorist groups and states of concern (like Iran and North Korea) building nuclear weapons. These concerns have caused many people to shun nuclear energy, and to seek alternatives that are not so intimately connected with nuclear weapons.

There are several hundred metric tons of weapons-grade plutonium worldwide. Storing plutonium is a security nightmare because it only takes several kilograms to make a nuclear bomb as powerful as the ones that destroyed the Japanese cities of Nagasaki and Hiroshima in World War II. However, since the 2001 terrorist attacks in the United States, the security of plutonium stockpiles and nuclear power plants has been increased substantially to reduce the chance terrorist groups could steal plutonium and enriched uranium and use them to make nuclear weapons.

Health consequences of Chernobyl
FIGURE 17.16

Shown is a 14-year-old who is recovering from thyroid cancer. A significant (25-fold) increase in thyroid cancer in children and adolescents occurred within a few years after the accident.

RADIOACTIVE WASTES

Four sites—located in Washington State, South Carolina, Nevada, and Utah—currently store most of the country's low-level radioactive wastes. Radioactive wastes are classified as either "low-level" or "high-level." **Low-level radioactive wastes** include glassware, tools, paper, clothing, and other items contaminated by radioactivity. The are produced by nuclear power plants, university research labs, nuclear medicine departments in hospitals, and industries.

High-level radioactive wastes produced during nuclear fission include the reactor metals (fuel rods and assemblies), coolant fluids, and air or other gases found in the reactor. High-level radioactive wastes are also generated during the reprocessing of spent fuel. Produced by nuclear power plants and nuclear weapons facilities, high-level radioactive wastes are among the most dangerous human-made hazardous wastes.

As the radioisotopes in spent fuel decay, they produce considerable heat, are extremely toxic to organisms, and remain radioactive for thousands of years. Their dangerous level of radioactivity requires special handling. Secure storage of these materials must be guaranteed for thousands of years, until the materials decay sufficiently to be safe. The safe disposal of radioactive wastes is one of the main difficulties that must be overcome if nuclear energy is to realize its potential in the 21st century.

What are the best sites for the long-term storage of high-level radioactive wastes? Many scientists recommend storing the wastes in stable rock formations deep in the ground. People's reluctance to have radioactive wastes stored near their homes complicates the selection of these sites. Meanwhile, radioactive wastes continue to accumulate. Many commercially operated nuclear power plants store their spent fuel in huge indoor pools of water or in storage casks on-site. None of these plants was designed for long-term storage of spent fuel (**FIGURE 17.17**).

low-level radioactive wastes Solids, liquids, or gases that give off small amounts of ionizing radiation.

high-level radioactive wastes Radioactive solids, liquids, or gases that initially give off large amounts of ionizing radiation.

Cask length: 17 feet

- Protective cover
- Lid with metallic seals
- Storage cask (neutron shield)
- Canister of steel
- Spent fuel rods

Storage casks for spent fuel FIGURE 17.17

A On-site storage casks at the Prairie Island nuclear power plant in Minnesota. Each cask holds 40 spent fuel assemblies (17.6 tons).

B Details of a storage cask. Each cask, designed to last at least 40 years, is monitored and will be replaced if leakage occurs.

Yucca Mountain In 1982 the passage of the **Nuclear Waste Policy Act** put the burden of developing permanent sites for civilian and military radioactive wastes on the federal government and required the first site to be operational by 1998. (The deadline has since been postponed to 2010 at the earliest.)

In a 1987 amendment to the Nuclear Waste Policy Act, Congress identified Yucca Mountain in Nevada as the only candidate for a permanent underground storage site for high-level radioactive wastes from commercially operated power plants (see What a Scientist Sees). Since 1983 the U.S. Department of Energy has spent billions of dollars conducting feasibility studies on Yucca Mountain's geology. Results indicate the site is safe, at least from volcanic eruptions and earthquakes. In 2002 Congress finally approved the choice of Yucca Mountain as the U.S. nuclear-waste repository, despite controversy and the state of Nevada's opposition.

Transporting high-level wastes from nuclear reactors and weapons sites by truck, rail, or air is a major concern of opponents of the Yucca Mountain site. The typical shipment would travel an average of 2,300 miles, and 43 states would have these dangerous materials passing through them on their way to Yucca Mountain.

Whether or not nuclear waste is eventually stored in Yucca Mountain, the scientific community generally agrees that storage of high-level radioactive waste in deep underground repositories is the best long-term option. An underground waste facility is far safer than storing high-level nuclear waste as we do now, at many different commercial nuclear reactors, which pose a greater risk of terrorist attacks, theft, and, possibly, human health problems.

What a Scientist Sees

A Yucca Mountain is an arid, sparsely populated area of Nevada.

B An environmental scientist looking at Yucca Mountain thinks about the huge complex of interconnected tunnels located in dense volcanic rock 305 m (1,000 ft) beneath the mountain crest. Canisters containing high-level radioactive waste may be stored in the tunnels.

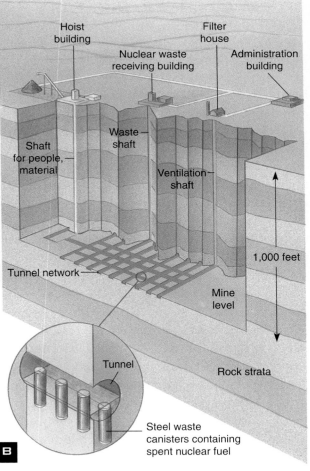

Over the past three decades, Soviet (and now Russian) practices for radioactive waste disposal have often violated international standards:

- Billions of gallons of liquid radioactive wastes were pumped directly underground, without being stored in protective containers. Russian officials claim that layers of clay and shale at the sites prevent leakage but admit that more leaks than expected have occurred.

- Highly radioactive wastes were dumped into the ocean in amounts double that of dumped wastes from 12 other nuclear nations combined.

- Both underground injection and underwater dumping of radioactive wastes continue because Russia lacks alternatives for nuclear waste processing. Potential health and environmental hazards associated with these wastes are unknown because so little data exist for these types of long-term storage.

DECOMMISSIONING NUCLEAR POWER PLANTS

decommissioning The dismantling of a nuclear power plant after it closes.

As nuclear power plants age, certain critical sections, like the reactor vessel, become brittle or corroded. At the end of their operational usefulness, nuclear power plants are not simply abandoned or demolished, because many parts have become contaminated with radioactivity.

Three options exist when a nuclear power plant is closed: storage, entombment, and decommissioning. If an old plant is put into **storage**, the utility company guards it for 50 to 100 years while some of the radioactive materials decay, making it safer to dismantle the plant later. Accidental leaks during the storage period are an ongoing concern.

Most experts do not consider **entombment**, permanently encasing the entire power plant in concrete, a viable option because the tomb would have to remain intact for at least 1,000 years. Accidental leaks would probably occur during that time, and we cannot guarantee that future generations would inspect and maintain the site.

The third option for the retirement of a nuclear power plant is to **decommission** the plant immediately after it closes. Advances in robotics may make it feasible to tear down sections of old plants that are too "hot" (radioactive) for workers to safely dismantle. As the plant is torn down, small sections of it are transported to a permanent storage site.

According to the International Atomic Energy Agency, 107 nuclear power plants worldwide were permanently retired as of 2004 (23 of these in the United States), and many nuclear power plants are nearing retirement age. During the 21st century, we may find we are paying more in our utility bills to decommission old plants than we are to construct new ones.

CONCEPT CHECK　STOP

How does a nuclear reactor produce electricity?

What are the environmental effects of generating electricity through coal combustion or through conventional nuclear fission?

What were some of the short-term effects of the nuclear power plant accident at Chernobyl in Ukraine? What are some long-term effects?

The Arctic National Wildlife Refuge

In 1960 Congress declared a section of northeastern Alaska protected because of its distinctive wildlife. In 1980 Congress expanded this wilderness area to form the Arctic National Wildlife Refuge (see FIGURE A). The proposed opening of the Arctic National Wildlife Refuge to oil exploration has been an ongoing environment-versus-economy conflict since the refuge's inception. On one side are those who seek to protect rare and fragile natural environments; on the other side are those whose higher priority is the development of some of the last major U.S. oil supplies.

The refuge, called "America's Serengeti," is home to many animal species, including polar bears, arctic foxes, peregrine falcons, musk oxen, Dall sheep, wolverines, and snow geese. It is the calving area for a large migrating herd of caribou (see FIGURE B). Although it is biologically rich, the tundra is an extremely fragile ecosystem, in part because of its harsh climate. Arctic organisms are particularly vulnerable to human activities.

In the mid-1990s, pro-development interests became more vocal, partly because in 1994, for the first time in its history, the United States imported more than half the oil it used. Although the Department of the Interior concluded that oil drilling in the refuge would harm the area's ecosystem, both the Senate and the House of Representatives passed measures to allow it. (President Clinton vetoed the bill.) In 2001, President George W. Bush announced his support for opening the refuge to oil drilling, but after a contentious debate in Congress, the Senate voted against doing so.

Supporters cite economic considerations as the main reason for drilling for oil in the refuge. Development of domestic oil would improve the balance of trade and make the United States less dependent on foreign countries for our oil.

Conservationists think oil exploration poses permanent threats to the delicate balance of nature in the Alaskan wilderness, in exchange for a temporary oil supply. They reason that the money spent drilling for oil would be better used for research into alternative, renewable energy sources and energy conservation—a more permanent solution to the energy problem.

Arctic National Wildlife Refuge

A Located in the northeastern part of Alaska, the refuge is situated close to the Trans-Alaska Pipeline, which begins at Prudhoe Bay and extends south to Valdez. The National Petroleum Reserve–Alaska is also shown.

B Members of the caribou herd whose calving grounds are on the Arctic National Wildlife Refuge.

CHAPTER SUMMARY

1 Energy Consumption

1. The per-capita energy consumption in highly developed nations is eight times higher than that in developing nations.

2 Coal

1. **Surface mining** is the extraction of mineral and energy resources near Earth's surface by first removing the soil, subsoil, and overlying rock strata. **Subsurface mining** is the extraction of mineral and energy resources from deep underground deposits. Surface mining is less expensive and safer but causes more serious environmental problems than subsurface mining.

2. Coal mining can lead to landslides and can pollute streams with sediment and **acid mine drainage**, pollution caused when sulfuric acid and dangerous dissolved materials such as lead, arsenic, and cadmium, wash from coal and metal mines into nearby lakes and streams. **Mountaintop removal** is a method of surface mining in which a huge shovel gradually removes an entire mountaintop to reach the coal located below. Burning coal releases more CO_2 and contributes more extensively to global climate warming than burning other fossil fuels. The combustion of coal contributes to **acid deposition**, a type of air pollution in which acid falls from the atmosphere to the surface as precipitation (acid precipitation) or as dry acid particles.

3. Power plants can make coal a cleaner fuel by installing **scrubbers** to clean the power plants' exhaust. **Fluidized-bed combustion** is a clean-coal technology in which crushed coal is mixed with limestone to neutralize the acidic sulfur compounds produced during combustion.

3 Oil and Natural Gas

1. More than half of the world's total estimated oil and natural gas reserves are located in the Persian Gulf region.

2. A serious spill along an oil or gas transportation route creates an environmental crisis, particularly in aquatic ecosystems. The burning of oil and natural gas produces CO_2 that can contribute to global climate warming. Burning oil also leads to acid deposition by producing nitrogen oxides. Natural gas contains almost no sulfur and produces much less CO_2 and other pollutants, as compared to oil and coal.

4 Nuclear Energy

1. **Nuclear energy** is the energy released by nuclear **fission** or **fusion**. A **nuclear reactor** is a device that initiates and maintains a controlled nuclear fission chain reaction to produce energy for electricity. A typical reactor contains a **reactor core**, where nuclear fission occurs; a **steam generator**; a **turbine**; and a **condenser**.

2. Generating electric power through nuclear energy emits few pollutants (such as CO_2) into the atmosphere, as compared to the combustion of coal, but generates highly dangerous radioactive waste, such as **spent fuel**, the used fuel elements that were irradiated in a nuclear reactor.

3. Accidents at nuclear power plants can release dangerous levels of radiation into the environment and result in human casualties. The safe storage of radioactive wastes is another concern associated with nuclear energy. **Low-level radioactive wastes** are radioactive solids, liquids, or gases that give off small amounts of ionizing radiation. **High-level radioactive wastes** are radioactive solids, liquids, or gases that initially give off large amounts of ionizing radiation. Radioactive wastes must be isolated securely for thousands of years. One option for the retirement of an aging nuclear power plant is to **decommission** it, which entails dismantling the old nuclear power plant after it closes.

KEY TERMS

- surface mining p. 409
- subsurface mining p. 409
- acid mine drainage p. 410
- fluidized-bed combustion p. 411
- nuclear energy p. 417
- fission p. 417
- fusion p. 417
- enrichment p. 418
- nuclear reactor p. 418
- spent fuel p. 420
- low-level radioactive wastes p. 422
- high-level radioactive wastes p. 422
- decommissioning p. 424

Passive solar heating FIGURE 18.2

A Behind the greenhouse glass of this passive solar home, rooms remain at steady temperatures during winter.

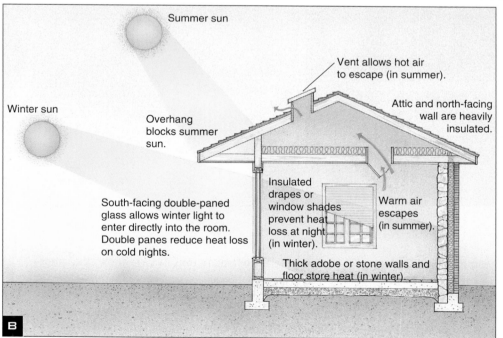

Summer sun

Winter sun

Overhang blocks summer sun.

Vent allows hot air to escape (in summer).

Attic and north-facing wall are heavily insulated.

South-facing double-paned glass allows winter light to enter directly into the room. Double panes reduce heat loss on cold nights.

Insulated drapes or window shades prevent heat loss at night (in winter).

Warm air escapes (in summer).

Thick adobe or stone walls and floor store heat (in winter).

B Several passive designs are incorporated into this home.

tank. Solar domestic water heating can provide a family's hot water needs year-round.

Active solar energy is not used for space heating as commonly as it is used for heating water, but it may become more important when diminishing supplies of fossil fuels force gas and oil prices higher.

In **passive solar heating**, solar energy heats buildings without the need for pumps or fans to distribute the heat.

Certain design features are incorporated into a passive solar heating system to warm buildings in winter and help them remain cool in summer (**FIGURE 18.2**). In the Northern Hemisphere, large south-facing windows receive more total sunlight during the day than windows facing other directions. Sunlight entering through the windows provides heat, which is then stored in floors and walls made of concrete or stone, or in containers of water. This stored heat is transmitted throughout the building naturally by **convection**, the circulation that occurs because warm air rises and cooler air sinks. Buildings with passive solar heating systems must be well insulated so that accumulated heat doesn't escape. Depending on the building's design and location, passive heating can save as much as 50 percent of heating costs. Currently, about 7 percent of new homes built in the United States have passive solar heating features.

> **passive solar heating** A system of putting the sun's energy to use without requiring mechanical devices to distribute the collected heat.

PHOTOVOLTAIC SOLAR CELLS

Photovoltaic (PV) solar cells can convert sunlight directly into electricity (see What a Scientist Sees). They are usually arranged on large panels that absorb sunlight even on cloudy or rainy days.

PV cells generate electricity with no pollution and minimal maintenance. They can be used on any scale, from small portable modules attached to camping lanterns to large, multimegawatt power plants, and can power satellites, uncrewed airplanes, highway signals, wristwatches, and calculators. The cells' widespread use to generate electricity is currently limited by their low efficiency at converting solar

■ photovoltaic solar cell (PV)
A wafer or thin film of solid state materials, such as silicon or gallium arsenide, that are treated with certain metals in such a way that the film generates electricity—that is, a flow of electrons—when solar energy is absorbed.

energy to electricity, and by the amount of land needed to hold the number of solar panels required for large-scale use.

In remote areas not served by electric power plants, like the rural areas of developing countries, it is more economical to use PV cells for electricity. Photovoltaics generate energy that can pump water, refrigerate vaccines, grind grain, charge batteries, and supply rural homes with lighting. According to the Institute for Sustainable Power, more than 1 million households in the developing countries of Asia, Latin America, and Africa have installed PV solar cells on the roofs of their homes. A PV panel the size of two pizza boxes supplies a rural household

What a Scientist Sees

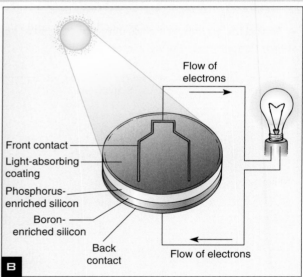

Photovoltaic Cells

A A student seeing the roof of the Intercultural Center of Georgetown University probably knows that it has arrays of photovoltaic (PV) cells to collect solar energy. The PV system supplies about 10 percent of the school's electricity.

B A scientist looking at those arrays knows that photovoltaic cells contain silicon and other materials. Sunlight excites electrons, which are ejected from silicon atoms. Useful electricity is generated when the ejected electrons flow out of the PV cells through a wire.

Solar shingles FIGURE 18.3

These thin-film solar cells look much like conventional roofing materials.

Future technological progress may make PVs economically competitive with electricity produced by conventional energy sources. The production of "thin-film" solar cells (FIGURE 18.3), which are much cheaper to manufacture than standard PVs, has decreased costs for PVs across the board. More than 120,000 Japanese homes have installed PV solar-energy roofing in the past few years. The Million Solar Roofs initiative, sponsored by the U.S. government, plans to have solar roofing on 1 million buildings by 2010. Another technological advance that shows promise is dye-sensitized solar cells, which can be produced at about one-fifth the cost of conventional silicon panels.

with enough electricity for five lights, a radio, and a television.

Utility companies can purchase PV devices in modular units, which can become operational in a short period. Rather than committing a billion dollars or more and a decade or more to build a new power plant, they can increase generating capacity in small increments. The PV units can provide the additional energy, for example, to power irrigation pumps on hot, sunny days.

The cost of manufacturing PV modules has steadily declined over the last 25 years, from an average factory price of almost $90 per watt in 1975 to about $3.10 per watt in 2003, according to PV Energy Systems, Inc. The cost of producing electricity from PVs has steadily declined from 1970 to the present. Despite this progress, the cost is still about $.15 to $.25 per kilowatt-hour. TABLE 18.1 compares the costs of generating electricity using different energy sources, including photovoltaics.

Generating costs of electric power plants
TABLE 18.1

Energy source	Generating costs (cents per kilowatt-hour)*
Hydropower	4–10
Biomass	6–8
Geothermal	3–8
Wind	4–5
Solar thermal	10–15
Photovoltaics	15–25
Natural gas	4–5
Coal	4–5
Nuclear power	10–15

*Electricity production and consumption are measured in kilowatt-hours (kWh). As an example, one 50-watt light bulb that is on for 20 hours uses one kilowatt-hour of electricity ($50 \times 20 = 1000$ watt-hours = 1 kWh).

SOLAR THERMAL ELECTRIC GENERATION

In **solar thermal electric generation**, electricity is produced by several different systems that collect sunlight and concentrate it using a combination of mirrors or lenses to heat a working fluid to high temperatures.

In one such system, computer-guided trough-shaped mirrors track the sun for optimum efficiency, center sunlight on nearby oil-filled pipes, and heat the oil to 390°C (735°F) (FIGURE 18.4). The hot oil is circulated to a water storage system and used to boil water into super-heated steam, which turns a turbine to generate electricity.

Solar thermal systems often have a backup—usually natural gas—to generate electricity at night and during cloudy days when solar power isn't operating. The world's largest solar thermal system of this type currently operates in the Mojave Desert in Southern California.

Solar thermal energy systems are inherently more efficient than other solar technologies because

■ **solar thermal electric generation** A means of producing electricity in which the sun's energy is concentrated by mirrors or lenses onto a fluid-filled pipe; the heated fluid is used to generate electricity.

they concentrate the sun's energy. With improved engineering, manufacturing, and construction methods, solar thermal energy may become cost-competitive with fossil fuels (see Table 18.1). In addition, the environmental benefits of solar thermal plants are significant: they don't produce air pollution or contribute to acid rain or global warming.

SOLAR-GENERATED HYDROGEN

Increasingly, people think of hydrogen as the fuel of the future, as it is abundant as well as easily produced. Electricity generated by photovoltaics or wind energy can split water into the gases oxygen and hydrogen, though this process isn't yet economical. Hydrogen can also be produced using conventional energy sources such as fossil fuels and nuclear power. However, using fossil fuels or nuclear energy to create hydrogen results in the same serious environmental and security problems we discussed in Chapter 17. We therefore limit our discussion to hydrogen fuel production using solar electricity, which is sustainable but not yet cost-efficient.

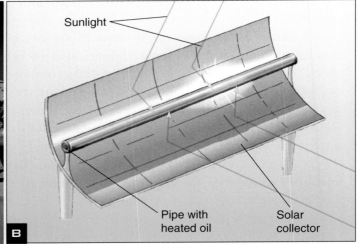

Solar thermal electric generation FIGURE 18.4

A A solar thermal plant in California uses troughs to focus sunlight on a fluid-filled tube, as shown in B. The heated oil is pumped to a water tank where it generates steam used to produce electricity. For simplicity, arrows show sunlight converging on several points; sunlight actually converges on the pipe throughout its length.

Hydrogen is a clean fuel; it produces water and heat as it burns and produces no sulfur oxides, carbon monoxide, hydrocarbon particulates, or CO_2 emissions. It does produce some nitrogen oxides, though in amounts fairly easy to control. Hydrogen has the potential to provide energy for transportation (in the form of hydrogen-powered automobiles) as well as for heating buildings and producing electricity.

It may seem wasteful to use electricity generated from solar energy to make hydrogen that will then be used to generate electricity. However, the electricity generated by existing photovoltaic cells must be used immediately, whereas hydrogen offers a convenient way to store solar energy as chemical energy. It can be transported by pipeline, possibly less expensively than electricity is transported by wire.

Production of hydrogen from PV electricity currently has a relatively low efficiency (perhaps 10 percent), which means that very little of the solar energy absorbed by the PV cells is actually converted into the chemical energy of hydrogen fuel. Low efficiency translates into high costs. Scientists are working to improve this efficiency because decreased costs could make solar-generated hydrogen fuel commercially viable.

Other challenges besides high costs face us if we are to replace gasoline with hydrogen as a transportation fuel. First, we would need to develop a complex infrastructure (like hydrogen pipelines) to provide hydrogen to service stations. Another challenge is developing **fuel cells** for motor vehicles that are inexpensive, safe, and can drive a long distance without the need to refuel. A fuel cell is an electrochemical cell similar to a battery (**FIGURE 18.5**).

Fuel cells produce power as long as they are supplied with fuel, whereas batteries store a fixed amount of energy. The major carmakers are now developing automobiles and buses powered by hydrogen fuel cells (see the chapter introduction).

> ■ **fuel cell** A device that directly converts chemical energy into electricity without producing steam that runs a turbine and generator; the fuel cell requires hydrogen and oxygen from the air.

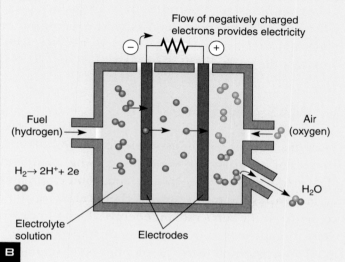

Fuel cells in a laboratory FIGURE 18.5

A These fuel cells combine hydrogen and oxygen to create electricity. B Cross-section of a fuel cell.

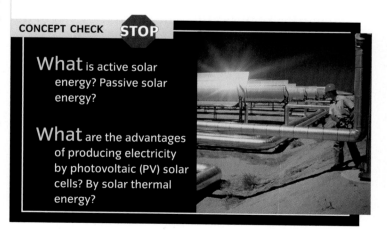

CONCEPT CHECK **STOP**

What is active solar energy? Passive solar energy?

What are the advantages of producing electricity by photovoltaic (PV) solar cells? By solar thermal energy?

Indirect Solar Energy

LEARNING OBJECTIVES

Define biomass and outline how it is used as a source of energy.

Compare the potential of wind energy and hydropower.

Some renewable energy sources indirectly use the sun's energy. Combustion of *biomass* (organic matter) is an example of indirect solar energy, because plants use solar energy for photosynthesis and store the energy in biomass. Windmills, or wind turbines, use *wind energy* to generate electricity. The damming of rivers and streams to generate electricity is a type of *hydropower*—the energy of flowing water.

BIOMASS ENERGY

Biomass, one of the oldest fuels known to humans, consists of materials like wood, fast-growing plant and algal crops, crop wastes, sawdust and wood chips, and animal wastes. Biomass contains chemical energy that comes from the sun's radiant energy, which photosynthetic organisms use to form organic molecules. Biomass is a renewable form of energy if managed properly.

Biomass fuel, which may be a solid, liquid, or gas, is burned to release its energy. Solid biomass fuels like wood, charcoal (wood turned into coal by partial burning), animal dung, and peat (partly decayed plant matter found in bogs and swamps)—supplies a substantial portion of the world's energy. At least half of the human population relies on it as their main source of energy. In developing countries, wood is the primary fuel for cooking and heat (**FIGURE 18.6**).

It is possible to convert biomass, particularly animal wastes, into **biogas**. Biogas, which is usually composed of a mixture of gases (mostly methane), is like natural gas. It is a clean fuel—its combustion produces fewer pollutants than either coal or biomass. In India and China, several million family-sized **biogas digesters** use microbial decomposition of household and agricultural wastes to produce biogas for cooking and lighting

(**FIGURE 18.7**). When biogas conversion is complete, the solid remains are removed from the digester and used as fertilizer.

> **biomass** Plant and animal material used as fuel.

Biogas has the potential to power fuel cells to generate electricity. A pilot program at Boston's main sewage treatment plant began producing electricity from biogas in 1997. Sewage sludge in large biogas digesters produces methane, which is then burned in a methane fuel cell to produce enough electricity for 150 homes. Like the hydrogen fuel cells discussed earlier in the chapter, methane fuel cells produce relatively few pollutants.

Nepal

Global Locator

Biomass FIGURE 18.6

Firewood is the major energy source for most of the developing world. Photographed in Nepal.

This small-scale biogas digester is being evaluated at a research center. Animal manure placed in the digester decomposes, releasing methane gas that can be used as cooking fuel.

Biogas digester in India FIGURE 18.7

Biomass can also be converted into liquid fuels, especially **methanol** (methyl alcohol) and **ethanol** (ethyl alcohol), which can be used in internal combustion engines. Mixing gasoline with 10 percent ethanol (usually produced from corn) produces a cleaner-burning mixture called *gasohol. Biodiesel,* made from plant or animal oils, is becoming more popular as an alternative fuel for diesel engines in trucks, farm equipment, and boats. The oil is often refined from waste oil produced at restaurants (like the oil used to make french fries); biodiesel burns much cleaner than diesel fuel.

Although some U.S. energy companies convert sugarcane, corn, or wood crops to alcohol, others are interested in the commercial conversion of agricultural and municipal wastes into ethanol. Currently, the profitability of ethanol is only possible because of government **subsidies** that reduce ethanol's cost. However, as more car companies introduce ethanol-friendly vehicles, these subsidies may cease to be necessary.

Biomass is attractive as a source of energy because it reduces dependence on fossil fuels and often makes use of waste products, thereby reducing our waste disposal problem. Biomass combustion is not completely free of the pollution problems of fossil fuels, but biomass combustion produces levels of sulfur and ash that are lower than those that coal produces.

Some problems associated with obtaining energy from biomass include the use of land and water that might otherwise be dedicated to agriculture. Shifting the agricultural balance toward energy production might decrease food production, contributing to higher food prices. Also, as mentioned earlier, at least half of the world's population relies on biomass as its main source of energy. Unfortunately, in many areas people burn wood faster than they replant trees. Intensive use of wood for energy has resulted in severe damage to the environment, including soil erosion, deforestation and desertification, air pollution, and degradation of water supplies.

Excessive use of crop biomass can also harm soil quality. Crop residues, such as cornstalks, are increasingly being used for energy. Crop residues left in and on the ground prevent erosion by holding the soil in place; their removal would eventually deplete the soil of minerals and reduce its productivity.

Indirect Solar Energy 437

Competition

compact development The design of cities in which tall, multiple-unit residential buildings are close to shopping and jobs, and all are connected by public transportation.

competition The interaction among organisms that vie for the same resources in an ecosystem (such as food or living space).

conservation biology The scientific study of how humans impact organisms and of the development of ways to protect biological diversity.

conservation easement A legal agreement that protects privately owned forest, rangeland, or other property from development for a specified number of years.

conservation tillage A method of cultivation in which residues from previous crops are left in the soil, partially covering it and helping to hold it in place until the newly planted seeds are established.

consumption overpopulation A situation that occurs when each individual in a population consumes too large a share of resources.

contour plowing Plowing that matches the natural contour of the land.

Coriolis effect The tendency of moving air or water to be deflected from its path and swerve to the right in the Northern Hemisphere and to the left in the Southern Hemisphere.

cost-benefit diagram A diagram that helps policymakers make decisions about costs of a particular action and benefits that would occur if that action were implemented.

crop rotation The planting of a series of different crops in the same field over a period of years.

decommissioning The dismantling of a nuclear power plant after it closes.

deep ecology worldview An understanding of our place in the world based on harmony with nature, a spiritual respect for life, and the belief that humans and all other species have an equal worth.

deforestation The temporary or permanent clearance of large expanses of forest for agriculture or other uses.

degradation (of land) Natural or human-induced processes that decrease the future ability of the land to support crops or livestock.

demographic transition The process whereby a country moves from relatively high birth and death rates to relatively low birth and death rates.

demographics The applied branch of sociology that deals with population statistics.

desert A biome in which the lack of precipitation limits plant growth; deserts are found in both temperate and tropical regions.

desertification Degradation of once-fertile rangeland or tropical dry forest into nonproductive desert.

dose–response curve In toxicology, a graph that shows the effect of different doses on a population of test organisms.

dust dome A dome of heated air that surrounds an urban area and contains a lot of air pollution.

ecological niche The totality of an organism's adaptations, its use of resources, and the lifestyle to which it is fitted.

ecological succession The process of community development over time, which involves species in one stage being replaced by different species.

ecology The study of the interactions among organisms and between organisms and their abiotic environment.

economic development An expansion in a government's economy, viewed by many as the best way to raise the standard of living.

ecosystem A community and its physical environment.

ecosystem services Important environmental benefits, such as clean air to breathe, clean water to drink, and fertile soil in which to grow crops, that the natural environment provides.

El Niño–Southern Oscillation (ENSO) A periodic, large-scale warming of surface waters of the tropical eastern Pacific Ocean that temporarily alters both ocean and atmospheric circulation patterns.

endangered species A species that faces threats that may cause it to become extinct within a short period.

endemic species Organisms that are native to or confined to a particular region.

El Niño

energy conservation Using less energy by reducing energy use and waste, for example.

energy efficiency Using less energy to accomplish a given task—by using new technology, for example.

energy flow The passage of energy in a one-way direction through an ecosystem.

enhanced greenhouse effect The additional warming that may be produced by human-increased levels of gases that absorb infrared radiation.

enrichment The process by which uranium ore is refined after mining to increase the concentration of fissionable U-235.

environmental chemistry A subdiscipline of chemistry in which commercially important chemical processes are redesigned to reduce environmental harm.

environmental ethics A field of applied ethics that considers the moral basis of environmental responsibility and how far this responsibility extends.

environmental justice The right of every citizen, regardless of age, race, gender, social class, or other factor, to adequate protection from environmental hazards.

environmental resistance   Unfavorable environmental conditions that prevent organisms from reproducing indefinitely at their biotic potential.

environmental science   The interdisciplinary study of humanity's relationship with other organisms and the nonliving physical environment.

environmental sustainability   The ability to meet humanity's current needs without compromising the ability of future generations to meet their needs.

environmental worldview   A worldview that helps us make sense of how the environment works, our place in the environment, and right and wrong environmental behaviors.

estuary   A coastal body of water, partly surrounded by land, with access to the open ocean and a large supply of fresh water from a river.

Estuary

evolution   The cumulative genetic changes in populations that occur during successive generations.

exponential population growth   The accelerating population growth that occurs when optimal conditions allow a constant reproductive rate.

external cost   A harmful environmental or social cost that is borne by people not directly involved in buying or selling a product.

first law of thermodynamics   Energy cannot be created or destroyed, although it can change from one form to another.

fission   The splitting of an atomic nucleus into two smaller fragments, accompanied by the release of a large amount of energy.

flowing-water ecosystem   A freshwater ecosystem such as a river or stream in which the water flows in a current.

fluidized-bed combustion   A clean-coal technology in which crushed coal is mixed with limestone to neutralize acidic compounds produced during combustion.

food insecurity   The condition in which people live with chronic hunger and malnutrition.

Flowing-water ecosystem

forest decline   A gradual deterioration and eventual death of many trees in a forest.

freshwater wetlands   Lands that shallow fresh water covers for at least part of the year; wetlands have a characteristic soil and water-tolerant vegetation.

fuel cell   A device that directly converts chemical energy into electricity without producing steam that runs a turbine and generator; the fuel cell requires hydrogen and oxygen from the air.

full cost accounting   The process of evaluating and presenting to decision makers the relative benefits and costs of various alternatives.

fusion   The joining of two lightweight atomic nuclei into a single, heavier nucleus, accompanied by the release of a large amount of energy.

genetic engineering   The manipulation of genes (for example, taking a specific gene from one species and placing it into an unrelated species) to produce a particular trait.

genetic resistance   Any inherited characteristic that decreases the effect of a given agent (like a pesticide) on an organism (like a pest).

geothermal energy   The use of energy from Earth's interior for either space heating or generation of electricity.

germplasm   Any plant or animal material that may be used in breeding.

greenhouse gases   The gases that absorb infrared radiation; include carbon dioxide, methane, nitrous oxide, chlorofluorocarbons, and tropospheric ozone.

groundwater   The supply of fresh water under Earth's surface that is stored in underground aquifers.

growth rate (r)   The rate of change (increase or decrease) of a population's size, expressed in percent per year.

Germplasm

gyres   Large, circular ocean current systems that often encompass an entire ocean basin.

habitat fragmentation   The breakup of large areas of habitat into small, isolated patches.

hazardous waste   Any discarded chemical that threatens human health or the environment.

high-level radioactive wastes   Radioactive solids, liquids, or gases that initially give off large amounts of ionizing radiation.

highly developed countries   Countries with complex industrialized bases, low rates of population growth, and high per-capita incomes.

hydropower   A form of renewable energy that relies on flowing or falling water to generate electricity.

incentive-based regulation   Pollution control laws that work by establishing emission targets and providing industries with incentives to reduce emissions.

industrialized agriculture Modern agricultural methods, which require a large capital input and less land and labor than traditional methods.

infant mortality rate The number of infant deaths under age 1 per 1000 live births.

infrared radiation Electromagnetic radiation with wavelengths longer than visible light but shorter than microwaves; perceived as invisible waves of heat energy.

integrated pest management (IPM) A combination of pest control methods that, if used in the proper order and at the proper times, keep the size of a pest population low enough to prevent substantial economic loss.

integrated waste management A combination of the best waste management techniques into a consolidated program to deal effectively with solid waste.

intertidal zone The area of shoreline between low and high tides.

invasive species Foreign species that spread rapidly in a new area if free of predators, parasites, or resource limitations that may have controlled their population in their native habitat.

low-level radioactive wastes Solids, liquids, or gases that give off small amounts of ionizing radiation.

malnutrition The impairment of health due either to an underconsumption of calories or specific nutrients or to an overconsumption of calories.

marginal cost of pollution abatement The added cost for all present and future members of society of reducing one unit of a given type of pollution.

marginal cost of pollution The added cost for all present and future members of society of an additional unit of pollution.

mass burn incinerator A large furnace that burns all solid waste except for unburnable items such as refrigerators.

microirrigation A type of irrigation that conserves water by piping it to crops through sealed systems.

minerals Elements or compounds of elements that occur naturally in Earth's crust.

moderately developed countries Developing countries with a medium level of industrialization and per-capita incomes lower than those of highly developed countries.

monoculture Ecological simplification in which only one type of plant is cultivated over a large area.

municipal solid waste Solid materials discarded by homes, offices, stores, restaurants, schools, hospitals, prisons, libraries, and other facilities.

national income accounts A measure of the total income of a nation's goods and services for a given year.

natural capital Earth's resources and processes that sustain living organisms, including humans; includes minerals, forests, soils, groundwater, clean air, wildlife, and fisheries.

Minerals

Neritic province

natural selection The tendency of better-adapted individuals—those with a combination of genetic traits better suited to environmental conditions—to survive and reproduce, increasing their proportion in the population.

neritic province The part of the pelagic environment that overlies the ocean floor from the shoreline to a depth of 200 m (650 ft).

nonmunicipal solid waste Solid waste generated by industry, agriculture, and mining.

nonpoint source pollution Pollutants that enter bodies of water over large areas rather than being concentrated at a single point of entry.

nonrenewable resources Natural resources that are present in limited supplies and are depleted as they are used.

nuclear energy The energy released by nuclear fission or fusion.

nuclear reactor A device that initiates and maintains a controlled nuclear fission chain reaction to produce energy for electricity.

nutrient cycling The pathway of various nutrient minerals or elements from the environment through organisms and back to the environment.

oceanic province The part of the pelagic environment that overlies the ocean floor at depths greater than 200 m (650 ft).

optimum amount of pollution The amount of pollution that is economically most desirable.

overburden Soil and rock overlying a useful mineral deposit.

overgrazing When too many grazing animals consume the plants in a particular area, leaving the vegetation destroyed and unable to recover.

overnutrition A serious overconsumption of calories that leaves the body susceptible to disease.

ozone thinning The removal of ozone from the stratosphere by human-produced chemicals or natural processes.

passive solar heating A system of putting the sun's energy to use without requiring mechanical devices to distribute the collected heat.

pathogen An agent (usually a microorganism) that causes disease.

people overpopulation A situation in which there are too many people in a given geographic area.

persistent organic pollutants (POPs) A group of persistent toxicants that bioaccumulate in organisms and travel thousands of kilometers through air and water to contaminate sites far removed from their source.

pesticide Any toxic chemical used to kill pests.

pheromone A natural substance produced by animals to stimulate a response in other members of the same species.

photochemical smog A brownish orange haze formed by chemical reactions involving sunlight, nitrogen oxides, and hydrocarbons.

photovoltaic solar cell (PV) A wafer or thin film of solid state materials, such as silicon or gallium arsenide, that are treated with certain metals in such a way that the film generates electricity—that is, a flow of electrons—when solar energy is absorbed.

plate tectonics The study of the processes by which the lithospheric plates move over the asthenosphere.

point source pollution Water pollution that can be traced to a specific spot.

population A group of organisms of the same species that live together in the same area at the same time.

population ecology The branch of biology that deals with the number of individuals of a particular species found in an area and how and why those numbers increase or decrease over time.

poverty A condition in which people are unable to meet their basic needs for adequate food, clothing, shelter, education, or health.

precautionary principle Making decisions about adopting a new technology or chemical product by assigning the burden of proof of its safety to the developers of that technology or product.

predation The consumption of one species (the prey) by another (the predator).

primary air pollutants Harmful chemicals that enter directly into the atmosphere from either human activities or natural processes.

primary treatment Treating wastewater by removing suspended and floating particles by mechanical processes.

rangeland Land that isn't intensively managed and is used for grazing livestock.

renewable resources Resources that are replaced by natural processes and that therefore can be used forever, provided they are not overexploited in the short term.

replacement-level fertility The number of children a couple must produce to "replace" themselves.

restoration ecology The study of the historical condition of a human-damaged ecosystem, with the goal of returning it as close as possible to its former state.

risk The probability of harm (such as injury, disease, death, or environmental damage) occurring under certain circumstances.

risk assessment The use of statistical methods to quantify the risks of an action so they can be compared and contrasted with other risks.

runoff The movement of fresh water from precipitation and snowmelt to rivers, lakes, wetlands, and, ultimately, the ocean.

salinization The gradual accumulation of salt in a soil, often as a result of improper irrigation methods.

Risk

saltwater intrusion The movement of seawater into a freshwater aquifer near the coast.

sanitary landfill The most common method of disposal of solid waste, by compacting it and burying it under a shallow layer of soil.

savanna A tropical grassland with widely scattered trees or clumps of trees.

scientific method The way a scientist approaches a problem, by formulating a hypothesis and then testing it by means of an experiment.

second law of thermodynamics When energy is converted from one form to another, some of it is degraded into heat, a less usable form that disperses into the environment.

secondary treatment Treating wastewater biologically to decompose suspended organic material; secondary treatment reduces the water's biochemical oxygen demand.

sewage Wastewater from drains or sewers (from toilets, washing machines, and showers); includes human wastes, soaps, and detergents.

shelterbelt A row of trees planted as a windbreak to reduce soil erosion of agricultural land.

sick building syndrome Eye irritations, nausea, headaches, respiratory infections, depression, and fatigue caused by indoor air pollution.

smelting The process in which ore is melted at high temperatures to separate impurities from the molten metal.

soil The uppermost layer of Earth's crust, which supports terrestrial plants, animals, and microorganisms.

soil erosion The wearing away or removal of soil from the land.

soil horizons The horizontal layers into which many soils are organized, from the surface to the underlying parent material.

solar thermal electric generation A means of producing electricity in which the sun's energy is concentrated by mirrors or lenses onto a fluid-filled pipe; the heated fluid is used to generate electricity.

Soil

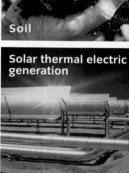
Solar thermal electric generation

source reduction An aspect of waste management in which products are designed and manufactured in ways that decrease the amount of solid and hazardous waste in the solid waste stream.

species richness The number of different species in a community.

spent fuel Used fuel elements that were irradiated in a nuclear reactor.

spoil bank A hill of loose rock created when the overburden from a new trench is put into the already excavated trench during strip mining.

standing-water ecosystem A body of fresh water surrounded by land and whose water does not flow; a lake or a pond.

subsistence agriculture Traditional agricultural methods, which are dependent on labor and a large amount of land to produce enough food to feed oneself and one's family.

Tundra

subsurface mining The extraction of mineral and energy resources from deep underground deposits.

surface mining The extraction of mineral and energy resources near Earth's surface by first removing the soil, subsoil, and overlying rock strata.

surface water Precipitation that remains on the surface of the land and does not seep down through the soil.

sustainable agriculture Agricultural methods that maintain soil productivity and a healthy ecological balance while having minimal long-term impacts.

sustainable consumption The use of goods and services that satisfy basic human needs and improve the quality of life but that also minimize the use of resources.

sustainable development Economic growth that meets the needs of the present without compromising the ability of future generations to meet their own economic needs.

sustainable forestry The use and management of forest ecosystems in an environmentally balanced and enduring way.

sustainable soil use The wise use of soil resources, without a reduction in the amount or fertility of soil, so it is productive for future generations.

sustainable water use The wise use of water resources, without harming the essential functioning of the hydrologic cycle or the ecosystems on which humans depend.

symbiosis Any intimate relationship or association between members of two or more species; includes mutualism, commensalism, and parasitism.

temperate deciduous forest A forest biome that occurs in temperate areas where annual precipitation ranges from about 75 cm to 126 cm (30 to 50 in).

temperate grassland A grassland with hot summers, cold winters, and less rainfall than is found in the temperate deciduous forest biome.

temperate rain forest A coniferous biome with cool weather, dense fog, and high precipitation.

temperature inversion A layer of cold air temporarily trapped near the ground by a warmer, upper layer.

tertiary treatment Advanced wastewater treatment methods that are sometimes employed after primary and secondary treatments.

threatened species A species whose population has declined to the point that it may be at risk of extinction.

tidal energy A form of renewable energy that relies on the ebb and flow of the tides to generate electricity.

total fertility rate (TFR) The average number of children born to each woman.

toxicology The study of toxicants, chemicals with adverse effects on health.

tropical rain forest A lush, species-rich forest biome that occurs where the climate is warm and moist throughout the year.

tundra The treeless biome in the far north that consists of boggy plains covered by lichens and small plants such as mosses; it has harsh, very cold winters and extremely short summers.

ultraviolet (UV) radiation That part of the electromagnetic spectrum with wavelengths just shorter than visible light; a high-energy form of radiation that can be lethal to organisms at high levels of exposure.

undernutrition A serious underconsumption of calories that leaves the body weakened and susceptible to disease.

urban heat island Local heat buildup in an area of high population.

urbanization The process whereby people move from rural areas to densely populated cities.

water pollution Any physical or chemical change in water that adversely affects the health of humans and other organisms.

Western worldview An understanding of our place in the world based on human superiority and dominance over nature, the unrestricted use of natural resources, and increased economic growth to manage an expanding industrial base.

wilderness A protected area of land in which no human development is permitted.

Urbanization

Water pollution

wildlife corridor A protected zone that connects isolated unlogged or undeveloped areas.

wind energy Electric energy obtained from surface air currents caused by the solar warming of air.

zero population growth The state in which population remains the same size because the birth rate equals the death rate.

PHOTO CREDITS

Chapter 1

Page 2: Adastra/Getty Images; page page 5: (top) Earth Imaging/Stone/ Getty Images; page 5: (bottom) Tomasz Tomaszewski/ NG Image Collection; page 6: (left) Peter Menzel; page 6: (right) Peter Menzel; page 6: (left) Peter Menzel; page 8: Reza Webistan/CORBIS; page 10: Macduff Everton/Iconica/Getty Images; page 13: (center) James P. Blair/NG Image Collection; page 13: John Anderson/NG Image Collection; page 13: (bottom) Sam Kittner/NG Image Collection; page 15: Mauro Fermariello/ Photo Researchers; page 16: (top) Steve Winter/NG Image Collection; page 16: (bottom) Courtesy of Matthew Price; page 18: Ben Osborne/ Stone/Getty Images; page 20: ALIX/Photo Researchers; page 21: Nick Nichols/NG Image Collection; page 23: (bottom) David H. Wells/CORBIS; page 23: (center) Joe Raedle/Getty Images

Chapter 2

Page 24: David Doubilet/NG Image Collection; page 28: Peter Menzel; page 27: (center) NG Image Collection; page 27: (top) Ira Block/NG Image Collection; page 29: Richard Nowitz/NG Image Collection; page 30: (center) Minnesota Historical Image Collection/CORBIS; page 30: (bottom) NG Image Collection; page 31: (bottom) Joel Sartore/NG Image Collection; page 31: (center) Tim Laman/NG Image Collection; page 32: Catherine Karnow/NG Image Collection; page 33: (top) William Thomas Cain/Getty Images; page 34: Jenny Hager/The Image Works; page 36: Donna DeCesare; page 37: Mike Goldwater - Network/ Matrix International, Inc.; page 38: (top) Nick Nichols/NG Image Collection; page 38: (bottom) Emory Kristof/NG Image Collection; page 39: Jose Azel/Aurora Photos; page 40: Wendy Stone/Corbis Images; page 41: (top) Roy Toft/NG Image Collection; page 42: John Liebenberg/Bloomberg News /Landov; page 43: (left) Terje Rakke/The Image Bank/Getty Images; page 45: Liane Cary/Age Fotostock

Chapter 3

Page 46: James P. Blair/NG Image Collection; page 48: USDA/NG Image Collection; page 49: Courtesy Library of Congress; page 50: Bettman/Corbis Images; page 51: (top) Courtesy National Archives; page 51: (center) Eugene Fisher/NG Image Collection; page 52: Photo by Tom Coleman/ Courtesy Aldo Leopold Foundation, Baraboo, WI.; page 53: AP/Wide World Photos; page 54: Jason Laure/Woodfin Camp & Associates; page 58: Marc Moritsch/NG Image Collection; page 59: Rick Poley/Visuals Unlimited; page 60: (bottom) Randy Olson/NG Image Collection; page 62: Herve Pelletier/SuperStock; page 66: (bottom) Gerd Ludwig/NG Image Collection; page 66: (top) Gerd Ludwig/NG Image Collection; page 56: (bottom) Peter Essick/Aurora Photos; page 69: Reuters Newmedia Inc./Corbis Images; page 63: (top) PhotoDisc, Inc./Getty Images; page 63: (bottom) PhotoDisc, Inc./Getty Images; page 65: Digital Vision; page 63: (bottom) PhotoDisc, Inc./Getty Images

Chapter 4

Page 74: (top) Les Stone/The Image Works; page 77: (top) Courtesy Millipore Corp; page 78: Karen Kasmauski/NG Image Collection; page 77: (bottom) Bruce Dale/NG Image Collection; page 79: (top) courtesy L. Beck, NASA Ames Research Center; page 80: Roy Toft/NG Image Collection; page 82: Darlyne A. Murawski/NG Image Collection; page 85: Courtesy Centers for Disease Control ; page 86: Daniel LeClair/Reuters/ Landov; page 88: Andy Levin/Photo Researchers; page 89: Raymond Gehman/NG Image Collection; page 90: Stuart Bauer; page 87: (top) Karen Kasmauski/ NG Image Collection; page 87: (bottom) James L. Amos/NG Image Collec-

tion; page 72: Image State; page 74: (bottom) PhotoDisc, Inc./Getty Images; page 75: (bottom) Corbis Digital Stock

Chapter 5

Page 94: Harmut Schwarzbach/Peter Arnold, Inc.; page 94: (bottom left inset) Robert Caputo/NG Image Collection; page 96: (bottom left) Corbis Digital Stock; page 96: (right) Digital Vision; page 97: (top) Norbert Rosing/ NG Image Collection; page 97: (center) Geroge F. Mobley/NG Image Collection; page 98: (top left) Jeremy Woodhouse/ PhotoDisc/Getty Images; page 98: (top right) Roy Toft/NG Image Collection; page 99: (left) Romeo Gacad/AFP/Getty Images; page 99: (right) Max Rossi/Reuters/ Corbis; page 100: Raymond Gehman/NG Image Collection; page 101: (top left) Paul Nicklen/NG Image Collection; page 101: (top right) Roy Taft/NG Image Collection; page 101: (top left) Michael Nichols/NG Image Collection; page 101: (bottom right) Norbert Rosing/NG Image Collection; page 108: (left) Dr. Robert Calentine/Visuals Unlimited; page 108: (right) Dr Jeremy Burgess/Photo Researchers; page 110: Jason Edwards/ NG Image Collection; page 112: Rob and Ann Simpson/Visuals Unlimited; page 114: Paul McCormick/Getty Images; page 115: (left) Michael Melford/ NG Image Collection; page 115: (right) Darlyne A. Murawski/NG Image Collection; page 116: Chris Johns/NG Image Collection; page 117: (left) Mattias Klum/NG Image Collection; page 117: (right) Robert Sisson/NG Image Collection; page 118: Macduff Everton/NG Image Collection; page 119: (left) Joel Sartore/NG Image Collection; page 120: ?AP/Wide World Photos; page 123: Darlyne A. Murawski/NG Image Collection; page 116: Corbis Digital Stock

Chapter 6

Page 124: Raymond Gehman/NG Image Collection; page 128: Joel Sartore/ NG Image Collection; page 129: Maria Stenzel/NG Image Collection; page 130: Sam Abell/NG Image Collection; page 131: James P. Blair/NG Image Collection; page 132: Robert Hynes/NG Image Collection; page 133: John Cunningham/Visuals Unlimited; page 135: Richard Olsenius/NG Image Collection; page 136: Mitsuaki Iwago/NG Image Collection; page 137: (top) Tom Bean/NG Image Collection; page 138: Kathleen Revis/NG Image Collection; page 140: Frans Lanting/NG Image Collection; page 141: Patricia Caulfield; page 142: (top) James P. Blair/NG Image Collection; page 142: (bottom) Joe Stancampiano/NG Image Collection; page 144: Nicole Duplaix/NG Image Collection; page 148: (top) Wolfgang Kaehler; page 148: (center left) Glenn N. Oliver/Visuals Unlimited; page 148: (center right) Wolfgang Kaehler; page 150: Raymond Gehman/NG Image Collection; page 153: (bottom) Jodi Cobb/NG Image Collection; page 153: (top) Joseph T. Collins/Photo Researchers

Chapter 7

Page 154: Bruce Dale/NG Image Collection; page 156: (left) Brian J. Skerry/NG Image Collection; page 156: (right) John Cancalosi/NG Image Collection; page 158: CNRI/Science Photo Library/Photo Researchers; page 159: Michael Abbey/Photo Researchers; page 160: Yvona Momatiuk and John Eastcott/Photo Researchers; page 162: Karen Kasmauski/NG Image Collection; page 163: James P. Blair/NG Image Collection; page 165: (left) Annie Griffiths Belt/NG Image Collection; page 165: (right) Karen Kasmauski/NG Image Collection; page 169: (left) Kenneth Garrett/NG Image Collection; page 169: (right) Larry C. Price; page 170: Jodi Cobb/ NG Image Collection; page 171: Robert Caputo/NG Image Collection; page 172: Karen Kasmauski/NG Image Collection; page 175: Richard Nowitz/ NG Image Collection; page 177: Stephanie Maze/NG Image

Collection; page 179: Stuart Franklin/NG Image Collection; page 180: (left) Getty Images; page 180: (right) Corbis Images; page 183: Lynn Johnson/NG Image Collection; page 172: Karen Kasmauski/NG Image Collection; page 174: Corbis Digital Stock

page 279: Peter W. Glynn/NG Image Collection; page 280: Science VU/Visuals Unlimited; page 282: Courtesy Dave Matilla/NOAA; page 283: Bill Curtsinger/NG Image Collection; page 285: Brian J. Skerry/NG Image Collection

Chapter 8

Page 184: Norbert Rosing/NG Image Collection; page 186: NASA/Corbis Images; page 187: (bottom) Kenneth Garrett/NG Image Collection; page 187: (top) Jack Finch/Photo Researchers; page 190: Digital Vision; page 192: Sarah Leen/NG Image Collection; page 194: Guy Starling/NG Image Collection; page 193: David Cavagnaro/Peter Arnold, Inc.; page 198: James P. Blair/NG Image Collection; page 200: John D. Cunningham/Visuals Unlimited; page 200: John D. Cunningham/Visuals Unlimited; page 202: Michael S. Yamashita/NG Image Collection; page 203: PhotoDisc/Getty; page 204: Ted Spiegal/NG Image Collection; page 206: Raymond Gehman/NG Image Collection; page 208: Peter Menzel Photography

Chapter 9

Page 210: James L. Stanfield/NG Image Collection; page 212: (left) Maria Stanzel/NG Image Collection; page 212: (right) Paul Nicklen/NG Image Collection; page 215: SuperStock, Inc.; page 217: James A. Sugar/NG Image Collection; page 219: Stock Trek/PhotoDisc Green/Getty; page 221: (left) Bill Curtsinger/NG Image Collection; page 221: (right) Joanna B. Pinneo/NG Image Collection; page 220: Michael Nichols/NG Image Collection; page 223: James P. Blair/NG Image Collection; page 225: Courtesy NASA; page 226: David Austen/NG Image Collection; page 228: (center) Robert Llewellyn; page 228: (bottom) Christian Mehr; page 228: (bottom) Christian Mehr; page 233: Thomas Nebbia/NG Image Collection

Chapter 10

Page 234: (opener and bottom left inset) Volkmar K. Wentzel/NG Image Collection; page 236: Jodi Cobb/NG Image Collection; page 239: ©AP/Wide World Photos; page 240: (left) Mark Wagner, London; page 240: (right) Courtesy USGS; page 242: Dan Lamont, Seattle, Washington; page 242: Courtesy of Worldsat International, Inc.; page 242: Courtesy of Worldsat International, Inc.; page 245: (left) Courtesy U. S. Dept. of Energy; page 245: (right) ©Raymond G. Barnes; page 247: James L. Stanfield/NG Image Collection; page 250: Steve Winter/NG Image Collection; page 256: Ted Shreshinsky/Corbis Images; page 259: (left) Mike Barlow/Dembinsky Photo Associates; page 250: (left) Rich Buzzelli/Tom Stack & Associates; page 250: (right) W.A. Banaszewski/Visuals Unlimited; page 258: Bruce Dale/NG Image Collection; page 260: Sandra Baker/Getty; page 262: NASA/Goddard Flight Center

Chapter 11

Page 262: (bottom left inset) B. Anthony Stewart/NG Image Collection; page 266: Washington University; land layer by the SeaWiFS Project, fire maps from ESA, sea surface temperature layer from JPL and cloud layer by SSEC, U. Wisconsin.; page 268: (top left) Burt Curtsinger/NG Image Collection; page 268: (top right) Alex Kerstitch/Visuals Unlimited; page 268: (center left) Stuart Westmorland/Photo Researchers; page 268: (center right) Paul Zahl/NG Image Collection; page 271: David Wrobel/Visuals Unlimited; page 272: (top) Norbert Wu/Peter Arnold, Inc.; page 270: Tim Layman/NG Image Collection; page 273: (left) Joe Stancampiano/NG Image Collection; page 273: (right) Courtesy of Bruce H. Robison, Monterey Bay Aquarium; page 276: Guy Hoppen/Accent Alaska; page 277: David Hiser/NG Image Collection; page 278: © Brandon Cole/Visuals Unlimited;

Chapter 12

Page 286: Pat Armstrong/Visuals Unlimited; page 286: (bottom left inset) Photo by Jack Waters www.telliquah.com; page 291: (top) Courtesy NOAA Laboratory for Satellite Altimetry; page 291: David Rydevick; page 292: (top) Dennis Drenner; page 292: (center) George Semple; page 292: (bottom) Charles D. Winters; page 293: (top left) Corbis Digital Stock; page 293: (second row, left) Digital Vision/Getty; page 293: (second row, center) PhotoDisc, Inc./Getty; page 293: (fourth row, far right) PhotoDisc, Inc./Getty; page 293: (fifth row, center) PhotoDisc, Inc./Getty; page 294: Bruce F. Molnia/ Terraphotographics/BPS; page 297: (left) David Hiser/Stone/Getty Images; page 297: (right) R. Ashley/Visuals Unlimited; page 298: (bottom) William Felger/Grant Heilman Photography; page 299: Leste Lefkowitz/Corbis Images; page 301: Courtesy U.S. Dept. of Agriculture; page 300: Photo by Ray Weil, courtesy Martin Rabenhorst; page 303: (left) Grant Heilman/Grant Heilman Photography; page 303: Courtesy U. S. Department of Agriculture; page 305: (left) Courtesy U. S. Department of Agriculture; page 305: (center) Brenda Tharp/Science Photo Library; page 305: (right) David Cavagnaro; page 306: Kevin Flaming/Corbis Images; page 307: Carl Pedersen/Alamy Images; page 309: Courtesy Charles Massie, Massie Reclamation Inc.; page 310: Age fotostock/Superstock

Chapter 13

Page 310: (bottom left inset) Age fotostock/Superstock; page 311: (far right, just above bottom) Raymond Gehman/NG Image Collection; page 313: Melissa Farlow/NG Image Collection; page 315: Kirtley-Perkins/Visuals Unlimited; page 317: Calvin Larsen/Photo Researchers, Inc.; page 321: (bottom left) Pegasus/Visuals Unlimited; page 321: (center left) NRSC/Photo Researchers, Inc.; page 321: (center right) Marin Hervey/Natural History Photographic Agency; page 318: Courtesy Forest Stewardship Council - US; page 319: Medford Taylor/NG Image Collection; page 323: Steve Smith/SUPERSTOCK; page 324: James Nachtwey/NG Image Collection; page 327: (top left) (c)Larry Ulrich; page 327: (top right) Tom Bean/Stone/Getty Images; page 328: Tom Murphy/NG Image Collection; page 329: (left) Chris Pease/Fish &Wildlife Service; page 329: (right) George F. Mobley/NG Image Collection; page 330: Dr. Carleton Ray/Photo Researchers, Inc.; page 332: (left) Bruce Dale/NG Image Collection; page 332: (right) James P. Blair/NG Image Collection; page 335: Corbis Sygma

Chapter 14

Page 336: Sisse Brimberg/NG Image Collection; page 336: (bottom left inset) H. Donnezan/Photo Researchers, Inc.; page 339: (left) Paul Souders; page 339: (right) Reuters Newmedia, Inc./Corbis Images; page 342: Reza/NG Image Collection; page 343: Sarah Leen/NG Image Collection; page 344: Photo by Agri-Graphics, Ltd,courtesy Winifred Hoffman; page 347: (left) ©AP/Wide World Photos; page 347: (right) Courtesy Peggy Greb/USDA; page 350: Jim Richardson/NG Image Collection; page 350: (top inset) Courtesy The Golden Rice Humanitarian Board; page 352: Charles D. Winters; page 353: Darlyne A. Murawski/NG Image Collection; page 354: (bottom) Courtesy Max Badgley; page 354: (top) Lawrence Migdale/Photo Researchers, Inc.; page 355: John D. Cunningham/Visuals Unlimited; page 357: Matthew Beck, Inverness, Florida; page 358: Cary Wolinsky/NG Image Collection; page 360: George Grall/NG Image Collection

Chapter 15

Page 360: (bottom left inset) Frans Lanting/Minden Pictures, Inc.; page 362: NG Image Collection; page 363: David Cavagnaro; page 364: Ed Darack/Visuals Unlimited; page 365: Doug Wechsler; page 368: (bottom) Courtesy Smithsonian Migratory Bird Center; page 369: Joel Sartore/NG Image Collection; page 371: (top left) Frans Lanting/Minden Pictures, Inc.; page 371: (top right) Courtesy USDA; page 371: (bottom) Chris Johns/NG Image Collection; page 373: Andrew Sacks/Time Life Pictures/Getty Images; page 374: (top) Jany Sauvanet/Natural History Photographic Agency; page 376: (left) University of Wisconsin - Madison Arboretum; page 376: (right) Virginia Kline/University of Wisconsin - Madison Arboretum; page 377: (left) Steve Nexbitt, Florida Fish and Wildlife Conservation Commission; page 377: (center) Richard O. Bierregaard, Jr.; page 377: (right) Courtesy Nordiska Genbanken, Alnarp, Sverige; page 379: Melissa Farlow/NG Image Collection; page 380: (top) Courtesy Steve Hillebrand/U.S. Fish &Wildlife Service; page 381: Courtesy Tracy Brooks, U.S. Fish &Wildlife Service; page 382: Michael Nichols/NG Image Collection

Chapter 16

Page 384: Joel Sartore/NG Image Collection; page 386: Alan L. Detrick/Photo Researchers, Inc.; page 388: Inga Spence/Visuals Unlimited; page 390: Jose Azel/Aurora Photos; page 393: ©Steve Greenberg; page 395: (top) Courtesy Baltimore Department of Public Works; page 395: (center) Shari Lewis/©AP/Wide World Photos; page 395: (bottom) Courtesy National Association for PET Container Resources [NAPCOR]; page 397: Andy Levin/Photo Researchers, Inc.; page 399: (left) Chromosohm/Photo Researchers, Inc.; page 399: (right) DOE/Science Source; page 401: (center) Courtesy USDA; page 401: (bottom) Gary Milburn/Tom Stack & Associates;

page 402: Peter Essick/NG Image Collection; page 404: Cary Wolinsky/ NG Image Collection; page 406: ©AP/Wide World Photos

Chapter 17

Page 406: (bottom left inset) Peter Cosgrove/©AP/Wide World Photos; page 408: Peter Essick/NG Image Collection; page 410: (top) Kristin Finnegan Photograhy; page 410: (bottom) John Shaw Photography; page 415: (top) Chris Wilkins/AFP/Getty Images; page 415: (bottom) Charles Mason/Black Star; page 418: Courtesy Westinghouse Electric Corp., Commercial Nuclear Fuel Division; page 420: (top) Novosti; page 420: (bottom) Caroline Penn/Corbis Images; page 422: (left) Peter Essick/Aurora Photos; page 423: Howells David/Gamma-Presse, Inc.; page 425: George F. Mobley/ NG Image Collection; page 427: Gerd Ludwig/NG Image Collection

Chapter 18

Page 428: (center and bottom left inset) Sarah Leen/NG Image Collection; page 432: Courtesy Solar Design Associates; page 431: Otis Imboden/ NG Image Collection; page 433: Courtesy of Uni-Solar. Energy Conservation Devices, Discover Magazine; inset, United Solar Systems, Discover Magazine.; page 434: L. Lefkowitz/Taxi/Getty Images; page 435: Pasquale Sorrentino/Photo Researchers, Inc.; page 436: Steve McCurry/NG Image Collection; page 437: Prof. David Hall/Photo Researchers, Inc.; page 439: Clean Water Action Alliance of Minnesota, Mark Frederickson/National Renewable Energy Laboratory; page 440: U.S. Army Corp of Engineers; page 443: Simon Fraser/Photo Researchers, Inc.; page 445: Courtesy Ontario Hydro; page 446: FOXTROT © 2006 Bill Amend/Reprinted with permission of Universal Press Syndicate. All rights reserved.; page 449: Barney Taxel, Cleveland, Ohio; page 451: Sarah Leen/NG Image Collection

TEXT, TABLE, AND LINE ART CREDITS

Chapter 1
Figure 1.2: Population Reference Bureau.

Chapter 2
First section, beginning on p. 26, concept of sustainability: *Our Common Future*. World Commission on Environment and Development. Cary, NC: Oxford University Press (1987), p. 22; second section, beginning on p. 31, list of 8 principles of deep ecology: Naess, A. and D. Rothenberg. *Ecology Community and Lifestyle*. Cambridge, UK: Cambridge University Press (2001). Reprinted with permission of Cambridge University Press; legend for Figure 2.8: PhillyCarShare web site, www.phillycarshare.org; fourth section, beginning on p. 35, principles for sustainable living: *Caring for the Earth*. Published jointly by the World Conservation Union, U.N. Environment Program, and World Wide Fund for Nature (1991).

Chapter 3
Page 52, excerpt from the "Wilderness Essay": a letter written by Wallace Stegner to David Pesonen of the U. of California's Wildland Research Center; fourth section, beginning on p. 60, Environmental Economics: Main source: Levin, J. "The Economy and the Environment: Revising the National Accounts." *IMF Survey* (June 4, 1990).

Chapter 4
Table 4.1: Data compiled from a variety of sources by Ropeil, D. and G.Gray. *Risk*. Boston: Houghton Miffliin Company (2002). Probabilities of risk calculated by author based on latest available data from U.S. Census Bureau; Table 4.2: Adapted from "Healthy Youth!" U.S. Centers for Disease Control and Prevention, National Center for Chronic Disease Prevention and Health Promotion, Division of Adolescent and School Health, Youth Risk Behavior Survey; Table 4.3: Complied by author; Table 4.5, Joesten, M.D. and J.L. Wood. *World of Chemistry*, 2nd edition. Philadelphia: Saunders College Publishing, (1996); Chapter 4 opener—inset, p. 71: Guillette, E.A., M.M. Meza, M.G. Aquilar, A.D. Soto, and I.E. Garcia. "An Anthropological Approach to the Evaluation of Preschool Children Exposed to Pesticides in Mexico." *Environmental Health Perspectives* (May 1998); Figure 4.1: Adapted from *Science and Judgment in Risk Assessment*. Washington, D.C.: National Academy Press (1994). Data in graph from "CDC's Second National Report on Human Exposure to Environmental Chemicals," published in 2003; Figure 4.7B: Grier, J.W. "Ban of DDT and Subsequent Recovery of Reproduction in Bald Eagles." *Science*, Vol. 218 (December 17, 1982). Redrawn by permission of the American Association for the Advancement of Science; Figure 4.8: Based on data from Woodwell, G.M., C.F. Worster, Jr., and P.A. Isaacson. "DDT Residues in an East Coast Estuary: A Case of Biological Concentration of a Persistent Insecticide." *Science*, Vol. 156 (May 12, 1967).

Chapter 5
Figures 5.8, 5.9, 5.10, and 5.12: Values are from Schlesinger, W.H. *Biogeochemistry: An Analysis of Global Change*, 2nd edition. Academic Press, San Diego (1997) and based on several sources; What a Scientist Sees, p. 112: Adapted from MacArthur, R.H. "Population Ecology of Some Warblers of Northeastern Coniferous Forests." *Ecology*, Vol. 39 (1958).

Chapter 6
Figure 6.1: Based on data from the World Wildlife Fund; Figure 6.2: Based on Holdridge, L. I, San Jose, Costa Rica (1967); Data for all climate graphs (in Figures 6.3-6.11) from www.worldclimate.com.

Chapter 7
Figure 7.5B: After V.C. Scheffer. "The Rise and Fall of a Reindeer Herd." *Sci. Month.*, Vol. 73 (1951); Figures 7.6, 7.12, and 7.13, Table 7.1, and bar graph for Critical and Creative Thinking Questions: Data from Population Reference Bureau; Figure 7.8: *World Population Prospects, The 2000 Revision*, United Nations Population Division; Figure 7.20: Map drawn using data from "Urban Agglomerations 2003," U.N. Department of Economic and Social Affairs, Population Division; What a Scientist Sees, page 173 (graph): Data from U.S. Census Bureau.

Chapter 8
Tables of major air pollutants (pp. 190-193), and Table 8.2: Compiled by author; What a Scientist Sees, p. 193 (graph), and Figure 8.7A: Environmental Protection Agency; Figure 8.13A: Adapted from Joesten, M.D., and J.L. Wood. *World of Chemistry*, Second Edition. Philadelphia: Saunders College Publishing (1996); Figure 8.14: Air Quality Planning and Standards, Office of Air and Radiation, EPA.

Chapter 9
Table 9.1: Carbon Dioxide Information Analysis Center, Environmental Sciences Division, Oak Ridge National Laboratory; Figure 9.2: Adapted from Figure 4.15 in Strahler, A. and A. Strahler. *Physical Geography*, 2nd edition. Hoboken, NJ: John Wiley and Sons, Inc. (2002) and based on data from J.T. Kiehl and K.E. Trenberth, "Earth's Annual Global Mean Energy Budget." *Bull. Am. Met. Soc.*, Vol. 78 (1997); Figure 9.5: Global Land-Ocean Temperature Index, Goddard Institute of Space Studies, NASA; Figure 9.6: Dave Keeling and Tim Whorf, Scripps Institution of Oceanography, La Jolla, California; Figure 9.13: Adapted from Davis, M.B., C. Zabinski, R.L. Peters, and T.E. Lovejoy, eds. "Changes in Geographical Range Resulting from Greenhouse Warming: Effects on Biodiversity in Forests." *Global Warming and Biological Diversity*. edited by R.L. Peters, and T.E. Lovejoy. New Haven, Connecticut: Yale University Press (1992); Case study map and graph: Calculations based on data from U.S. Energy Information Administration and Population Reference Bureau; map in Critical and Creative Thinking Questions: U.S. Global Change Research Program public archives.

Chapter 10
Tables 10.1, 10.2, and 10.3: Compiled by author; Figure 10.2: Adapted from Figure 2.8 on p. 28 in Strahler, A. and A. Strahler. *Physical Geography: Science and Systems of the Human Environment*. Hoboken, NJ: John Wiley & Sons, Inc. (2002); Figure 10.5: *Control of Water Pollution from Urban Runoff*. Paris: Organization for Economic Cooperation and Development (1986); Figure 10.9: U.S. Geological Survey; Figure 10.16: Adapted from Joesten, M.D., and J.L. Wood. *World of Chemistry*, 2nd edition. Philadelphia: Saunders College Publishing (1996).

Chapter 11
Figure 11.2: After Broecker, W.S. "The Great Ocean Conveyor." Oceanography, Vol. 4 (1991); Figure 11.3A: Data adapted from the Climate Analysis Center, National Weather Service, and NOAA, Camp Springs, MD; Figure 11.6: Adapted from Figure 14-1 in Karleskint, G. *Introduction to Marine Biology*. Philadelphia: Harcourt College Publishers (1998); Figure 11.7A: Adapted from Marangos, J.E., M.P. Crosby, and J.W. McManus. "Coral Reefs and Biodiversity: A Critical and Threatened Relationship." *Oceanography*, Vol. 9 (1996); What a Scientist Sees, p. 279, Figure B: Adapted from National Assessment Synthesis Team, *Climate Change Impacts on the United States: The Potential Consequences of Climate Variability and Change*, (Report for

the U.S. Global Change Research Program). Cambridge, UK: Cambridge University Press (2001); graph in Critical and Creative Thinking Questions: Adapted from Vital Signs 2006-2007, page 27, Worldwatch Institute (2006), and based on data from the FAO.

Chapter 12
Table 12.1: Compiled by author; p. 290, text on Indian Ocean earthquake and tsunami: Adapted from USGS data and NOAA news reports; Figure 12.6: Parts b, c, d based on Colliers Encyclopedia, 1969, Crowell–Collier Educational Corporation, Vol. 16, p. 304; Figure 12.7: Adapted from Joesten, M.D., and J.L. Wood. *World of Chemistry*, Second Edition. Philadelphia: Saunders College Publishing (1996); Figure 12.17: *One Planet, Many People: Atlas of Our Changing Environment*. United Nations Environment Program (2005)—Part A based on graph on page 30 (from FAO 2000), Part B based on table on page 29 (from *World Atlas of Desertification*).

Chapter 13
Table 13.1: U.S. Dept. of Interior, U.S. Dept. of Agriculture, and U.S. Dept. of Defense; Table 13.2: National Park Service; Table 13.3: From Box 1.1 in Noss, R.F., M.A. O'Connell, and D.D. Murphy. *The Science of Conservation Planning: Habitat Conservation Under the Endangered Species Act*, Island Press: World Wildlife Fund (1997). Reprinted with permission; Figure 13.1: Adapted from the map, "Federal Lands in the Fifty States," produced by the Cartographic Division of the National Geographic Society (October 1996); Figure 13.6: Based on Goode Base Map (S.R. Eyre 1968); Figure 13.10: Data obtained from Marsh, W.M. and J.M. Grossa, Jr., *Environmental Geography*, 2nd edition, John Wiley & Sons, Inc. (2002); graph in Critical and Creative Thinking Questions: World Resources Institute and U.N. Food and Agricultural Organization.

Chapter 14
Figure 14.2: Based on data from U.N. Food and Agricultural Organization and UNICEF; Figure 14.3: Adapted from USDA and FAO data, as reported in Brown, L.R. et al. *Vital Signs 2003*. New York: W.W. Norton & Company (2002); Figure 14.4: Adapted from G.H. Heichel, "Agricultural Production and Energy Resources." *American Scientist*, Vol. 64 (January/February 1976); Figure 14.8: Data from USDA; Figure 14.16: From "International Travel and Health. Vaccination Requirements and Health Advice: Situation as of January 1, 1992." Geneva: World Health Organization (1992). Reproduced by permission of the World Health Organization; Figure 14.19: Adapted from Gardner, G., "IPM and the War on Pests." *World Watch*, Vol. 9, No. 2 (March/April 1996). Redrawn by permission of Worldwatch Institute; What a Scientist Sees, p. 354 (graph): Adapted from Debach, P. *Biological Control by Natural Enemies*. New York: Cambridge University Press (1974); graph in Critical and Creative Thinking Questions: Data from UNESCO/FAO.

Chapter 15
Table 15.1: Adapted from page 527 of *Climate Change Impacts on the United States*, A report of the National Assessment Synthesis Team, U.S. Global

Change Research Program, Cambridge University Press (2001); Table 15.2: U.S. Fish and Wildlife Service; Data in first section on number of species: Data from Reaka-Kudla, M.L., D.E. Wilson, and E.O. Wilson. *Biodiversity II*. Washington, D.C.: Joseph Henry Press (1997); Figure 15.1B: After M.L. Cody and J.M. Diamond, eds., *Ecology and Evolution of Communities*. Harvard University, Cambridge, (1975); Figure 15.11: Compiled by author from various sources; What a Scientist Sees, p. 369: Map by Conservation International; quote in Critical and Creative Thinking Questions: Aldo Leopold, *A Sand County Almanac*. Oxford University Press, Inc. (1991); graph in Critical and Creative Thinking Questions: Data from The Nature Conservancy.

Chapter 16
Table 16.1: Compiled by author from various sources; Figures 16.2 and 16.3: Data from EPA; Figure 16.4: Data from Ecobalance, Inc., and Integrated Waste Services Association; Figures 16.9-16.11: Data from EPA; Figure 16.17: Adapted from Rocky Mountain Arsenal Remediation Venture Office; p. 405, graph in Critical and Creative Thinking Questions: Data from EPA.

Chapter 17
Unless noted otherwise, all energy facts cited in this chapter were obtained from the Energy Information Administration (EIA), the statistical agency of the U.S. Department of Energy (DOE).Table 17.1: Integral Consulting. "2005 Assessment of Lingering Oil and Resource Injuries from the *Exxon Valdez* Oil Spill (Restoration Project 040776)"; Table 17.2: Adapted from two sources: Hinrichs, R.A. and M. Kleinbach. *Energy: Its Use and the Environment*, 3rd edition. Philadelphia: Harcourt College Publishers (2002) and *Science for Democratic Action*, Vol. 8, No. 3 (May 2000); Figures 17.1A, 17.6, and 17.8: EIA, U.S. Department of Energy; Figure 17.1B: Adapted from Figure 13-3a in Harris, J.M. *Environmental and Natural Resource Economics: A Contemporary Approach*, second edition. Houghton Mifflin (2006) and based on data from the World Bank; p. 427, graph in Critical and Creative Thinking Questions: EIA, U.S. Department of Energy.

Chapter 18
Unless noted otherwise, all energy facts cited in this chapter were obtained from the Energy Information Administration (EIA), the statistical agency of the U.S. Department of Energy (DOE). Table 18.1: Princeton Energy Resources International; Figure 18.5B: Adapted from Hinrichs, R.A. *Energy: Its Use and the Environment*, 2nd ed. Philadelphia: Saunders College Publishing (1996); Figure 18.8C: From *Vital Signs 2005*, Worldwatch Institute, W.W. Norton (2005) and based on data from BTM Consult, American Wind Energy Association, and European Wind Energy Association; p. 451, map in Critical and Creative Thinking Questions: U.S. Department of Energy.

INDEX

In page references, *f* denotes a figure and *t* denotes a table.

A

Abiotic environment, 96, 126
Abortion, 174
Acadia National Park, 327*t*
Acid deposition
    coal and, 410, 410*f*
    in Copper Basin, Tennessee, 286
    defined, 227
    development of, 228, 229*f*
    effects of, 228, 229*f*
    global warming and, 227
    oil, natural gas, and, 414
    ozone depletion and, 227
    politics of, 228
    recovery from, 228, 230–231
    smelting and, 296
    species endangerment and, 370
Acid mine drainage, 296, 297*f*, 410
Acid rain, 59, 63, 227
    See also Acid deposition
Acids, sources of, 398*t*
Active solar heating, 430–431, 430*f*
Acute toxicity, 75
Adam Joseph Lewis Center for Environmental
    Studies, 449, 449*f*
Adaptation
    evolution and, 143
    to global warming, 222–223
Additivity, 86
Adélé penguins, global warming and, 220, 221*f*
Aerosols, 218
Afghanistan
    polio in, 162*f*
    refugees in, 8
Africa
    deforestation in, 318–319
    elephants of, 370, 371*f*, 380
    gender roles in, 169*f*
    hunger in, 338, 339*f*, 340
    Lake Victoria, 94
    locust problem in, 349
    poaching in, 373
    savannas in, 136
    tsunami in, 290
Agenda 21, 42
Age structure diagram, 166, 167*f*
Age structure of countries, 166–168, 167*f*,
    168*f*
Agriculture, 336–359
    energy production by, 437
    environmental impacts of, 346–347
    genetic diversity and, 344, 365
    genetic engineering in, 345, 345*f*, 350–351,
        350f, 351*f*
    global warming and, 220, 222–223

    industrialized, 341–342, 341*f*, 346, 347*f*, 347*t*
    intercropping, 342
    irrigation (*see* Irrigation)
    Lake Victoria, 94
    nomadic herding, 342, 342*f*
    organic, 349
    pest control (*see* Pesticides)
    prime farmland loss and, 343
    salinization, 242, 303, 303*f*, 346, 347
    shifting cultivation, 342
    slash-and-burn, 21, 320, 342
    soil and, 304–306, 304*f*
    subsistence, 320, 342
    suburban spread and, 343, 343*f*
    sustainable, 304–306, 348–349, 348*f*, 356
    tropical rain forests versus, 21
    types of, 341–342
    water pollution from, 251–252
    water waste in, 246
    yields
        defined, 341
        increasing crop, 344–345
        increasing livestock, 346
    See also Food resources
Agroecosystem, 349
Air pollution, 59, 190–206
    agricultural practices and, 346–347
    in Chattanooga, 206
    children and, 86–87
    classes of pollutants, 190–193
    coal and, 410–411
    controlling pollutants, 200–203, 206
    defined, 190
    in developing countries, 202–203
    dust domes, 198, 199*f*
    effects of, 194–199
    human health and, 195, 195*t*
    human sources of, 194
    indoor, 203–205, 205*f*
    in Mexico City, 177*f*
    mining and, 296
    primary, 190, 191*f*, 194, 194*f*
    radon, 204–205, 204*f*
    secondary, 190, 191f
    sources of outdoor, 193–194, 194*f*
    topography effects on, 197–198, 197*f*
    transport of, long distance, 184
    urban, 195–196
    urban heat islands, 198–199, 199*f*
    weather effects on, 197–198, 197*f*
Air toxics, 192–193
Alaska
    Arctic National Wildlife Refuge, 425
    deforestation and, 319
    Delta River, 313*f*
    Fort Yukon, 128*f*
    Glacier Bay, 148*f*
    Izembek Lagoon, 271
    oil spill in, 415–416, 415*f*

    Pribilof Islands, 160*f*
    sea otters off coast of, 272
    Tongass National Forest, 332
    wilderness areas in, 328, 329
Aldrin, 82*t*
Alligators, role of, in ecosystem, 364*f*
Alpine tundra, 128
Alternative agriculture. *See* Sustainable
    agriculture
Aluminum recycling, 394, 395*f*
Amazon rain forest, 77*f*, 78, 163*f*, 320, 321*f*
Amazon River, 251*f*
American Council for an Energy-Efficient
    Economy, 447
American Dust Bowl, 52
American Farmland Trust, 343
American Forest Resource Alliance, 330
American Forestry Association, 50
American warbler study, 112
Ammonification, 108
Amoebic dysentery, 76*t*
Amphibians, declines and deformities in, 360
Ancylostomiasis, 76*t*
Androgens, 90
Anopheles mosquitoes, 352
Antagonism, 86
Antarctica
    food web of, 283
    global warming and, 210, 220, 221*f*
    ozone hole in, 224, 225*f*, 226, 227, 283
Anthropocentric worldview, 32
Antibiotics used in livestock, 346
Antiquities Act, 51
Aquaculture, 275*f*, 277, 277*f*
Aquatic ecosystems, 138
    brackish (estuaries), 142–143
    freshwater, 138–140, 139*f*, 140*f*
    freshwater wetlands, 138, 141
    See also Ocean
Aquifers
    defined, 237
    depletion of, 240–241
    Ogallala, 241, 241*f*
Aral Sea, 243, 243*f*
Arcata Marsh and Wildlife Sanctuary, 256
Archaea, 146, 146*f*
Architecture, green, 443, 449, 449*f*
Arctic, ozone thinning over, 224
Arctic National Wildlife Refuge, 425
Arctic tundra, 128, 128*f*
Arid lands, 240
Aristotle, 143
Army Corps of Engineers, 234
Arsenic, sources of, 398*t*
Artificial eutrophication, 251
Asia
    air pollution in, 203
    hunger in, 338, 339*f*, 340
Aspirin, lethal dose of, 83*t*

Assimilation, 108
Asthenosphere, 288
Aswan Dam, 441
Atlanta, Georgia, 193
Atmosphere
    circulation in, 188–189, 188*f*
    climate and, 212–215
    composition of, 186*t*
    defined, 98, 186
    interaction with ocean, 266–267
    layers of, 187*f*
    pressure in, 188
Atmospheric aerosols, 218
Atomic structure, 416–417, 416*f*
Audubon, John James, 49
Audubon's warblers, 49*f*
Aurora, 187*f*
Australia
    coal reserves in, 409
    Mediterranean climate in, 133
    savannas in, 136
Automobiles
    biomass fuels for, 437
    cleaner, 428
    fuel cells for, 435
    fuel efficiency of, 446, 446*f*
    recycling and reuse, 384
    sharing, 33, 33*f*

B

Background extinction, 366
Bald eagle, 80*f*, 357
Baleen whales, 283
Ballast water, 372
Bangladesh
    disease and population concentration, 78*f*
    fertility rates in, 171
    global warming and, 210
    infant mortality rates in, 165*f*
    malaria in, 78
    polio in, 162*f*
    population growth in, 166*t*
    reproductive rights in, 174
Bayou Sauvage National Wildlife Refuge, 142*f*
Bear Lake, Colorado, 139
Beech trees, global warming and, 221*f*
Belem, Brazil, 132*f*
Belem-Brasilia Highway, 320
Bellwether species, 360
Benthic environment, 268, 269*f*, 270–272
Benthos, 138
Bioaccumulation, 79–80, 81*f*, 353
Biocentric preservationist, 50
Biocentric worldview, 32
Biochemical oxygen demand (BOD), 248, 248*f*
Biodegradable, 388
Biodiesel, 437
Biodiversity. *See* Biological diversity

Biodiversity hotspots, 369
Biogas, 436
Biogas digesters, 436, 437*f*
Biogeochemical cycles
    carbon cycle, 104–105, 105*f*, 120
    defined, 104
    hydrologic cycle, 106–107, 106*f*, 138
    nitrogen cycle, 106–107, 106*f*, 107*f*
    phosphorus cycle, 108, 109*f*, 110
Biogeography, 144
Biological controls, 355, 355*t*
Biological diversity, 37–39
    agricultural practices and, 344, 346–347
    areas of decline in, 368–369
    defined, 37, 363
    ecosystem diversity, 363
    genetic diversity, 363, 365–366
    hotspots for, 369
    need for, 363–364
    species richness and, 362–363, 362*f*
    See also Conservation biology; Conservation
        policies and laws; Endangered species
Biological magnification, 79, 80, 81*f*, 184, 353
Biological oxygen demand, 248, 248*f*
Biomass, 436, 436*f*
Biomass energy, 436–437
    electric generating costs of, 433*t*
Biomes
    boreal forest, 129, 129*f*
    chaparral, 133–134, 133*f*
    defined, 126
    desert, 137, 137*f*
    major terrestrial, 126*f*
    savanna, 136, 136*f*
    temperate deciduous forest, 131, 131*f*
    temperate grassland, 135, 135*f*
    temperate rain forest, 130, 130*f*, 332
    temperature and precipitation, 127*f*
    tropical rain forest, 132–133, 132*f*
    tundra, 128, 128*f*
Biosphere, 97–98
Biotic environment, 96
Biotic potential, 157–158
Birds of America (The), 49*f*
Birth rate, 157, 161, 164–165
    See also Fertility
Black Triangle, 66
Blackbeard Island Wilderness, Georgia, 329*f*
Blast furnace, 295*f*
Bleaching, coral, 220, 279
Blue Ridge Mountains, 193
Board of National Ministries of the American
    Baptist Churches, 34
Bonneville Dam, 245*f*
Boreal forest, 129, 129*f*
    deforestation of, 319
Borneo, 132*f*, 223*f*
Botswana, 144*f*

Bottom ash, 390
Brackish ecosystems, 142–143
Brazil
    Amazon rain forest, 77*f*, 78, 163*f*, 320, 377*f*
    Amazon River, 251*f*
    Belem, 132*f*
    Belem-Brasilia Highway, 320
    biological diversity in, 38*f*
    Curitiba, 180
    deforestation and, 320, 321*f*
    environmental degradation in, 163*f*
    hydropower in, 440
    malaria in, 220*f*
    population growth in, 166*t*
    tropical rain forests in, 21
    urban planning in, 180
Broad-spectrum pesticide, 352
Bronchitis, 195
Brownfields, 176
Bulgaria
    environmental problems in, 66
    population growth in, 168
Bureau of Land Management (BLM), 312,
    313*t*, 322, 325, 328
Bush, George W., 231, 282, 332, 425, 428
Bushmeat, 373
Butterflies, global warming and, 220, 221*f*
Bycatch, 275*f*, 276

C

Cadmium, sources of, 398*t*
Caffeine, lethal dose of, 83*t*
California
    air pollution in, 87, 196–198, 198*f*
    Arcata, 256, 256*f*
    biodiversity decline in, 368
    Culver City, 133*f*
    flooding in, 239*f*
    geothermal power plant in, 442
    Hetch Hetchy Valley, 51, 51*f*
    Humboldt County, 32*f*
    Joshua Tree National Park, 137*f*
    Northridge earthquake, 290
    old-growth forests in, 46
    San Francisco, 289
    San Joaquin Valley, 240*f*
    Santa Barbara oil spill, 57
    Sequoia National Park, 50
    solar thermal plant in, 434, 434*f*
    Superfund sites in, 401
    tire incineration in, 390*f*
    wildfires in, 133, 134
    wind turbines in, 438
    Yosemite National Park, 50, 51*f*, 326, 327*f*,
      327*t*
Cameroon, Korup National Park, 310
Canada
    acid deposition and, 228

deforestation in, 319
E. coli outbreak in, 75
energy consumption by, 408*f*
Fort Smith, 129*f*
global warming and, 221*f*
Great Lakes pollution and, 258
hydropower in, 440
livestock practices in, 346
Ontario and surface runoff, 239*f*
recycling in, 394
solid waste generation in, 386
superinsulated building in, 445*f*
tidal energy in, 443
Vancouver, 175*f*
Yukon air pollution, 184
Cancer-causing substances, risk assessment of, 84–85
Capital, natural, 60, 61*f*
Captive breeding, 377, 377*f*, 378
Car sharing, 33, 33*f*
Carbon cycle, 104–105, 105*f*, 120
Carbon dioxide
    coal and, 411
    concentrations of, 1958 to present, 217*f*
    global warming and, 120, 217–219, 218*f*, 218*t*
    managing emissions of, 222
    oil and natural gas and, 414
    per-capita emissions, 231*f*
    waste-to-energy incineration and, 390*f*
Carbon management, 222
Carbon monoxide, 195*t*
Carbon oxides, 191
Carbon tetrachloride, 224, 226
Carcinogens, 84–85
Caring for the Earth, 35
Carnivores, 101
Carrying capacity (*K*)
    defined, 36
    environmental resistance and, 158–160, 159*f*
    population projections and, 163
    rangeland, 324
Carson, Rachel, 52–53, 53*f*
Cascade Range, 215
Cause of death probabilities, 72*t*
Central Europe, environmental problems in, 66
Chaparral, 133–134, 133*f*, 362*f*
Chemicals
    accidents involving, 400
    children and exposure to, 83, 86–87
    as endocrine disrupters, 90
    lethal dose values, 83, 83*t*
    risk assessment of mixtures, 86
Chemistry, environmental, 401
Chernobyl, Ukraine, 421, 421*f*
Children
    chemical exposure of, 83, 86–87
    hunger and, 339*f*
    as labor source, 170, 170*f*
    pesticides and, 70, 86–87

Chile
    Mediterranean climate in, 133
    species richness in, 362*f*
China
    acid deposition and, 227, 228
    air pollution in, 202, 202*f*, 203
    biogas digesters in, 436
    coal reserves and consumption in, 409, 409*f*
    energy consumption by, 408
    hydropower in, 440
    population growth in, 154, 166*t*
    Three Gorges Project, 441
    tidal energy in, 443
    water problems in, 243
    wheat production by, 336
Chlordane, 82*t*
Chlorine, in drinking water, 254
Chlorofluorocarbons (CFCs)
    changes in, preindustrial to present, 218*f*
    controlling, 226
    defined, 224
    ozone depletion and, 217, 224
    sources of, 398*t*
Cholera, 76*t*
Chronic bronchitis, 195
Chronic toxicity, 75
Circulation
    atmospheric, 188–189, 188*f*
    oceanic, 264, 265*f*
Civilian Conservation Corps, 52
Clean Air Act, 201–202, 206
Clean Air Amendments of 1990, 65, 192–193, 201–202, 411
Clean coal technologies, 411
Clean Water Act, 256–257
Clear-cutting, 316–317
    in Olympic National Park, 330*f*
    in Tongass National Forest, 332
Cleveland, Grover, 50
Climate
    in biomes, 126, 127*f*
    defined, 212
    Mediterranean, 133
    precipitation and, 214–215
    solar radiation and, 213–214, 213*f*
    See also Biomes
Climate change. *See* Global warming
Climax community, 147
Clinton, Bill, 46, 332, 425
Coal, 409
    cleaning up, 411, 411*f*
    as commercial energy source, 412*f*
    electric generating costs of, 433*t*
    environmental impacts of, 410–411, 410*f*
    mining of, 409
    nuclear energy compared to, 418, 419*t*, 420
Coastal development, 275*f*, 278
Cocaine, lethal dose of, 83*t*

Coevolution, 113–114, 113*f*
Coffee plantations, bird decline and, 368
Cogeneration, 413, 446, 446*f*
Coke, in mineral processing, 295
Cold deserts, 137
Colorado
    Bear Lake, 139
    Leadville, 297*f*
    soil erosion in, 303*f*
Colorado potato beetle, 347*f*
Colorado River, 242, 242*f*
Colorado River Compact, 242
Columbia River, 245, 245*f*, 403, 441
Combined heat and power (CHP), 446
Combustion, 105
Command and control regulation, 65
Commensalism, 114–115, 115*f*
Commercial harvest of live organisms, 374
Commons, global, 24
Communism, pollution legacy of, 66
Communities
    climax, 147
    defined, 96
    pioneer, 148
Comoro Islands, 369
Compact development, 176, 180
Competition, 118–119, 119*f*
Composting, 391–392, 391*f*
Comprehensive Environmental Response, Compensation, and Liability Act (CERCLA), 400
Condenser, in nuclear power plant, 418, 419*f*
Conservation
    defined, 48
    energy, 444, 448, 448*f*
    ex situ, 374
    in situ, 374
    land resource, 331
    at local, national, and global levels, 39–40
    mid-20th century, 52–53
    soil, 48, 48*f*, 59, 304–306, 304*f*
    species, 376–378, 377*f*
    water, 246, 247*f*, 247*t*
Conservation biology
    challenges in, 375*f*
    defined, 374
    habitat protection, 375
    habitat restoration, 376
    species conservation, 376–378, 377*f*
    See also Biological diversity; Endangered species
Conservation easement, 325, 343
Conservation International, 369
Conservation policies and laws
    Endangered Species Act, 46, 357, 366, 378–379
    international, 380
Conservation Reserve Program (CRP), 306
Conservation tillage, 304, 305*f*

Conservationists, utilitarian, 50
Consumers, in ecosystems, 100–102, 101*f*
Consumption
    defined, 4
    energy, 408, 408*f*, 444–445
    of natural resources, 6–12, 6*f*, 408
    sustainable, 27–28
Consumption overpopulation, 9–10, 10*f*, 11*f*, 27–28
Contaminant level, maximum, 256
Continental shelves, oil deposits under, 414
Contour plowing, 48*f*, 305
Contraceptive use and fertility, 172, 172*f*, 174
Control group, 17
Convection, 198, 431
Convention on Biological Diversity, 380
Convention on International Trade in Endangered Species of Wild Flora and Fauna (CITES), 380
Convergent plate boundary, 288*f*, 289*f*
Copper Basin, Tennessee, 286
Copper ore, 286
Copper ore tailings, 297*f*
Coral, 270
Coral bleaching, 220, 279
Coral reefs, 270–271, 270*f*
    human impacts on, 279
Coriolis effect, 189, 189*f*, 264, 265*f*
Cost-benefit diagram, 64, 65*f*
Côte d'Ivoire, fertility rates in, 172*f*
Council on Environmental Quality, 57
Crater Lake, 250
Crop rotation, 305, 305*f*
Crops
    genetically engineered, 345, 345*f*, 350–351, 350*f*, 351*f*
    genetic diversity of, 344, 363, 363*f*
    increasing yields of, 344–345
Crude oil, 412
Cryptosporidiosis, 76*t*
Cullet, 394
Culture, fertility and, 169–170
Curitiba, Brazil, 180
Currents, ocean, 264, 265*f*
Cuyahoga River pollutant fire, 59
Cuyahoga Valley National Park, 327*t*
Cyanide heap leaching, 297
Cyanides, sources of, 398*t*
Cyanobacteria, 108
Czech Republic, environmental problems in, 66

D
Dams
    Aswan, 441
    Bonneville, 245*f*
    Columbia River and, 245, 245*f*
    in Dinosaur National Monument, 51
    Grand Coulee, 245*f*

    hydropower and, 440–441
    Little Goose, 440*f*
    Three Gorges Project, 441
    tidal energy and, 443–444
Darwin, Charles, 143, 145*f*
Darwin's finches, 145*f*
Data, defined, 15
DDT (dichlorodiphenyltrichloroethane), 82*t*
    bald eagle and, 357
    effects on birds, 79–80, 80*f*, 81*f*
    as endocrine disrupter, 90
    exception to global ban, 82
    in Great Lakes, 258
    red scale infestation caused by, 354*f*
    Silent Spring, 53
Death causes and probabilities, 72*t*
Death rate, 157, 161–162
Death Valley National Park, 58*f*
Deciduous forest, temperate, 131, 131*f*
Decommissioning nuclear power plants, 424
Decomposers, in ecosystems, 100, 101*f*, 102
Decomposition, nutrient cycling and, 302
Deep ecology worldview, 31–32, 31*f*
Defenders of Wildlife, 331
Deforestation, 318–319
    of boreal forests, 319
    defined, 318
    results of, 319
    of tropical forests, 163*f*, 320, 321*f*, 368
Degradation
    environmental, 8–9, 163*f*
    habitat, 370
    land, 324, 346, 347*t*
    rangeland, 323–324
    resource, 60
    soil, 304*f*
Dematerialization, 393
Demographics, 164
Demographics of countries, 164
    age structure, 166–168, 167*f*, 168*f*
    demographic transition, 164, 165*f*, 166, 166*t*
    fertility (*see* Fertility)
    infant mortality rates, 164, 165*f*
    most populous countries, 165*f*
    See also Population growth
Demography, 14
Denitrification, 108
Denmark
    industrial ecosystem in, 307
    wind energy in, 438
Density of seawater, 264
Derelict lands, 298
Desert, 137, 137*f*, 240
Desertification, 324*f*
    defined, 324
    rangeland, 323–324
Detritivores, 102
Detritus, 102
    in tundra, 128

Devall, Bill, 31
Developed countries. *See* Highly developed countries; Moderately developed countries
Developing countries
    air pollution in, 202–203
    biological diversity in, 37–38
    consumption of animal products by, 336
    consumption of energy by, 408, 408*f*, 444–445
    consumption of resources and, 7, 9, 10, 27
    global warming and, 230–231
    mineral deposits of, 296
    ozone layer protection in, 226
    quality of life and, 36–37
    smoking in, 87
    solar power in, 432
    urbanization in, 177, 178*f*
    water pollution in, 259
    See also Less developed countries (LDCs)
Development
    coastal, 275*f*, 278
    compact, 176, 180
    economic, 340
    at local, national, and global levels, 39–40
    quality of life and, 36–37
    sustainable, 26, 26*f*
    world summit on, 42
Dieldrin, 82*t*
Diesel fuel, 428
Differential reproductive success, 144, 145*f*
Dinosaur National Monument, dam construction in, 51
Dioxins, 82*t*, 85, 90, 398, 398*t*, 399*f*
Disease-causing agents
    in environment, 75–76
    in water, 249*t*
Diseases
    emerging, 76–79
    global warming and, 219–220
Dispersal, 157
Divergent plate boundary, 288*f*, 289*f*
Diversity. *See* Biological diversity
Domains of life, 146–147, 146*f*
Dose, 83
Dose–response curve, 84, 84*f*
Drainage basin, 237
Drift nets, 276
Drinking water
    disease-causing agents in, 75–76, 76*t*, 77*f*
    herbicides and, 80
    legislation on, 90, 256
    purification of, 254, 254*f*
    See also Water; Water pollution
Drip irrigation, 246, 247*f*
Dumping, ocean, 278
Dust Bowl, American, 52, 325
Dust domes, 198, 199*f*
Dutch Belted cow, 344*f*

Dye-sensitized solar cells, 433
Dysentery, 76*t*, 77

E

E. coli (Escherichia coli), 75–76, 77*f*
Earth
    layers of, 288–289, 288*f*
    seasonal progression of, 213–214, 214*f*
Earth Day, 53–54, 54*f*, 57
Earth Summit, 230, 380
Earthquakes, 290, 291*f*, 441
Easement, conservation, 325, 343
Eastern Europe
    air pollution in, 203
    environmental problems in, 66
Eco-justice, 34
Ecological footprints, 10, 11*f*
Ecological niches, 110–111
Ecological succession, 147–149, 148*f*, 149*f*, 322
Ecologically certified wood, 318
Ecologically sustainable forest management,
    316, 318
Ecology
    defined, 14, 96–98
    industrial, 307
    landscape, 97
    population, 156–160
    restoration, 376
Ecology, Community, and Lifestyle, 31
Economic development, 340
Economic growth, 4
Economics, 60
    See also Environmental economics
Ecosystem-based approached to ocean
    management, 281–282
Ecosystem services, 364*t*
    of atmosphere, 186
    defined, 37, 141, 364
    of government-owned land, 312–313
    of soil organisms, 301–302
    species richness and, 364
Ecosystems
    agricultural, 349
    cycling of matter in, 104
        carbon cycle, 104–105, 105*f*, 120
        hydrologic cycle, 106–107, 106*f*, 138
        nitrogen cycle, 107–108, 107*f*, 108*f*
        phosphorus cycle, 108, 109*f*, 110
    defined, 96
    diversity of, 363
    endangered, 331, 331*t*
    energy flow through, 99
        path of, 102, 102*f*, 103*f*, 104
        producers, consumers, and decom-
            posers, 100–102, 101*f*
        thermodynamics, 99–100
    Florida Everglades, 124
    global warming and, 220, 221*f*
    imbalance in Lake Victoria, 94

industrial, 206, 307
minimum critical size of, 376, 377*f*
niches, 110–112
organism interactions
    competition, 118–119
    keystone species, 119
    predation, 115–117
    symbiosis, 113–115
ozone depletion and, 224
See also Aquatic ecosystems; Biomes
Ecotone, 363
Ecotourism, 21
Ecuador, deforestation in, 321*f*
Edge effect, 363
Edith's Checkerspot butterfly, global warming
    and, 221*f*
Education, fertility and, 171–173
Effective dose–50 percent (ED$_{50}$), 84
Egypt
    air pollution in, 203
    energy consumption by, 408*f*
    family planning in, 36*f*
    global warming and, 210
    schistosomiasis in, 441
Ehrlich, Paul R., 11, 53
El Niño–Southern Oscillation (ENSO),
    266–267, 266f, 267*f*, 280
Electricity
    generating costs of, 433*t*
    net metering of, 447
    power companies and energy efficiency, 447
    from waste-to-energy incineration, 390,
        390*f*, 391*f*
    See also Nuclear energy; Renewable energy
        resources; Solar energy
Electrostatic precipitator, 200*f*, 391*f*
Elephant
    habitat loss and, 370, 371*f*
    protection of, 380
Elevation, 126, 128
Emigration, 157
Emission limitations, national, 257
Emphysema, 195
Endangered species, 366–367, 367*f*
    defined, 366
    extinction and, 366
    habitat loss and, 370
    international protection of, 380
    invasive species and, 372–373, 372*f*
    legislation and, 59
    list of, 379*t*
    overexploitation and, 373–374
    pollution and, 370
    reintroduction of, 377
    See also Biological diversity; Conservation
        biology
Endangered Species Act, 46, 357, 366,
    378–379
Endemic species, 368

Endocrine disrupters, 90, 249*t*, 258
Endrin, 82*t*
Energy, 406–451
    defined, 99
    kinetic, 99, 99*f*
    potential, 99, 99*f*
    thermodynamics, 99–100
Energy conservation
    defined, 444
    at home, 448, 448*f*
Energy efficiency
    defined, 444
    electric power companies and, 447
    green architecture, 449, 449*f*
    superinsulated buildings, 445*f*
    technologies for, 445–446
Energy flow
    defined, 102
    path of, through ecosystems, 102, 102*f*,
        103*f*, 104
Energy Policy Act, 428
Energy resources
    agricultural use of, 341, 341*f*
    consumption of, 408, 408*f*, 444–445
    mining use of, 298
    offshore extraction of, 280
    renewable (*see* Renewable energy
        resources)
    world commercial, 412*f*
    See also Coal; Nuclear energy; Oil and
        natural gas
Enhanced greenhouse effect, 218, 218*f*
Enrichment
    uranium, 417, 418
    water, 248, 250–251, 347
ENSO (El Niño–Southern Oscillation),
    266–267, 266*f*, 267*f*, 280
Enteritis, 76*t*
Entombment of closed nuclear power plants,
    424
Entropy, 100
Environmental changes, emerging diseases
    and, 76–79
Environmental chemistry, 401
Environmental degradation
    in Brazil, 163*f*
    habitat, 370
    land, 324, 346, 347*t*
    poverty and, 9
    rangeland, 323–324
    soil, 304*f*
    resource, 60
    refugees and, 8
Environmental economics, 60, 61*f*
    national income accounts, 60–62
    pollution and, 63–64, 64*f*, 65*f*
    pollution control
        cost and benefits, 62
        strategies, 65, 65*f*

Environmental ethics, 29, 366
Environmental health hazards, 75
   disease-causing agents, 75–76
   emerging diseases, 76–79
Environmental history
   conservation in mid-20th century,
     52–53
   environmental movement, 53–54
   forest protection, 49–50
   frontier attitude, 49
   national parks and monuments, 50–51
   timeline of events, 55*f*
Environmental impacts
   of agriculture, 346–347
   of coal, 410–411, 410*f*
   of humans, 4–12, 26–28
   of livestock, 346–347
   of mineral use, 296–298, 297*f*
   of nuclear energy vs. coal, 418, 419*t*, 420
   of oil and natural gas, 414–416, 425
   of pesticides, 346–347
Environmental impact statements (EISs),
   57, 57*f*
Environmental justice, 34–35
Environmental legislation, 57–58, 65
   See also individual laws
Environmental literacy, 56
Environmental movement, 53–54, 55*f*, 330
Environmental problems
   in Central and Eastern Europe, 66
   framework for addressing, 18–19, 19*f*
   religion and, 40
   of urban areas, 176
   world summit on, 42
Environmental Protection Agency (EPA)
   air pollution tests by, 190
   Clean Air Act oversight by, 201
   creation of, 57
   "Draft Report on the Environment 2003,"
     59
   endocrine disrupter tests by, 90
   regulation by, 58
   toxic waste cleanup oversight by, 400–401
   water quality oversight by, 256–257
Environmental resistance, 158–160
Environmental science
   defined, 14
   goals of, 14–15
   prediction, 17
   as process, 15
   scientific method, 16*f*, 17
   theories, 17–18
Environmental sustainability, 12–14, 13*f*, 24,
   97, 206
   See also entries under Sustainable
Environmental Working Group, 80
Environmental worldviews, 30
Environmentalists, 53

Epicenter, 290
Epidemiologists, 76
Epiphytes, 114–115
Equatorial uplift, 214
Erosion, soil, 303, 303f
   agricultural practices and, 346, 347
   mining and, 296, 297
Escherichia coli (E. coli), 75–76, 77*f*
Estrogens, 90
Estuaries
   defined, 142
   as ecosystems, 142–143
   hydrologic cycle and, 106
   overdrawing surface waters and, 241
   tidal energy and, 443
Ethanol
   biomass converted to, 437
   lethal dose of, 83*t*
Ethics
   defined, 29
   environmental, 29, 366
   environmental justice and, 35
Ethiopia, poverty in, 36, 37*f*
Eubacteria, 146, 146*f*
Euglena, 146
Eukarya, 146f, 147
Eukaryotes, 146–147, 146*f*
Euphotic zone, 269*f*, 273
Europe, ozone levels over, 224
European Union
   automobile recycling policy of, 384
   genetic engineering policy of, 351
   hormone-treated beef restrictions by, 346
Eutrophic lake, 250, 251*f*
Eutrophication, 250–251, 251*f*
   artificial, 251
Evaporation, hydrologic cycle and, 106, 106*f*,
   236
Evolution
   defined, 143
   diversity and, 365
   domains and kingdoms of life, 146–147,
     146*f*
   natural selection, 144, 145*f*
Ex situ conservation, 374
Exosphere, 186, 187*f*
Expansionist worldview, 30
Experimental group, 17
Explosive waste, 398*t*
Exponential population growth, 158, 158*f*
External cost, 63
Extinction, 367*f*
   background, 366
   defined, 366
   mass, 366
   See also Conservation biology
Extrapolation, 85
Exxon Valdez oil spill, 415–416, 415*f*

F
Fallow, 342
Family planning services, 172, 174
Farm Bill (Food Security Act) of 1985, 306,
   343
Farmland Protection Program, 343
Faults, 290
Fecal coliform test, 76, 77*f*
Federal land management, 312–313, 330, 332
Federal Land Policy and Management Act, 325
Fertility
   culture and, 169–170
   education and, 172–173
   family planning services and, 172, 172*f*, 174
   gender inequality and, 171–172
   government policies and, 174
   replacement-level, 164
   in most populous countries, 166*t*
   total rate, 164
Fibrous root system, 323
First law of thermodynamics, 99–100
Fish and Wildlife Service (FWS), 312, 313*t*,
   328, 357, 375, 378, 381
Fisheries
   El Niño effects on, 266–267
   Georges Bank, 262, 280
Fish ladder, 245*f*
Fishing industry
   aquaculture, 275*f*, 277, 277*f*, 283
   bycatch, 275*f*, 276
   marine harvest by, 274–275
   methods used by, 276
   open management by, 277
   overfishing by, 262, 275–277, 275*f*
   subsidies for, 281
Fission, 417–418, 417*f*
Flood plain, 238
Floods, 238–239, 239*f*
Florida
   Big Cypress Swamp, 141*f*
   Everglades, 124
   Fort Meade, 298*f*
   Lake Apopka, 90
   panthers in, 379*f*
Flowing-water ecosystems, 138, 140
Fluidized-bed combustion of coal, 411, 411*f*
Fly ash, 390
Focus, of earthquakes, 290
Food chains, 102, 102*f*
Food insecurity, 338
Food Quality Protection Act, 90
Food resources
   grain production, 340*f*
   grain stockpiles, 336
   hunger and, 338, 339*f*, 340
   poverty and, 340
   See also Agriculture
Food security, 336

Food Security Act (Farm Bill) of 1985, 306, 343
Food web, 103*f*, 104
   Antarctic, 283
Forbs, 323
Forest decline, 228, 229*f*
Forest Reserve Act, 50
Forest Stewardship Council, 318
Forests
   boreal, 129, 129*f*, 319
   deforestation, 318–319
      of boreal forests, 319
      results of, 319
      of tropical forests, 320, 321*f*
   ecologically certified wood, 318
   global warming and, 315, 319
   harvesting, 316–317
   hydrologic cycle and, 314*f*, 315
   importance of, 314–315
   management of, 330, 332
      sustainable, 316
      traditional, 315–316
   mangrove, 143
   national, 312*f*, 313*t*, 322, 332
   old-growth, 46
   protecting, 49–50
   temperate deciduous, 131, 131*f*
   temperate rain, 130, 130*f*, 332
   tropical dry, 320, 321*f*
   tropical rain, 21, 77*f*, 78, 132–133, 132*f*,
      163*f*, 320, 321*f*, 368
Fort Meade, Florida, 298*f*
Fossil fuels
   defined, 105, 213
   global warming and, 120, 217, 222,
      230–231
   See also Coal; Oil and natural gas
Framework Convention on Tobacco Control,
   87
France, tidal energy in, 443
Fresh water. *See* Drinking water; Water
Freshwater ecosystems, 138–140, 139*f*, 140*f*
Freshwater wetlands, 138, 141
Frogs, as bellwether species, 360
Frontier attitude, 30, 49
Fuel assemblies, 418, 418*f*
Fuel cells, 435, 435*f*
   methane, 436
Fuel rods, 418, 418*f*
Full cost accounting, 58
Fungicides, 352
Furans (dibenzofurans), 82*t*
Fusion, 417

G

Galápagos Islands, 145*f*
Game Management, 52
Ganges River, 259*f*
Gasohol, 437

Gause, G.F., 159*f*
Gender discrimination, 171–172, 171*f*
Gender roles, 169, 169*f*
Gene banks, 377–378, 377*f*
General Land Law Revision Act, 50
Genetic diversity
   in corn, 363*f*
   defined, 363
   importance of, 365–366
   loss of, in agriculture, 344
Genetic engineering
   in agriculture, 345, 345*f*, 350–351, 350*f*,
      351*f*
   concerns about, 351
   defined, 350
   importance of, 365
Genetic pest controls, 355*t*
Genetic resistance, 353
Genetically modified, 351
Georges Bank fishery, 262, 280
Georgia
   Blackbeard Island Wilderness, 329*f*
   Marietta, 56*f*
Geothermal energy, 442–443, 443*f*
   electric generating costs of, 433*t*
Geothermal heat pumps, 443
Germany
   acid deposition damage in, 228
   birth rates in, 164
   coal reserves in, 409
   environmental problems in, 66
   population growth in, 167*f*, 168
   reproductive rights in, 174
   wind energy in, 438
Germplasm, 344
Geysers power plant, California, 442
Gifford Pinchot National Forest, clear-cut
   forest in, 13*f*
Glacier ice melt, 210
Glass recycling, 394, 395*f*
Global commons, 24
Global distillation effect, 184
Global Polio Eradication Initiative, 162*f*
Global warming
   acid deposition and, 227
   Antarctic food web and, 283
   as carbon cycle imbalance, 120
   causes of, 217–219
   coal and, 411
   dealing with, 222–223
   effects of, 210, 219–220, 221*f*, 222
   evidence of, 216, 216*f*
   forests and, 315, 319
   international implications of, 230–231
   nuclear energy and, 420
   ocean and, 275*f*, 280
   oil and natural gas and, 414
   ozone depletion and, 227
   species endangerment and, 370

Gold, wastes from mining, 297
Government policies and fertility, 174
Grain production, 340*f*
Grain stockpiles, 336
Grand Banks, Hibernia oil platform, 61*f*
Grand Canyon, 326, 327*t*
Grand Coulee Dam, 245*f*
Grand Teton National Park, 327*t*
Grasses, sea, 271, 271*f*
Grasslands, national, 312*f*
   See also Rangelands
Gray water, 246, 247*f*
Great Barrier Reef, 271
Great Lakes, 258*f*
   invasive species in, 372–373, 373*f*
   water pollution in, 258
Great Smoky Mountains National Park, 327*t*
Green architecture, 443, 449, 449*f*
Green Belt Movement in Kenya, 40*f*
Green chemistry, 401
Green Muscle, 349
Green revolution, 344–345
   second, 349
Green wood, 318
Greenhouse effect, enhanced, 218, 218*f*
Greenhouse gases, 217–219, 227
   changes in, preindustrial to present,
      218*t*
   coal and, 411
   defined, 218
   in developing and highly developed
      countries, 230–231
   oil and natural gas and, 414
   See also Global warming
Gross domestic product (GDP), 60
Gross national income (GNI), 164
Groundwater, 237, 237*f*
   pollution of, 253, 253*f*
Growth hormone used in cattle, 89
Growth rate of population.
   See Population growth
Gulf of Mexico, oil drilling in, 414, 414*f*
Gyres, 264

H

Habitat, 110
Habitat degradation, species endangerment
   and, 370
Habitat destruction, species endangerment
   and, 370
Habitat fragmentation, 346, 347
   species endangerment and, 370, 371*f*
Haeckel, Ernst, 96
Haiti, hunger in, 339*f*
Halons, 224
Hanford Nuclear Reservation, Washington
   State, 403, 403*f*
Hardin, Garrett, 24

Harrison, Benjamin, 50
Hawaii
    biodiversity decline in, 368
    formation of, 289
    Mauna Loa Observatory, 217f
    as national monument, 282
Hayes, Denis, 53
Hazardous air pollutants, 192–193
Hazardous waste, 397–403
    chemical accidents, 400
    cleanup of, 400–401, 401f
    defined, 397
    dioxins, 82t, 85, 90, 398, 398t, 399f
    landfill, 402f
    managing, 400–402
    PCBs (polychlorinated biphenyls), 59, 82t,
        90, 184, 258, 398t, 399, 399f
    production of, 401–402
    radioactive, 398t, 403, 420, 422–424
    types of, 398–399, 398t
    See also Solid waste
Health effects
    of air pollution, 195, 195t
    of global warming, 219–220
    of ozone depletion, 226, 226f
    of pollutants
        cancer-causing substances, 84–85
        chemical mixtures, 86
        children and chemical exposure, 86–87
        measuring, 83–84
Health hazards, environmental, 75–79
Heat energy, 213, 218
Heat exchange, atmospheric circulation and,
    188–189, 188f
Heat islands, urban, 198–199, 199f
Hepatitis, infectious, 76t
Heptachlor, 82t
Herbicides, 352
    drinking water study, 80
Herbivores, 101
Heroin, lethal dose of, 83t
Hetch Hetchy Valley, California, 51, 51f
Hexachlorobenzene, 82t
Hibernia oil platform, 61f
High-grade ores, 294
High-level radioactive wastes, 422
Highly developed countries
    birth rates in, 164–165
    consumption of animal products by, 336
    consumption of energy by, 408, 408f, 445
    consumption overpopulation in, 9–10,
        27–28
    defined, 6
    global warming and, 230–231
    infant mortality rates in, 164
    life expectancies in, 164
    overnutrition in, 338
    ozone layer protection and, 226
    resource consumption and, 6–7

solid waste generation in, 386
    urbanization in, 177
Himalayas, 289
Holdren, John P., 11
Honduras, air pollution and respiratory
    disease in, 86f
Hormones
    endocrine disrupters and, 90
    growth, in cattle, 89, 346
    as pest controls, 355t
Host, 115
Hot spot, 289
Hot springs, as energy source, 442
Human health. See Health effects
Human impacts on environment, 4–5
    developed and developing countries, 6
    population and consumption, 6–12
        equation for assessing, 10
    sustainable consumption, 27–28
    sustainable development, 26
    See also Environmental impacts
Human population. See Population;
    Population growth
Human trafficking, 8
Human values
    defined, 29
    environment and, 29
    voluntary simplicity, 33
    worldviews, 30–32
Humane Society, 380
Humus, 300, 301f
Hungary
    environmental problems in, 66
    reproductive rights in, 174
Hunger, world, 338, 339f, 340
Hybrid cars, 428
Hydrocarbons, 192, 193
Hydrochlorofluorocarbons (HCFCs), 226
Hydrofluorocarbons (HFCs), 226
Hydrogen bond of water, 238
Hydrogen fuels, 428
    solar-generated, 434–435
Hydrogen sulfide, 442
Hydrologic cycle, 106–107, 106f
    forests and, 314f, 315
    fresh water and, 138, 236–237, 236f
Hydropower, 436, 440–441, 440f
    as commercial energy source, 412f
    defined, 440
    electric generating costs of, 433t
Hydrosphere, 98
Hydrothermal reservoir, 442
Hypothesis, 17

I
Iceland
    formation of, 289
    geothermal power in, 442, 443f

Idaho
    Salmon River, 323f
    Selway-Bitterroot Wilderness, 329f
Immigration, 157
In situ conservation, 374
Incentive-based regulation, 65
Incineration, 390–391, 391f, 401–402
India
    air pollution in, 202–203
    biogas digesters in, 436, 437f
    coal reserves and consumption in, 409
    deforestation in, 321f
    ecological footprint of, 10, 11f
    energy consumption by, 408f
    polio in, 162f
    population growth in, 166t
    poverty in, 27f
    reproductive rights in, 174
    savannas in, 136
    tsunami in, 290
    wind energy in, 438
Indonesia
    El Niño effects on, 267
    gold mining in, 297
    population growth in, 166t
    rice production in, 345
    tsunami in, 290, 291f
Indoor air pollution, 203–205, 205f
Industrial ecology, 307
Industrial ecosystems, 206, 307
Industrial importance of organisms, 365
Industrial smog, 195
Industrial water pollution, 251–252, 251f
Industrial water waste, 246
Industrialized agriculture, 341–342, 341f
    environmental effects of, 346, 347f, 347t
Infant mortality rates, 164, 165f
Infectious diseases, 75–76, 76t
Infectious hepatitis, 76t
Infectious waste, 398t
Infrared radiation
    defined, 213
    global warming and, 218
Inorganic chemicals, in polluted water, 249t
Inorganic plant and algal nutrients, in
    polluted water, 249t
Insecticides, 352
Institute for Sustainable Power, 432
Institute of Nuclear Power Operations, 421
Institute of Scrap Recycling Industries, 394
Integrated pest management, 349, 356, 356f
Integrated waste management, 396, 396f
Intercropping, 342
International Food Policy Research Institute, 345
International Labor Organization, 170
Interspecific competition, 118
Intertidal zone, 268, 269f
Intraspecific competition, 118
Invasive species, 329, 372–373, 372f

IPAT model, 10–12
Iran
    air pollution in, 203
    natural gas reserves in, 413
Irradiation, food, 355*t*
Irrigation
    aquifer depletion and, 241
    of arid and semiarid lands, 240, 240*f*
    from Colorado River, 242
    global issues, 243
    microirrigation, 246, 247*f*
    salinization and, 242, 303, 303*f*, 346, 347
Islands
    habitat, 370, 371*f*
    urban heat, 198–199, 199*f*
Israel, infant mortality rates in, 165*f*
Izembek Lagoon, Alaska, 271

J

J curve, 158, 158*f*, 162
Japan
    acid deposition and, 228
    Kobe earthquake, 290
    PCBs in, 399
    population growth in, 166*t*
    resource consumption and, 6*f*
    solar energy in, 433
Joshua Tree National Park, 137*f*

K

Kalundborg, Denmark, 307
Kansas, Lawrence, 135*f*
Kelps, 272, 272*f*
Kentucky, mountaintop removal in, 410
Kenya
    Green Belt Movement in, 40*f*
    infant mortality rates in, 172
Keystone species, 119
Kinetic energy, 99, 99*f*
Kingdoms of life, 146–147, 146*f*
Kobe, Japan, 290
Korup National Park, Cameroon, 310
Krill, 283
Kuwait, oil spill cleanup in, 416
Kwashiorkor, 339*f*
Kyoto Protocol, 120, 231

L

La Niña, 268, 280
Lake Apopka, chemical spill in, 90
Lake Erie, pollution in, 53
Lake Roosevelt, 245*f*
Lake Victoria, Africa, 94
Lakes
    eutrophic, 250
    oligotrophic, 250
    zonation in, 138–139
Land degradation, 324, 346, 347*t*

Land use
    conservation of land resources, 331
    federal land management, 312–313, 330, 332
    forests, 314–315
        deforestation, 318–321
        management of, 315–318, 330, 332
        in U.S., 322
    Korup National Park, Cameroon, 310
    national parks, 326–328, 332
    rangelands, 323–325
    in United States, 312–313, 312*f*, 313*t*
    wilderness areas, 328–329
Landfills
    hazardous waste, 402*f*
    sanitary, 388–389
Landscape, 96, 97
Landscape ecology, 97
Landslides, 290
Latin America
    air pollution in, 203
    gender roles in, 169*f*
    hunger in, 340
    reproductive rights in, 174
Latitude
    precipitation and, 214
    solar intensity and, 213*f*
    temperature changes and, 213–214
Lava, 289
Leachate, 388, 402, 402*f*
Leaching
    cyanide heap, 297
    soil, 300
Lead
    as endocrine disrupter, 90
    lethal dose of, 83*t*
    sources of, 398*t*
Leadville, Colorado, 297*f*
Legislation, environmental, 57–58, 65
    See also individual laws
Lemurs, 369
Leopold, Aldo, 52, 52*f*
Less developed countries (LDCs)
    birth rates in, 164
    defined, 6, 7
    infant mortality rates in, 164, 165*f*
    life expectancies in, 164
    per capita GNI PPP in, 164
    See also Developing countries
Lethal dose, 83
Lethal dose–50 percent (LD$_{50}$), 83, 83*t*
Libya, 342*f*
Life history characteristics, 157
Lifestyle, consumption and, 27, 28
Limnetic zone, 138–139
Liquefied natural gas, 413
Liquefied petroleum gas, 413
Liquid hydrogen fuel, 428
Lithosphere, 98, 288, 288*f*

Litter, 300
Little Goose Dam, 440*f*
Littoral zone, 138–139
Livestock
    environmental impacts of, 346–347
    genetic diversity of, 344
    increasing yields of, 346
Long Island DDT study, 80, 81*f*
Longlines, 276
Los Angeles, air pollution in, 87, 196–198, 198*f*
Louisiana
    air pollution in, 13*f*
    Bayou Sauvage National Wildlife Refuge, 142*f*
Love Canal, New York, 397, 397*f*
Low-grade ores, 294
Low-input agriculture. *See* Sustainable agriculture
Low-level radioactive wastes, 422

M

MacArthur, Robert, 112
Madagascar, 369
Magma, 289
Magnuson-Stevens Fishery Conservation and Management Act, 281
Malaria, 78–79, 219, 220*f*, 352, 353*f*
Maldives, global warming and, 210
Mali, resource consumption and, 7*f*
Malnutrition, 338, 339*f*
Malthus, Thomas, 161
Man and Nature, 50
Manganese nodules, 280, 280*f*, 281
Mangrove forests, 143
Maple Ridge Wind Farm, New York, 438
Marasmus, 339*t*
Marginal cost, 63
Marginal cost of pollution, 63–64, 64*f*, 65*f*
Marginal cost of pollution abatement, 64, 64*f*, 65*f*
Mariculture, 277
Marine pollution, 274, 275*f*
Marine reserves, 282, 282*f*
Marine sanctuaries, 312*f*
Marine snow, 274
Marsh, George Perkins, 50
Marshes, 141
    salt, 143
Massachusetts, wind farm proposed for, 438
Mass burn incinerators, 390–391, 391*f*
Mass extinction, 366
Mauna Loa Observatory, Hawaii, 217*f*
Maximum contaminant level, 256
Mead, Margaret, 41
Medicines, genetic diversity and, 365, 365*f*
Mediterranean climates, 133
Meltdown, 420

Mercury
   as endocrine disrupter, 90
   in fish, 59
   in Great Lakes, 258
   sources of, 398*t*
Mesosphere, 186, 187*f*
Mesquite Flat Dunes, 58*f*
Metals, 292, 292f, 293*f*
   heavy, 398*t*
   recycling, 394, 395*f*
Methane, 217, 217*f*, 218*t*, 388, 436
Methanol, 437
Methyl bromide, 224, 226
Methyl chloroform, 224, 226
Mexico
   acid deposition damage in, 228, 229*f*
   air pollution in, 177*f*, 203
   energy consumption by, 408*f*
   green revolution in, 345
   malaria in, 78–79, 79*f*
   resource consumption and, 6*f*
   sustainable forestry in, 316, 318
   water problems in, 242, 243
Michigan, Superfund sites in, 401
Microirrigation, 246
Micronesia
   Caroline Islands, 142*f*
   clam farm, 277*f*
Million Solar Roofs initiative, 433
Mineral extraction offshore, 280
Minerals, 292–298
   defined, 292
   environmental impacts of, 296–298
   extracting, 294–295, 294*f*
   processing, 295, 295*f*
   in soil, 300
   types of, 292, 293*f*, 294
   uses of, 293*f*
Mining
   coal, 409, 410, 410*f*
   environment and, 296, 297*f*, 410
   mountaintop removal, 410
   open-pit surface, 294, 294*f*
   restoration of lands, 298, 298*f*
   shaft, 294*f*, 295
   slope, 294*f*, 295
   strip, 294–295, 294*f*, 409
   subsurface, 295, 409
   surface, 294, 294*f*, 409, 410, 410*f*
Minnesota
   logging in, 30*f*
   radioactive waste storage in, 422*f*
Mirex™, 82*t*
Missouri River, 234
Missouri River Basin Association, 234
Mitigation of global warming, 222, 223*f*
Mobile sources of air pollution, 194
Moderately developed countries
   birth rates in, 164

defined, 6
   infant mortality rates in, 164
   urbanization in, 180
Modular incinerators, 390–391
Mojave Desert, California, 434
Monoculture, 315, 342, 347
Monongahela National Forest, 131*f*
Montreal Protocol, 226, 227
Morphine, lethal dose of, 83*t*
Mount Pinatubo, Philippines, 216*f*, 219*f*, 289
Mount Saint Helens, Washington, 289
Mountaintop removal mining, 410
Mozambique, global warming and, 210
Muir, John, 50, 50*f*, 51*f*
Municipal sewage treatment, 255–256, 255*f*
Municipal solid waste, 386–387, 387*f*
Municipal water waste, 246, 247*f*, 247*t*
Mutualism, 114, 114*f*
Myers, Norman, 369

N

Naess, Arne, 31
Nantucket Sound wind farm, 438
Narrow-spectrum pesticide, 352
National Academy of Science, 120
National Air Pollution Control Association, 206
National Appliance Energy Conservation Act (NAECA), 445
National conservation strategy, 380
National Council of Churches, 40
National emission limitations, 257
National Environmental Policy Act (NEPA), 57–58
National forests, 312*f*, 313*t*, 322, 332
National grasslands, 312*f*
National income accounts
   defined, 60
   natural resource depletion, 61
   pollution control cost and benefits, 62
National Marine Fisheries Service, 281
National marine sanctuaries, 312*f*
National monuments, creation of, 50–51
National Oceanic and Atmospheric Administration, 267, 277, 281
National Park System (NPS), 326
   creation of, 50–51, 326
   land administered by, 312, 313*t*
   locations of parks, 312*f*
   most popular parks, 327*t*
   overcrowding in, 327*f*
   threats to, 326–328
   wilderness areas and, 328
National Research Council, 86, 204
National Response Center (NRC), 400
National Wetlands Coalition, 330
National Wilderness Preservation System, 58, 328, 329

National wilderness refuges, 312*f*, 313*t*
National Wildlife Federation, 53
National Wildlife Refuge System, 375
National wildlife refuges, 312*f*, 313*t*
Natural capital, 60, 61*f*
Natural gas. *See* Oil and natural gas
Natural increase in human populations, 157
Natural resources. *See* Resources
Natural selection, 144, 145*f*
Nekton, 138, 273
Nelson, Gaylord, 53
Nepal, air pollution in, 202
Nertic province, 268, 269*f*, 273, 273*f*
Nerve gas, 398*t*
Net domestic product (NDP), 60
Net metering of electricity, 447
Nevada
   Reno, 137*f*
   Yucca Mountain, 423
New Jersey, Superfund sites in, 401
New York
   Earth Day, 54*f*
   Love Canal, 397, 397*f*
   pollution in, 53, 88*f*
   Rikers Island composting, 392
   Superfund sites in, 401
   wind farm in, 438
New Zealand, shelterbelts in, 306*f*
Nicotine, lethal dose of, 83*t*
Nigeria
   ecological footprint of, 10
   energy consumption by, 408*f*
   polio in, 162*f*
   population growth in, 166*t*, 167*f*
   urban problems in, 179*f*
NIMBY (not in my backyard), 20
NIMTOO (not in my term of office), 20
Nitric acid, 228
Nitrification, 108
Nitrogen cycle, 107–108, 107*f*, 108*f*
Nitrogen fixation, 108, 108*f*
Nitrogen oxides, 191, 195*t*, 231, 414
Nitrous acid, 228
Nitrous oxide, 217, 218*f*, 224, 228, 229*f*
Nomadic herding, 342, 342*f*
Noninfectious diseases, 75
Nonmetallic minerals, 292, 292*f*, 293*f*
Nonmunicipal solid waste, 387
Nonpoint source pollution, 251–252
Nonrenewable resources, 7
   See also Coal; Nuclear energy; Oil and natural gas
North American Association for Environmental Education, 56
North Carolina, acid fog in, 410*f*
Northern spotted owl, 46
North Korea, acid deposition and, 228
Northridge, California, 290
Northwest Forest Plan, 46

Norway
    acid deposition and, 228
    seed bank in, 377*f*
Nuclear energy
    atomic structure and, 416–417, 416*f*
    coal compared to, 418, 419*t*, 420
    as commercial energy source, 412*f*
    defined, 417
    electric generating costs of, 433*t*
    fission, 417–418, 417*f*, 419*f*
    fusion, 417
    oil dependence and, 420
    weapons linked to, 421
Nuclear power plants
    accidents in, 420–421
        Chernobyl, 421, 421*f*
        meltdown, 420
        Three Mile Island, 420–421
    decommissioning, 424
    entombment of, 424
    parts of, 418, 419*f*
    storage of, 424
    wastes from, 20, 20*f*, 422–424
Nuclear reactor, 418, 419*f*
Nuclear Waste Policy Act, 423
Nuclear weapons, 421
Nutrient cycling, 302, 302*f*

O
Oberlin College, 449, 449*f*
Ocean
    circulation patterns of, 264, 265*f*
    conveyer belt, 264, 265*f*, 280
    coral reefs, 270–271, 270*f*, 279
    currents, 264, 265f
    density of seawater, 264
    El Niño–Southern Oscillation, 266–267,
        266*f*, 267*f*, 280
    global, 264
    human impacts on
        Antarctic food web, 283
        aquaculture, 275f, 277, 277*f*
        climate change, 275*f*, 280
        coastal development, 275*f*, 278
        coral reefs, 279
        dumping, 278
        fisheries, 274–277, 275*f*
        habitat deterioration, 274, 275*f*
        plastic debris, 278, 278*f*
        pollution, 274, 275*f*
        resource extraction, 280, 281
        shipping, 278
    interaction with atmosphere, 266–267
    kelps, 272, 272*f*
    La Niña, 268, 280
    life zones, 268, 269*f*
        benthic environment, 268, 269*f*,
            270–272

    euphotic zone, 269*f*, 273
    intertidal zone, 268, 269*f*
    neritic province, 268, 269*f*, 273, 273*f*
    oceanic province, 268, 269*f*, 273–274,
        273*f*
    pelagic environment, 268, 273–274
marine snow, 274
protecting, 281–282
sea grasses, 271, 271*f*
upwells in, 266, 267*f*
vertical mixing of, 264
Ocean Dumping Ban Act, 278
Oceanic province, 268, 269*f*, 273–274, 273*f*
Ogallala Aquifer, 241, 241*f*
Ohio
    Cuyahoga River pollutant fire, 59
    green architecture in, 449, 449*f*
Oil and natural gas
    Arctic National Wildlife Refuge and, 425
    cogeneration, 413
    crude oil, 412
    electric generating costs of, 433*t*
    embargo by OPEC, 406
    environmental impacts of, 414–416
    importance as energy source, 412–413, 412*f*
    nuclear energy as alternative to, 420
    offshore drilling, 414, 414*f*
    prices and supplies of, 406
    refining, 412, 412*f*
    reserves of, 413–414, 413*f*
    spills, 414–416, 415*f*
Oil Pollution Act, 416
Old-growth forests, Pacific Northwest, 46
Oligotrophic lake, 250
Olympic National Park, 327*t*, 330*f*
Omnivores, 101
Ontario, Canada, 239*f*
OPEC (Organization of Petroleum Exporting
    Countries), 406
Open management in fishing industry, 277
Open-pit surface mining, 294, 294*f*, 296
Optimum amount of pollution, 64
Oregon
    Estacada, 130*f*
    McKenzie River, 41*f*
    old-growth forests in, 46
Ores, 286, 292, 294
Organic agriculture, 349
Organic compounds, in polluted water, 249*t*
Organization of Petroleum Exporting
    Countries (OPEC), 406
Origin of Species by Means of Natural
    Selection (The), 143
Otters, sea, 272
Our Common Future, 26
Overburden, 294, 294*f*
Overfishing, 262, 275–277, 275*f*
Overgrazing, 324
Overnutrition, 338

Overpopulation
    consumption, 9–10, 10*f*, 11*f*, 27–28
    people, 9
Overproduction, natural selection and, 144,
    144*f*
Oxides, 292
Oxygen demand, biochemical, 248, 248*f*
Ozone
    defined, 192
    in drinking water, 254
    health effects of, 195*t*
    stratospheric, 224, 225*f*
    tropospheric, 192, 217
Ozone depletion, 224, 225*f*
    acid deposition and, 227
    causes of, 224
    effects of, 224, 226
    global warming and, 227
    recovery from, 226
    species endangerment and, 370
Ozone hole, 224, 225*f*, 226, 227
Ozone thinning, 224, 283

P
Pacific Northwest
    Audubon's warblers, 49*f*
    old-growth forests in, 46
Pacific Tsunami Warning System, 290
Pakistan
    earthquake in, 290
    polio in, 162*f*
    population growth in, 166*t*
Paper recycling, 394
Paramecium, 159*f*
Parasite, 115
Parasitism, 115, 115*f*
Parent material, 299–300, 300*f*
Particulate matter, 190, 195*t*
Passive smoking, 87
Passive solar heating, 431, 431*f*
Pathogens, 76, 352
PCBs (polychlorinated biphenyls), 59, 82*t*, 90,
    184, 258, 398*t*, 399, 399*f*
Pelagic environment, 268, 269*f*, 273–274
Pennsylvania
    mountaintop removal in, 410
    PhillyCarShare, 33*f*
    pollution in, 53
    Superfund sites in, 401
    Three Mile Island, 420–421
People overpopulation, 9
Per-capita GNI PPP, 164
Permafrost, 128
Persian Gulf
    oil reserves in, 413, 413*f*
    oil spill in, 416
Persistence, 79
Persistent compounds, as air pollution, 184

Persistent organic pollutants (POPs), 82, 82t
Peru
    El Niño effects on, 266–267
    malaria in, 78
    poverty in, 27f
    Tambopata River, 140f
    Yungay landslide, 290
Pest, 352
Pest management, integrated, 349, 356, 356f
Pesticides, 352–357
    alternatives to, 355, 355t, 356
    benefits of, 352–353
    broad-spectrum, 352
    children and, 70, 86–87
    defined, 346
    as endocrine disrupters, 90
    environmental impacts of, 346–347
    environmentally safe, 349
    movement in environment, 80, 81f, 354f
    narrow-spectrum, 352
    problems with, 353–354
    Silent Spring, 53
    types of, 352
    See also Chemicals; DDT (dichlorodiphenyl-
        trichloroethane)
Pesticides in the Diets of Infants and Children, 86
Petrochemicals, 412
Petroleum, 412
    See also Oil and natural gas
Petroleum gas, liquefied, 413
Pew Oceans Commission, 274
Pheromones, 355, 355f, 355t
Philippines
    Mount Pinatubo, 216f, 219f, 289
    terracing in, 305f
PhillyCarShare, 33f
Phosphorus cycle, 108, 109f, 110
Photochemical smog, 191, 192, 193, 196, 196f,
    198f
Photodegradable, 388
Photosynthesis, 100, 100f, 104–105
Photovoltaic (PV) solar cells, 432–433, 432f
    electric generating costs of, 433t
Phytoplankton, 273
Pinchot, Gifford, 50
Pioneer community, 148
Plankton, 138, 270
Plasma torch, 402
Plastic
    disposal of, 388
    as ocean pollution, 278, 278f
    recycling, 394, 395f
Plate boundaries, 288–289, 288f, 289f
Plate tectonics, 288–289, 288f
Plutonium, 421
Poaching, 373
Point source pollution, 251, 251f
Poland, environmental problems in, 66
Polar easterlies, 188f, 189

Polar property of water, 238, 238f
Poliomyelitis, 76t
Pollutants
    air, 190–193
    health effects determination, 83–87
    persistent organic, 82
Polluted runoff, 251
Pollution
    in Central and Eastern Europe, 66
    children and, 86–87
    economic definition of, 60
    economist's view of, 63–65
    marginal cost of, 63–64, 64f, 65f
    optimum amount of, 64
    soil, 303
    species endangerment and, 370
    See also Air pollution; Water pollution
Pollution abatement, marginal cost of, 64, 64f,
    65f
Pollution control
    cost and benefits of, 62, 62f
    economic strategies for, 65, 65f
Polychlorinated biphenyls (PCBs), 59, 82t, 90,
    184, 258, 398t, 399, 399f
Polyculture, 342
Polyethylene terphthalate (PET), 394, 395f
Population
    carrying capacity, 36, 158–160, 159f, 163,
        324
    defined, 96, 156
    demographics of countries, 164
        age structure, 166–168, 167f, 168f
        demographic transition, 164, 165f,
            166, 166t
        fertility (*see* Fertility)
        infant mortality rates, 164, 165f
        most populous countries, 165f
    environmental resistance and, 158–160
    growth of (*see* Population growth)
    size, change in, 157, 157f
Population Bomb (The), 53
Population crash, 160, 160f
Population ecology, 156–160
Population growth, 4, 5f
    in China, 154
    deforestation and, 320
    exponential, 158, 158f
    fertility and (*see* Fertility)
    food problems and, 338, 339f, 340
    maximum, 157–158, 158f
    natural selection and, 144
    to present, 161, 161f
    projections, 161, 162–163, 162f
    rate of, 157
    stabilizing, 169–174
    sustainable living and, 36–37
    urbanization and, 175–180
    zero, 162
Population growth momentum, 167

Portugal, reproductive rights in, 174
Potato beetle, 347f
Potential energy, 99, 99f
Poverty, 5f
    defined, 4
    food and, 340
    resource consumption and, 9, 27
Prairies, 135, 376f
Precautionary principle, 88–89
Precipitation
    in biomes, 126, 127f
    climate and, 212
    defined, 214
    factors affecting, 214–215
    forests and, 314f, 315
    global warming and, 219
    hydrologic cycle and, 106f, 107, 236
Predation, 115–117, 116f, 117f
Prediction, scientific, 17
Prescribed burning, 150
Preservation, 48
Preservationist, biocentric, 50
Pressurized water reactor, 419f
Prevailing winds, 189, 264
Primary air pollutants, 190, 191f, 194, 194f
Primary consumers, 101
Primary sludge, 255, 255f
Primary succession, 148, 148f
Primary treatment, 255, 255f
Prime farmland, 343
Prince William Sound, Alaska, 415f
Principle of inherent safety, 400
Producers, in ecosystems, 100–101
Product stewardship, 384
Profundal zone, 138–139
Prokaryotes, 146, 146f
Proxy Falls, 215
Public Rangelands Improvement Act, 325
Purchasing power parity (PPP), 164
Purification of drinking water, 254, 254f
Purse-seine net, 276

Q
Quality of life, sustainable living and, 36–37
Quarry, 294

R
Radiation
    infrared, 213, 218
    solar, 213–214
    ultraviolet (UV), 224
Radioactive substances, in polluted water, 249t
Radioactive waste
    Hanford Nuclear Reservation, 403
    high-level, 422
    low-level, 422
    sources of, 398t
    storage of, 20, 420, 422, 422f

Russia, 424
 Yucca Mountain, 423
Radon, 204–205, 204*f*
Rain forests
 agriculture versus, 21
 human activity and disease in, 77*f*, 78
 temperate, 130, 130*f*, 332
 tropical, 132–133, 132*f*, 163*f*, 320, 321*f*, 368
Rain shadow, 215, 215*f*
Range, 366
Rangelands
 carrying capacity of, 324
 defined, 323
 degradation and desertification of,
  323–324
 issues involving, 325
 overgrazed, 324
 trends in U.S., 325
Reactor core, 418, 419*f*
Recycling, 386*f*
 automobile, 384
 glass, 394, 395*f*
 materials, 393–394, 395*f*
 metals, 394, 395*f*
 paper, 394
 plastic, 386*f*, 394, 395*f*
 tires, 394
 water, 246, 247*f*
Red tides, 249*t*
Refining, petroleum, 412, 412*f*
Refugees, environmental degradation and, 8
Refuse-derived fuel incinerators, 391
Regulation
 command and control, 65
 environmental, 58
 incentive-based, 65
Religion
 environment and, 40, 54
 fertility rates and, 170
Renewable energy resources
 biomass energy, 436–437, 436*f*
 electric generating costs of, 433*t*
 geothermal energy, 442–443, 443*f*
 green architecture, 449, 449*f*
 home electricity production, 447
 hydropower, 440–441, 440*f*
 solar energy (*see* Solar energy)
 tidal energy, 443–444
 wind energy, 438, 439*f*
 See also Energy conservation; Energy
  efficiency
Renewable resources, 7
Replacement-level fertility, 164
Reproductive pest controls, 355*t*
Reproductive rights in different countries, 174
Reproductive success, differential, 144, 145*f*
Resource Conservation and Recovery Act
 (RCRA), 400
Resource degradation, 60

Resource partitioning, 110–112
Resources
 consumption of, 6–12, 6*f*, 408
 defined, 48
 depletion of, 61, 61*f*
 nonrenewable, 7
 renewable, 7
 See also Coal; Nuclear energy; Oil and
  natural gas; Renewable energy resources
Response to toxicant, 83
Restoration
 habitat, 376
 mining land, 298, 298*f*
 prairie, 376*f*
 soil, 306
Restoration ecology, 376
Reuse
 automobile, 384
 product, 393
 water, 246
Rhine River, 244, 244*f*
Rhizobium, 108
Richter scale, 290
Rikers Island composting, 392
Ring of fire, 289
Risk, 72–74
Risk assessment, 73–74
 of cancer-causing substances, 84–85
 of chemical mixtures, 86
 of children exposed to chemicals, 83,
  86–87
 defined, 73
 precautionary principle of, 88–89
 steps in, 73*f*
Risk management, 73
Rivers, features of, 140*f*
Roadless Area Conservation Rule, 332
Rocks, 292
Rocky Mountain National Park, 327*t*, 329
Rodenticides, 352
Romania
 environmental problems in, 66
 poverty in, 5*f*
Roosevelt, Franklin Delano, 36, 52
Roosevelt, Theodore, 50, 50*f*, 375
Root system, fibrous, 323
Runoff
 defined, 106, 237
 polluted, 251
 urban, 252, 252*f*
Russia
 birth defects in, 66*f*
 coal reserves in, 409
 deforestation in, 319
 environmental problems in, 66
 natural gas reserves in, 413
 population growth in, 166*t*, 168
 reproductive rights in, 174
 tidal energy in, 443

S
S curve, 159, 159*f*, 161-162
Safe Drinking Water Act, 90, 256
Salinity, 138
Salinization, 242, 303, 303*f*, 346, 347
Salmon River, Idaho, 323*f*
Salt marshes, 143
Saltwater intrusion, 222, 240
Sand County Almanac, 52, 52*f*
San Francisco, California, 289
Sanitary landfills, 388, 389*f*
Saudi Arabia, reproductive rights in, 174
Savanna, 136, 136*f*
Schistosomiasis, 76*t*, 441
Science, process of, 15–18
Scientific method, 16*f*, 17
Scotland, reforestation in, 38*f*
Scrubbers, 391f, 411
Sea grasses, 271, 271*f*
Sea level, global warming and, 210, 220, 222,
 275*f*, 280
Sea otters, 272
Seasons, progression of, 213–214, 214*f*
Seaweeds, 272
Secondary air pollutants, 190, 191*f*
Secondary consumers, 101
Secondary succession, 149, 149*f*, 322
Secondary treatment, 255, 255*f*
Second green revolution, 349
Second law of thermodynamics, 100
Sediment pollution, 249*t*
Seed banks, 377–378, 377*f*
Seed tree cutting, 316, 317
Seismic waves, 290
Seismograph, 290
Seismologists, 290
Selective cutting, 316, 317
Selway-Bitterroot Wilderness, Idaho, 329*f*
Semiarid lands, 240
Sentinel species, 360
Sequoia National Park, 50
Serengeti National Park, Tanzania, 136*f*
Sessions, George, 31
Sewage, 248, 248*f*, 249*t*
 biogas electric generation from, 436
 livestock and, 346–347
 treatment, 255–256, 255*f*
Shade plantations, 368
Shaft mine, 294*f*, 295
Shelterbelts, 306, 306*f*
Shelterwood cutting, 316, 317
Shifting cultivation, 342
Shipping, effects of, on ocean, 278
Short-grass prairies, 135
Sick building syndrome, 203
Sierra Club, 50, 51*f*, 53
Silent Spring, 53, 53*f*
Simplicity, voluntary, 33

Sinks for waste products, environment as, 60, 61*f*
Slag, 295, 390
Slash-and-burn agriculture, 21, 320, 342
Slope mine, 294*f*, 295
Sludge, 255–256, 255*f*
Smelting, 286, 295, 295*f*, 296
Smith, Robert Angus, 227
Smithsonian Migratory Bird Center, 368
Smog
    defined, 195
    industrial, 195
    photochemical, 191, 192, 193, 196, 196*f*, 198*f*, 414
Smoking
    cancer risk of, 73–74, 74*f*, 74*t*
    disease risk of, 87
Sodium cyanide, lethal dose of, 83*t*
Soil, 299–306, 299*f*, 300*f*
    composition of, 300
    defined, 299
    formation of, 299
    organisms in, 301–302, 301*f*
    sustainable use of, 302
Soil conservation, 48, 48*f*, 59, 304–306, 304*f*
Soil Conservation Service, 52
Soil degradation, 303*f*, 304*f*
Soil erosion, 303, 303*f*
    agricultural practices and, 346, 347
    mining and, 296, 297
Soil horizons, 300, 300*f*
Soil pollution, 303
Soil profile, 300, 300*f*
Soil reclamation, 306
Solar energy
    active solar heating, 430–431, 430*f*
    electric generating costs of, 433*t*
    heating buildings and water, 430–431
    indirect, 436–441
    passive solar heating, 431, 431*f*
    photovoltaic solar cells, 432–433
    solar-generated hydrogen, 434–435
    solar thermal electric generation, 434
Solar-generated hydrogen, 434–435
Solar hydrogen fuel, 428
Solar radiation
    absorption of, 213*f*
    climate and, 213–214
    latitude and, 213*f*
Solar shingles, 433*f*
Solar thermal electric generation, 434, 434*f*
    costs of, 433*t*
Solid waste, 384–396
    amount of, 39f, 386
    composting, 391–392, 391*f*
    disposal of, 387–392, 387*f*
    incineration, 390–391, 391*f*
    integrated waste management of, 396, 396*f*
    open dumps, 387

recycling, 393–394, 395*f*
    automobiles, 384
    glass, 394, 395*f*
    metals, 394, 395*f*
    paper, 394
    plastic, 394, 395*f*
    tires, 394
    reducing, 392–396
    reusing products, 393
    sanitary landfills, 388–389
    source reduction, 393
    types of, 386–387, 387*f*
    See also Hazardous waste
Solvent
    organic, 398*t*
    water as, 238
Somalia, hunger in, 339*f*
Source reduction, 393, 401
Sources of raw materials, environment as, 60, 61*f*
South Africa
    coal reserves in, 409
    Mediterranean climate in, 133
South America
    deforestation in, 318–319
    savannas in, 136
    tropical rain forest in, 132
South Carolina, radioactive waste in, 422
South Korea, acid deposition and, 228
Southeast Asia, tropical rain forest in, 132
Spain
    global warming and, 221*f*
    wind energy in, 438
Species
    bellwether, 360
    conservation of, 376–378, 377*f*
    defined, 96, 362
    endangered (*see* Endangered species)
    endemic, 368
    extinct, 366–367, 367*f*
    invasive, 329, 372–373, 372*f*
    keystone, 119
    number of, 362–363
    sentinel, 360
    threatened, 366–367, 379*t*
Species richness, 362–363, 362*f*
    ecosystem services and, 364
Spent fuel
    defined, 420
    storage of, 422, 422*f*
Spoil bank, 295
Spotted owl, 46
Sri Lanka
    fertility rates in, 171
    tsunami in, 290
Standing-water ecosystem, 138–139
Stationary sources of air pollution, 194
Steam generator, in nuclear power plant, 418, 419*f*

Stegner, Wallace, 52
Stewardship, 24
Stockholm Convention on Persistent Organic Pollutants, 82
Storage casks for spent fuel, 422*f*
Storage of closed nuclear power plants, 424
Stratosphere, 186, 187*f*
    ozone depletion in, 224–227, 225*f*
Streptococcus, 158*f*
Strip cropping, 305
Strip mining, 294–295, 294*f*, 409
Strychnine, lethal dose of, 83*t*
Subduction, 289, 289*f*
Sub-lethal dose, 83
Subsidence, 240, 240*f*
Subsidy
    ethanol, 437
    fishing, 281
Subsistence agriculture, 320, 342, 342*f*
Subsurface mining, 295, 409
Succession, ecological, 147–149, 148*f*, 149*f*, 322
Sudan
    desertification in, 324*f*
    gender discrimination in, 171*f*
Sulfides, 292
Sulfur dioxide, 228, 229*f*
    marginal cost of, 63–64
    from smelting, 286
Sulfur emissions, 218–219, 219*f*
Sulfur haze, 218
Sulfuric acid, 228
Sulfur oxides, 191, 195*t*
Sun plantations, 368
Superfund Act, 397, 400, 401
Superfund National Priorities List, 59, 401
Superfund Program, 400–401
Superinsulated buildings, 445*f*
Surface mining, 294, 294*f*, 296, 409
Surface Mining Control and Reclamation Act (SMCRA), 298, 410
Surface water, 237
Sustainability, 12–14, 13*f*, 24, 97, 206
Sustainable agriculture, 304–306, 348–349, 348*f*, 356
Sustainable consumption, 27–28
Sustainable development, 26*f*
    defined, 26
    world summit on, 42
Sustainable forestry, 316, 318
Sustainable living
    biological diversity and, 37–39
    development and conservation, 39–40
    plan for, 35
    population growth and quality of life, 36–37
    social change, 40–41
Sustainable soil use, 302
Sustainable water use, 244–247
Swamps, 141

Sweden, acid deposition and, 228
Symbiosis, 113–115
Synergy, 86

T

Tailings, 296, 297f
Taiwan
    acid deposition and, 228
    PCBs in, 399
Tallgrass prairies, 135
Tambopata River, Peru, 140f
Tanzania
    elephant habitat in, 371f
    Serengeti National Park, 136f
TAO/TRITON array, 267
Taxpayers for Common Sense, 325
Taylor Grazing Act, 325
Temperate deciduous forest, 131, 131f
Temperate grassland, 135, 135f
Temperate rain forest, 130, 130f, 332
Temperature
    in biomes, 126, 127f
    climate and, 212
    latitude and, 213–214
    mean annual, 1960 to 2004, 216f
Temperature inversion, 197, 197f
Tennessee
    Chattanooga, 206, 206f
    Copper Basin, 286
    Kingsport, 34f
    mountaintop removal in, 410
    Nashville, 131f
Terracing, 305, 305f
Tertiary consumers, 101
Tertiary treatment, 256
Texas, chemical accidents in, 400
Thailand, tsunami in, 290, 291f
Theories, 17–18
Thermal pollution, 249t
Thermal stratification, 138
Thermodynamics
    defined, 99
    first law of, 99–100
    second law of, 100
Thermosphere, 186, 187f
Thin-film solar cells, 433, 433f
Thoreau, Henry David, 49–50
Threatened species, 366–367, 379t
Three Gorges Project, China, 441
Three Mile Island, Pennsylvania, 420–421
Threshold, 84
Thunderstorm, 187f
Thyroid hormones, 90
Tidal energy, 442, 443–444
Tides, 443
Tires
    incineration of, 390, 390f
    recycling, 394

Tongall Timber Reform Act, 332
Tongass National Forest, Alaska, 332
Topography
    air pollution and, 197–198, 197f
    soil formation and, 299
Topsoil, 300, 300f
Total fertility rate, 164, 166t
    See also Fertility
Toxicity, 75
Toxaphene™, 82t
Toxicants
    defined, 75
    dose of, 83
    global ban of persistent organic
        pollutants, 82
    movement in environment, 80, 81f
    persistent, 79–80
    response to, 83
Toxic waste. See Hazardous waste
Toxicology, 75
Trade winds, 188f, 189, 264, 266–267
Transform plate boundary, 288f, 289f
Transpiration, 314f, 315
Trawl bag, 276
Tree plantation, 315f
Trickle irrigation, 246, 247f
Trophic level, 102
Tropical dry forests, deforestation of, 320,
    321f
Tropical rain forests
    agriculture versus, 21
    biodiversity decline in, 368
    as biome, 132–133, 132f
    deforestation of, 163f, 320, 321f
    distribution of, 321f
    human activity and disease in, 77f, 78
Troposphere, 186, 187f
Tropospheric ozone, 217
Tsunami, 290, 291f
Tundra, 128, 128f
Turbines
    geothermal, 443f
    nuclear power, 418, 419f
    water, 440–441, 440f
    wind, 439, 439f
Typhoid, 76t

U

Ukraine, Chernobyl nuclear accident, 421,
    421f
Ultraviolet (UV) radiation
    defined, 224
    in drinking water, 254
    ozone depletion and, 224, 225f, 226
U.N. Climate Change Convention, 230–231
U.N. Conference on Environment and
    Development, 42
U.N. Conference on Human Settlements, 179

U.N. Convention on the Law of the Sea
    (UNCLOS), 281
U.N. Environment Programme, 35, 40, 380
U.N. Fish Stocks Agreement, 281
U.N. Food and Agricultural Organization
    (FAO), 274–275, 277, 316, 318, 338, 346,
    351
U.N. Framework Convention on Climate
    Change, 230
U.N. Intergovernmental Panel on Climate
    Change (IPCC), 120, 216
U.N. International Convention for the
    Prevention of Pollution from Ships
    (MARPOL), 278
U.N. Millennium World Peace Summit of
    Religious and Spiritual Leaders, 40
U.N. World Commission on Environment and
    Development, 26
U.N. World Summit on Sustainable
    Development, 42
Undernutrition, 338, 339f
United Nations
    population projections by, 161, 162–163,
        162f
    sustainability and, 40
United States
    acid deposition and, 227, 228, 230
    air pollution in, 184, 203, 204, 410, 418
    amphibian declines in, 360
    biodiversity decline in, 368
    chaparral in, 133
    coal in, 409–410, 411
    conservation policies and laws in, 378–379
    consumption overpopulation in, 9–10, 10f,
        11f, 28f
    DDT ban in, 357
    ecological footprint of, 10, 11f
    emissions in, 1970 and 2000, 201f
    endangered ecosystems in, 331t
    energy consumption by, 408, 408f, 444
    environmental history in, 46, 49–55, 55f
    environmental legislation in, 57–59,
        201–202
    farmland loss in, 343
    fertility rates in, 172, 173
    forests in, 322, 322f
    geothermal power in, 442
    global warming and, 221f, 230–231
    Great Lakes, 258, 372–373, 373f
    hazardous waste in, 397, 399–403
    hydropower in, 440
    infant mortality rates in, 164
    invasive species in, 372–373, 373f
    Kyoto Protocol and, 120, 231
    land use in, 312–313, 312f, 313t
    life expectancies in, 164
    livestock practices in, 346, 347f
    nuclear energy in, 419f, 420, 420–421, 424

oil and natural gas in, 412, 413–414, 425
oil spill in, 415–416, 415*f*
ozone layer protection in, 226
per capita GNI PPP in, 164
population growth in, 166*t*, 167*f*
prairies in, 135
precautionary principle in laws of, 89
rangeland trends in, 325
recycling in, 393–394, 395*f*
secondary succession in, 149
smoking risk in, 74*t*, 87
soil conservation in, 304, 306
solar energy in, 431
solid waste in, 39*f*, 386, 388, 390, 392
species overexploitation in, 373
temperate deciduous forest in, 131
urbanization in, 175–178
water problems in, 239–242, 246, 252
water quality legislation in, 256–257
wildfires in, 150
wildlife refuges in, 312*f*, 313*t*, 375
wind energy in, 438
Upwells in ocean, 266, 267*f*
Uranium
enrichment of, 417, 418
fuel, 418, 418*f*
ore, 417
Urban heat islands, 198–199, 199*f*
Urban runoff, 252, 252*f*
Urbanization, 166, 175–176
defined, 175
environmental benefits of, 176
environmental problems and, 176
planning, in Brazil, 180
trends in, 177, 178*f*, 179
U.S. Commission on Ocean Policy, 281
U.S. Department of Energy, 423
U.S. Environmental Protection Agency.
See Environmental Protection Agency
(EPA)
U.S. Forest Service (USFS), 312, 313t, 322,
325, 328, 332
Utah, radioactive waste in, 422
Utilitarian conservationists, 50

V

Values
defined, 29
environment and, 29
voluntary simplicity, 33
worldviews, 30–32
Vancouver, Canada, 175*f*
Variable, 17
Variation, natural selection and, 144
Vietnam, global warming and, 210
Virginia
Falls Church, 29*f*
mountaintop removal in, 410

Volcanoes, 289
Voluntary simplicity, 33

W

Wackernagel, Mathis, 10
Warm deserts, 137
Washington State
Gifford Pinchot National Forest, 13*f*
Hanford Nuclear Reservation, 403
Little Goose Dam, 440*f*
Mount Saint Helens, 289
Olympic National Park, 130*f*
radioactive waste in, 422
Waste. *See* Hazardous waste; Solid waste
Waste-to-energy incinerator, 390, 390*f*, 391*f*
Water
aquifers, 237, 237*f*
depletion of, 240–241
conservation, 246, 247*f*, 247*t*
dams and reservoirs, 245, 245*f*
development and flow changes, 239, 239*f*
drinking (*see* Drinking water)
floods, 238–239, 239*f*
global issues, 243–244
gray, 246, 247*f*
groundwater, 237, 237*f*
hydrologic cycle, 106–107, 106*f*, 138,
236–237, 236*f*
importance of, 236–238
industrial waste, 246
managing, 244–247
properties of, 238, 238*f*
quality
drinking water purification, 254, 254*f*
pollution control, 256–259
sewage treatment, 255–256, 255*f*
resource problems of, 238–244
salinization, 242, 303, 303*f*, 346, 347
surface, 237
overdrawing, 241–242
sustainable use, 244–247
See also Ocean; Water pollution
Water pollution, 59, 248–253
agricultural practices and, 346–347
controlling, 256–259
at home, 257, 257*t*
defined, 248
in developing countries, 259
diseases transmitted by, 75–76, 76*t*, 77*f*
eutrophication, 250–251
in Great Lakes, 258
groundwater, 253, 253*f*
marine, 274, 275*f*, 278
mining and, 296, 297
sewage, 248, 248*f*, 249*t*
sources of, 251–252
types of, 248, 249*t*, 250–251
See also Drinking water

Water Pollution Control Act, 256–257
Water power. *See* Hydropower
Water rights, 234
Water table, 240
Water treatment, 254, 254*f*
Water treatment plant, 59*f*
Watersheds, 21, 106, 237
Weather
air pollution and, 197–198, 197*f*
defined, 212
Weathering processes, 299
West Virginia
Monongahela National Forest, 131*f*
mountaintop removal in, 410
Westerlies, 188*f*, 189, 264
Western worldview, 30, 30*f*
Wetlands, freshwater, 138, 141, 237
Whales, baleen, 283
Wheat yields, 344*f*
Whitman, Christine Todd, 89
Whooping crane, captive breeding of, 377*f*
Wild and Scenic River Act, 441
Wilderness, 328–329
Wilderness Act, 52, 328
"Wilderness Essay," 52
Wilderness refuges, 312*f*, 313*t*
Wildfires, 150
using goats to fight, 134
Wildlife corridor, 316
Wildlife refuges, 312*f*, 313*t*, 375
Wind energy, 436, 438, 439*f*
electric generating costs of, 433*t*
global growth of, 439*f*
Winds
atmospheric circulation and, 188–189
ocean-atmosphere interaction and,
266–267
ocean circulation and, 264, 265*f*
Wise-use movement, 330
Wolves, reintroducing to Yellowstone, 381
Women
culture and fertility of, 169–170
education and fertility of, 173
family planning services and, 172, 174
government policies and fertility of, 174
refugees, 8
reproductive rights of, 174
social and economic status of, 171–172
World Bank, 4, 42, 162
World Conservation Strategy, 380
World Conservation Union (IUCN), 35, 380
World food problems
grain production, 340*f*
grain stockpiles, 336
hunger, 338, 339*f*, 340
poverty and, 340
World Health Organization, 76–77, 87, 338,
346

World Resources Institute, 274, 409
World Wildlife Fund, 310, 377*f*, 380
World Wide Fund for Nature, 35
Worldviews
    anthropocentric, 32
    biocentric, 32
    deep ecology, 31–32
    defined, 30
    environmental, 30
    expansionist, 30
    Western, 30, 30*f*
Worldwatch Institute, 9
Wyoming
    coal mining in, 410*f*

    grassland in, 135*f*
    wolves returned to, 381

Y
Yaqui Indian pesticide study, 70
Yellow River, 243
Yellowstone National Park, 50, 326, 327*t*, 328, 381
Yields
    crop, 344–345
    defined, 341
    livestock, 346
Yosemite National Park, 50, 51*f*, 326, 327*f*, 327*t*

Yosemite National Park Bill, 50
Yucca Mountain, Nevada, 423
Yukon, air pollution in, 184
Yungay, Peru, 290

Z
Zambia, Lusaka, 136*f*
Zebra mussel, 372–373, 373*f*
Zero population growth, 162
Zion National Park, 327*f*, 327*t*
Zonation, 138–139
    of ocean (*see* Ocean, life zones)
Zooplankton, 273
Zooxanthellae, 270, 279

Units of Measure: Some Useful Conversions

SOME COMMON PREFIXES

Prefix and Symbol	Meaning	Example
Giga- (G)	Billion	1 gigaton = 1,000,000,000 tons
Mega- (M)	Million	1 megawatt = 1,000,000 watts
Kilo- (k)	Thousand	1 kilojoule = 1000 joules
Centi- (c)	Hundredth	1 centimeter = 0.01 meter
Milli- (m)	Thousandth	1 milliliter = 0.001 liter
Micro- (μ)	Millionth	1 micrometer = 0.000001 meter
Nano- (n)	Billionth	1 nanometer = 0.000000001 meter
Pico- (p)	Trillionth	1 picocurie = 0.000000000001 curie

LENGTH   Standard Unit = Meter

1 meter (m) = 39.37 in. = 3.28 ft
1 in. = 2.54 cm
1 km = 0.621 mi
1 mi = 1.609 km = 1609 m
1 nautical mile = 1.15 mi = 1.85 km

VOLUME   Standard Unit = Liter

1 liter (L) = 1,000 cm^3 = 1.057 qt (U.S.)
1 gallon (U.S.) = 3.785 L
1 mi^3 = 4.166 km^3

ENERGY   Standard Unit = Joule

1 joule (J) = 0.24 cal
1 calorie = 4.184 J
1 Calorie = 1000 calories = 1 kcal
1 kilocalorie = 4.184 kJ
1 British thermal unit = 252 cal
1 kilowatt-hour = 3,600,000 J

PRESSURE   Standard Unit = Pascal

1 bar = 10^5 Pa
1 atm = 1.01 bar = 1.01×10^5 Pa
1 millibar = 0.0145 lb/in^2

AREA   Standard Unit = Square Meter (m^2)

1 hectare = 10,000 m^2 = 0.01 km^2 = 2.471 acres
1 acre = 0.405 hectare
1 km^2 = 100 hectares = 0.386 mi^2
1 mi^2 = 640 acres = 259 hectares = 2.59 km^2

MASS   Standard Unit = Kilogram

1 kilogram (kg) = 2.205 lb = 35.3 oz
1 ton = 2000 lb
1 metric ton = 1000 kg = 1.103 ton = 2204.6 lb
1 short ton = 907 kg
1 lb = 453.6 g

ELECTRICAL POWER   Standard Unit = Watt

1 watt (W) = 1 J/second

TEMPERATURE   Standard Unit = Celsius

°C = (°F − 32) × $^5/_9$
°F = °C × $^9/_5$ + 32
1°C = 1.8°F